Advances in Wastewater Research

Series Editor

Maulin P. Shah, Environmental Microbiology Consultant, Ankleshwar, Gujarat, India

The Advances in Wastewater Research highlights the prospective future development of wastewater treatment techniques, research and field applications. The series also focuses on the limitations and challenges associated with some generally accepted wastewater strategies and evaluates the possible applications of these corrective strategies to eliminate toxic pollutants from the environment through integrated technologies in wastewater treatment. The series covers the research areas including, but not limited to, toxic pollutants removal from wastewater, methods and procedures, materials, management of wastewater, valorisation possibilities of waste resulting from the removal of toxic pollutants from wastewater, wastewater recycling, environmental impact, and wastewater policies. The series contribute to remove, reduce, or neutralize water contaminants that threaten human health, and ecosystem productivity and integrity. The series includes peer-reviewed monographs, edited volumes, textbooks, and conference proceedings.

Harshita Jain · Prangya Rath ·
Laxmi Kant Bhardwaj · Maulin P. Shah
Editors

Emerging Contaminants in the Aquatic Environment

Global Overview

Springer

Editors
Harshita Jain
Amity Institute of Environmental Sciences
Amity University
Noida, Uttar Pradesh, India

Laxmi Kant Bhardwaj
Amity Institute of Environmental
Toxicology, Safety and Management
Amity University
Noida, Uttar Pradesh, India

Prangya Rath
Amity Institute of Environmental Sciences
Amity University
Noida, Uttar Pradesh, India

Maulin P. Shah
Enviro Technology Limited
Ankleshwar, Gujarat, India

ISSN 3004-9725 ISSN 3004-9733 (electronic)
Advances in Wastewater Research
ISBN 978-981-96-8547-9 ISBN 978-981-96-8548-6 (eBook)
https://doi.org/10.1007/978-981-96-8548-6

This Springer imprint is published by the registered company Springer Nature Singapore Pte Ltd.
The registered company address is: 152 Beach Road, #21-01/04 Gateway East, Singapore 189721, Singapore

Contents

1 Understanding of Emerging Contaminants 1
Laxmi Kant Bhardwaj, Prangya Rath, Amit Kumar, Ayushi Suratramka, and Vartika Singh
1.1 Introduction 3
1.1.1 Classification/Types of Emerging Contaminants (ECs) 3
1.2 Sources of Emerging Contaminants 6
1.3 Harmful Effects of Emerging Contaminants (ECs) 9
1.4 Detection Methods of Emerging Contaminants (ECs) 9
1.5 Treatment of Emerging Contaminants (ECs) 10
1.6 Challenges in the Treatment of Emerging Contaminants (ECs) 12
1.7 Recommendations 12
1.8 Conclusions 13
References 14

2 Sources of Emerging Contaminants 19
Dolly Sharma, Abhiram Kumar, Amaresh Mishra, Jyoti Chaprana, Rashmi Mishra, and Bhaswati Banerjee
2.1 Introduction 21
2.2 Classification of Emerging Contaminants (ECs) 22
2.2.1 Biological Contaminants 22
2.2.2 Chemical Contaminants 23
2.2.3 Other Classifications Based on Origin and Impact 25
2.3 Environmental Concerns and Risks 27
2.3.1 Impact on Human Health 27
2.3.2 Ecosystem Disruptions 27
2.3.3 Global Distribution Patterns 28

2.4 Sources of Emerging Contaminants (ECs) 28
2.4.1 Anthropogenic Sources 28
2.4.2 Natural Sources 32
2.5 Challenges and Future Prospects 32
2.6 Implications for Environmental Policy and Resource Management 33
2.7 Recommendation 34
2.8 Conclusions 35
References 35

3 Transport and Fate of Emerging Contaminants in Aquatic Systems 41
Laxmi Kant Bhardwaj, Rupam Priya, Moharana Choudhury, Amit Kumar, and Prangya Rath
3.1 Introduction 43
3.2 Transport Mechanisms of Emerging Contaminants (ECs) 44
3.2.1 Hydrodynamic Transport (Advection and Dispersion) 44
3.2.2 Sorption and Desorption Processes 45
3.2.3 Bioaccumulation and Biomagnification 45
3.2.4 Role of Colloids and Particulates 46
3.3 Fate of Emerging Contaminants (ECs) in Aquatic Systems 47
3.3.1 Chemical and Biological Degradation 47
3.3.2 Photodegradation and Hydrolysis 48
3.3.3 Persistence and Transformation Products 49
3.4 Key Contaminants and Their Behavior 49
3.5 Environmental and Human Health Implications 50
3.6 Analytical Approaches for Detection and Monitoring of Emerging Contaminants (ECs) 51
3.7 Remediation and Mitigation Strategies 53
3.8 Environmental Regulations and Management Strategies 54
3.9 Case Studies and Practical Implications 54
3.10 Future Perspectives and Research Directions 55
3.11 Recommendations 56
3.12 Conclusions 56
References 57

4 Health and Environmental Hazards 61
Erukulla Priyanka, Abhiram Kumar, and Kumar Pranav Narayan
4.1 Introduction 63
4.1.1 Heat Stress 64
4.1.2 The Impact of Air Pollution on Health 65
4.2 Emerging Contaminants (ECs) 66
4.2.1 Pesticides 69
4.2.2 Pharmaceuticals 69
4.2.3 Lifestyle and Personal Care Compounds (PCCs) 69

4.2.4 Industrial Additives and By-Products 69
4.2.5 Food Additives 70
4.2.6 Flame Retardants 70
4.2.7 Plasticizers 70
4.2.8 Perfluoroalkyl Substances (PFAS) 70
4.2.9 Surfactants 71
4.2.10 Quaternary Ammonium Compounds 71
4.2.11 Nanomaterials (NMs) 72
4.2.12 Microplastics (MPs) 72
4.2.13 Endocrine-Disrupting Chemicals (EDCs) 73
4.3 Sources and Transport Pathways 73
4.4 Impact on Human Health and Environment 74
4.5 Importance of Addressing Health and Environmental Hazards 75
4.6 Detection and Monitoring 76
4.6.1 Conventional Methods 77
4.6.2 New Methods 78
4.7 Mitigation and Remediation Strategies 78
4.8 Future Directions 78
4.9 Case Studies 79
4.10 Strategies for Efficient Mitigation and Prevention 80
4.11 Recommendations 80
4.12 Conclusions 81
References 82

5 Emerging Contaminants from Agricultural Waste and Their Impact on Aquatic Ecosystems 87
Shreya Singh, Laxmi Kant Bhardwaj, Prangya Rath, and Shivendra Singh
5.1 Introduction 89
5.2 Agriculture Waste, Types, and Their Sources 90
5.3 Sources of Emerging Contaminants (ECs) in Agriculture 91
5.3.1 Treated Wastewater and Effluents 92
5.3.2 Livestock and Manure 93
5.3.3 Pesticides, Agrochemicals (AgCs), and Fertilizers 93
5.3.4 Irrigation and Runoff from Contaminated Water Bodies 93
5.3.5 Plastic Mulch and Agricultural Plastics 94
5.3.6 Atmospheric Deposition 94

5.4 Transport Pathways of Emerging Contaminants (ECs) 95
5.5 Impacts of Emerging Contaminants (ECs) on Aquatic Ecosystems 95
5.6 Detection Methods/Techniques for Emerging Contaminants (ECs) 97
5.7 Mitigation and Management Strategies 98
5.8 Challenges and Future Directions 99
5.9 Recommendations 100
5.10 Conclusions 100
References 101

6 Microplastics Are Contaminants of Emerging Concern in Aquatic Environments: An Overview 105
Prangya Rath, Laxmi Kant Bhardwaj, Abhiram Kumar, and Kumar Pranav Narayan
6.1 Introduction 106
6.1.1 Classification of Microplastics (MPs) 107
6.2 Sources and Transport of Microplastics (MPs) 108
6.2.1 Wastewater and Industrial Effluent 114
6.2.2 Run-Off from Land-Based and Combined Sewer Overflows 114
6.2.3 Atmospheric Deposition 115
6.2.4 Drinking Water Production and Distribution 115
6.2.5 Fragmentation and Degradation of Macroplastics 115
6.3 Health Impact of Microplastics (MPs) 116
6.3.1 Physical Hazard 117
6.3.2 Chemical Hazard 117
6.3.3 Microbiological Hazard 117
6.4 Detection of Microplastics (MPs) 118
6.4.1 Visual Identification 118
6.4.2 Density Separation 118
6.4.3 Spectroscopic Techniques 118
6.4.4 Thermo-Analytical Technique 119
6.4.5 Chemical Method 119
6.4.6 Hyperspectral Imaging (HSI) Technique 119
6.4.7 Combined Methods 119
6.5 Treatment of Microplastics (MPs) 119
6.5.1 Sorption and Filtration Methods 120
6.5.2 Chemical Methods 120
6.5.3 Biological Methods 121
6.6 Recommendations 122
6.7 Conclusions 122
References 123

7 Advanced Monitoring and Detection Techniques for Emerging Contaminants in Aquatic Environments 129
Arittrabha Sengupta, Jeniva Jasmine, Bhavya Chaudhary, Kritika Singh, Sahil Sharma, Harsh Sable, and Rohan Samir Kumar Sachan
7.1 Introduction 130
7.1.1 Environmental Occurrence and Distribution of Emerging Contaminants (ECs) 131
7.1.2 Sources of Emerging Contaminants (ECs) 131
7.1.3 Pathways of Emerging Contaminants (ECs) into the Environment 132
7.2 Effect-Based Measures for Emerging Contaminant (EC) Detection 134
7.2.1 Detection and Quantification of Emerging Contaminants (ECs) 135
7.2.2 Technological Advances in Detection 135
7.3 High-Resolution Mass Spectrometry (HRMS) Techniques 139
7.3.1 Types of High-Resolution Mass Spectrometry (HRMS) 139
7.4 Integrating Advanced Analytical Methods 140
7.4.1 Advanced Analytical Methods 141
7.4.2 Benefits of Advanced Analytical Techniques in Aquatic Contaminant Monitoring 144
7.4.3 Challenges and Considerations 146
7.5 Recommendations 148
7.6 Conclusion and Future Directions 148
References 150

8 Applications of Graphene Nanocomposites in Advanced Detection and Removal of Emerging Water Contaminants 157
Souravi Bardhan and Dipak Kr Chanda
8.1 Introduction 159
8.2 Graphene Nanocomposites as a Promising Candidate in Water Treatment 162
8.2.1 Graphene Oxide (GO) 165
8.2.2 Reduced Graphene Oxide (rGO) 166
8.2.3 Graphene Quantum Dots (GQDs) 166
8.2.4 Graphene-Based Membranes 167
8.3 Applications of Graphene Nanocomposites in the Detection of Emerging Contaminants (ECs) 167
8.3.1 Detection of Antibiotic Residues 170
8.3.2 Detection of Personal Care Products (PCPs) 174
8.3.3 Detection of Pharmaceutical Compounds 174
8.3.4 Detection of Microplastics (MPs) 175
8.3.5 Detection of Plasticizers 176

8.3.6 Detection of Perfluorinated Organic Compounds (PFOCs) 177
8.4 Removal of Emerging Contaminants (ECs) from Water Using Graphene Nanocomposites 178
8.4.1 Adsorption 179
8.4.2 Photocatalysis 179
8.4.3 Membrane Filtration (MF) 180
8.4.4 Electrochemical Degradation 181
8.5 Recommendations 182
8.6 Conclusion and Future Perspectives 183
References 184

9 Emerging Trends in Challenges in Technological Innovation for Sustainable Water Management 193
Qudrat Ullah, Muhammad Waqar, Muhammad Sajad, Muhammad Qasim, and Thanet Khomphet
9.1 Introduction to Sustainable Water Management (SWM) 194
9.2 Technological Innovations in Water Resource Management 196
9.3 Water Quality Management (WQM) 198
9.4 Integrated Water Resource Management (IWRM) 202
9.5 Problems in Practicing Integrated Water Resources Management (IWRM) 204
9.6 Learned and Best Practices from Integrated Water Resource Management (IWRM) Implementation Cases 205
9.7 Climate Change and Water Resources 206
9.8 Policy Responses to Climate Change 208
9.9 The Nexus Between Water, Food, and Animal Sciences 209
9.10 Water Scarcity and Demand Management 211
9.11 Innovative Financing for Water Projects 214
9.12 Community-Based Water Management 217
9.12.1 Examples of Community Mobilization Programmed 217
9.12.2 Difficulties Experiences by Communes 218
9.13 Trends in New Water Management 218
9.13.1 Opportunities Regarding Artificial Intelligence (AI) and Machine Learning (ML) 218
9.13.2 New and Creative Ways of Reusing Water 219
9.14 Recommendations 219
9.15 Conclusions 220
References 220

10 Trends of Emerging Contaminants in the Aquatic Environment—Risk Assessment and Management Strategies 227
Preeti Mehta Kakkar, Ayushi Aggarwal, and Natasha Nair
10.1 Introduction ... 229
10.2 Emerging Contaminants (ECs) Explanation and Signification ... 232
10.2.1 Personal Care Products (PPCs) ... 233
10.2.2 Microplastics (MPs) ... 234
10.2.3 Nanomaterials ... 235
10.2.4 Industrial Chemicals and Unconventional Sources ... 235
10.2.5 Pesticides ... 236
10.3 Factors Influencing the Presence and Transport of Contaminants ... 236
10.4 Spatial and Temporal Variations in Contamination Levels: Implications for Environmental Monitoring ... 237
10.5 Bioaccumulation and Biomagnification in Aquatic Food Webs: Mechanisms and Ecological Implications ... 238
10.6 Risk Assessment Methodologies ... 239
10.6.1 Emerging Pollutants in Aquatic Environments: Screening, Sampling, and Preconcentration Techniques ... 239
10.6.2 Emerging Methods for Detection and Analysis of Emerging Contaminants (ECs) ... 241
10.6.3 Risk Quotient (RQ) Method ... 242
10.7 Novel Considerations in Risk Assessment of Emerging Contaminants (ECs) ... 243
10.8 Management and Remediation of Emerging Contaminants (ECs) ... 244
10.8.1 Conventional Wastewater Treatment Technologies ... 244
10.8.2 Secondary Wastewater Treatment ... 245
10.9 Remediation of Emerging Contaminants (ECs) ... 245
10.9.1 Membrane Filtration ... 245
10.9.2 Adsorption ... 246
10.9.3 Nanotechnology and Nanomaterials ... 247
10.9.4 Advanced Oxidation Process (AOP) ... 249
10.9.5 Phytoremediation ... 249
10.10 Public Policy and Government Regulation ... 250
10.11 Future Directions and Conclusions ... 250
References ... 251

11 Regulatory Framework and Policy Responses 257
Anamika Shrivastava and Harshita Jain
11.1 Introduction 258
11.1.1 Importance of Regulatory and Policy Measures 260
11.1.2 Global Challenges in Regulating Emerging Contaminants (ECs) 261
11.2 International Frameworks and Conventions 262
11.2.1 Key Conventions on Emerging Contaminants (ECs) 263
11.2.2 Role of the United Nations Environment Program (UNEP) and Contributions of Other International Organizations 264
11.2.3 Limitations and Gaps in International Agreements 265
11.3 National-Level Policies in Emerging Contaminants (ECs) 266
11.3.1 Public versus Private Sector Roles in Regulation of Emerging Contaminants (ECs) 266
11.3.2 Policy Gaps in Low-Income Countries 267
11.4 Regulatory Challenges 268
11.4.1 Lack of Standardized Monitoring Protocols 269
11.4.2 Challenges in Risk Assessment for Emerging Contaminants (ECs) 270
11.4.3 Resource Constraints and Enforcement Issues 272
11.5 Policy Responses and Innovations 272
11.5.1 Polluter Pays Principle (PPP) 272
11.5.2 Extended Producer Responsibility (EPR) 274
11.5.3 Economic Instruments: Taxes, Subsidies, and Incentives 274
11.5.4 Public-Private Partnerships in Policy Implementation 275
11.6 Stakeholder Engagement 276
11.6.1 Role of Non-governmental Organizations (NGOs) and Civil Society in Policy Advocacy 276
11.6.2 Community Participation in Regulatory Processes 276
11.6.3 Awareness Campaigns and Educational Initiatives 277
11.7 Future Directions in Regulatory Frameworks 278
11.7.1 Integration of Science-Based Evidence in Policymaking 278
11.7.2 The Role of Big Data and Artificial Intelligence (AI) in Policy Development 279
11.7.3 Strengthening Global Cooperation and Harmonization 279

11.7.4 Linking Regulations to Sustainable Development Goals (SDGs) ... 280
11.8 Conclusion ... 280
References ... 281

12 Emerging Contaminants in Mumbai's Untreated Sewage to Rhine's Industrial Runoff—A Comparative Case Study in India and Europe ... 285
Niketa Bhati, Laxmi Kant Bhardwaj, and Harshita Jain
12.1 Introduction ... 286
12.2 Mumbai, India: A Megacity's Challenge with Untreated Wastewater ... 288
12.2.1 Potential Sources of Emerging Contaminants (ECs) in Mumbai's Wastewater ... 289
12.2.2 Ecological Impacts of Emerging Contaminants (ECs) in Mumbai's Water Bodies ... 290
12.3 Rhine River Basin, Europe: A Legacy of Industrial Activity ... 291
12.3.1 Types of Emerging Contaminants (ECs) Likely Present in the Rhine ... 292
12.3.2 Ecological Consequences of Emerging Contaminants (ECs) in the Rhine Basin ... 294
12.4 Case Comparisons: Mumbai versus Rhine ... 295
12.4.1 Mitigating Emerging Contaminants (EC) Pollution: Tailored Approaches for Different Challenges ... 297
12.5 Strategies for Mumbai: Wastewater Treatment and Disposal Practices ... 298
12.6 Strategies for the Rhine: Regulation, Remediation, and Ongoing Monitoring ... 301
12.7 Discussion and Future Prospects ... 302
12.8 Conclusion ... 303
References ... 303

List of Figures

Fig. 1.1 Diagrammatic representation of the study 7
Fig. 1.2 Origins and pathways of emerging contaminants (ECs) in the environment 7
Fig. 2.1 Occurrence, impacts, sources, challenges and future prospects, implications for environmental policy, and resource management for ECs 22
Fig. 2.2 The evolution of emerging contaminants (ECs) about advancements in detection and monitoring of trace-level toxic chemicals: Since the early 2000s, the term "emerging contaminants" (ECs) has been used to describe newly identified classes of pollutants. Substances such as polychlorinated biphenyls (PCBs), per- and poly-fluoroalkyl substances (PFAS), and plastics serve as notable examples of chemicals that have been widely used for decades (represented by gray bars). Over time, these substances were recognized as contaminants (marked by pins) and subsequently regulated or phased out (indicated by the faded-out shadow), though the timelines for restriction varied. The lower panel illustrates how certain ECs emerged as substitutes for previously banned or restricted pollutants, as indicated by the arrows (Wang et al. 2024). (https://doi.org/10.1016/j.xinn.2024.100612) CC by 4.0. e (http://creativecommons.org/licenses/by-nc-nd/4.0/) 25

Fig. 2.3 Schematic representation of the diverse pathways of EC generation, use, and environmental release: Various sectors, including industries, agriculture, households, hospitals, and wastewater treatment plants, contribute to the distribution of emerging contaminants (ECs). These contaminants enter the environment through multiple pathways, such as industrial emissions, agricultural runoff, household waste, medical discharges, and effluent from treatment facilities. Collectively, these sources play a significant role in the widespread dissemination of ECs, impacting ecosystems and public health (Wang et al. 2024). (https://doi.org/10.1016/j.xinn.2024.100612) CC by 4.0. e (http://creativecommons.org/licenses/by-nc-nd/4.0/) ... 31
Fig. 3.1 Transport of emerging contaminants (ECs) ... 44
Fig. 4.1 Climate change-related factors like pollen, air pollution, and heat stress set off complex immune responses that can either promote inflammation or tolerance ... 65
Fig. 5.1 Different types of agricultural waste ... 91
Fig. 6.1 Classification of microplastics (MPs) based on their origin ... 108
Fig. 6.2 Sources and transport pathways of microplastics (MPs) ... 116
Fig. 7.1 Source of emerging contaminants (ECs) ... 133
Fig. 7.2 Challenges to the advanced analytical techniques utilized in pollutant monitoring ... 146
Fig. 8.1 Pictorial representation depicting various categories of emerging contaminants (ECs) and the pathways through which they enter aquatic systems ... 160
Fig. 8.2 **a** Various forms of carbon materials; **b** various derivatives of graphene obtained by conversion of graphene-to-graphene oxide and to reduced graphene oxide (GO). Reproduced from Memisoglu et al. (2023). Open access, **c** various synthesis routes of development of graphene ... 163
Fig. 8.3 **a** Illustration of removal of various emerging contaminants (ECs) in water using graphene NMs (Assad et al. 2024). Open access. **b** Magnetic composite based on GO, chitosan, and organoclay for the removal of ECs like BPA, 17α-ethinylestradiol, and triclosan (de Almeida 2022). Reproduced with permission from Elsevier ... 169
Fig. 10.1 Mechanism of absorption ... 247
Fig. 10.2 Important nanomaterial characteristics for water purification ... 248
Fig. 10.3 Phytoremediation and its various processes ... 250
Fig. 11.1 Policy responses and innovations ... 273
Fig. 11.2 Stakeholder engagement in regulatory processes ... 277

Fig. 12.1 Impact of emerging contaminants (ECs) and their management 287
Fig. 12.2 Comparative analysis of emerging contaminant (EC) pollution in Mumbai and the Rhine River Basin 289
Fig. 12.3 Process steps to treat emerging contaminants (ECs) in Mumbai wastewater 298
Fig. 12.4 Process steps to treat emerging contaminants (ECs) in Rhine River Basin 299

List of Tables

Table 1.1 Different types of emerging contaminants (ECs) with their common sources ... 6
Table 2.1 Contribution of various sources to emerging contaminant (EC) pollution ... 29
Table 3.1 Transport process of emerging contaminants (ECs) with description and key factors ... 47
Table 4.1 Categories of emerging contaminants (ECs) and their effects on the environment ... 67
Table 4.2 Comparison between conventional methods and new methods for detecting environmental hazards, along with their advantages and disadvantages ... 77
Table 5.1 Agricultural waste with their environmental impact and management practices ... 92
Table 5.2 Different types of emerging contaminants (ECs) with their sources and pathways ... 96
Table 6.1 Occurrence of different types of microplastics (MPs) in aquatic environment ... 109
Table 7.1 Different analytical techniques for particular emerging contaminants (ECs) ... 136
Table 8.1 Graphene nanocomposite-based detection of various emerging contaminants (ECs) ... 171
Table 9.1 Summary of technological innovations ... 199
Table 9.2 Comparison of water treatment technologies ... 202
Table 9.3 Climate change impact assessment ... 210
Table 9.4 Water demand management techniques ... 215
Table 10.1 Various emerging contaminants (ECs) and their certain characteristics ... 231

Table 10.2 Detection techniques for pharmaceutical compounds in water and wastewater, covering a wide array of commonly found substance 242
Table 12.1 Challenge and management strategies in Mumbai and the Rhine River Basin 296

Chapter 1
Understanding of Emerging Contaminants

Laxmi Kant Bhardwaj, Prangya Rath, Amit Kumar, Ayushi Suratramka, and Vartika Singh

Abstract Emerging contaminants (ECs) are toxic ingredients that have been newly discovered in the atmosphere. They have negative effects on human health and the environment even at insignificant levels. These harmful substances are consumed daily in modern civilization. These pollutants are mainly reported in wastewater and arrive at different water reservoirs and pollute them. Water pollution has been a universal crisis for the past few years. These compounds mostly occur from pharmaceuticals, personal care products (PCPs), pesticides, and synthetic dye manufacturing. The intensity of these chemicals is increasing because established wastewater treatment methods are not effective in recognizing and removing them. As their intensity has risen in the environment and proved to be detrimental to both the environment and human health, they have captured the attention of researchers and scientists in recent times. Consequently, there is a need for enhanced equipment to mitigate or reduce these pollutants. This chapter outlines conclusions that focus on how ECs are generated, where they come from, the negative effects they have on humans and on the environment, and methods for their identification.

Keywords Emerging contaminants (ECs) · Pesticides · Personal care products (PCPs) · Microplastics (MPs) · Environment · Human health

---

L. K. Bhardwaj (✉)
Amity Institute of Environmental Toxicology, Safety and Management, Amity University, Noida, Uttar Pradesh, India
e-mail: bhardwaj.laxmikant@gmail.com

P. Rath
Amity Institute of Environmental Sciences, Amity University, Noida, Uttar Pradesh, India

A. Kumar
Molecular Biology and Biotechnology, CSIR-National Botanical Research Institute, Lucknow, Uttar Pradesh, India

A. Suratramka
Institute of Allied Medical Science and Technology, NIMS University, Jaipur, Rajasthan, India

V. Singh
Amity Institute of Global Warming and Ecological Study, Amity University, Noida, Uttar Pradesh, India

H. Jain et al. (eds.), *Emerging Contaminants in the Aquatic Environment*, Advances in Wastewater Research, https://doi.org/10.1007/978-981-96-8548-6_1

Abbreviations

Ag-NPs	Silver nanoparticles
AOPs	Advanced oxidation processes
As	Arsenic
BFRs	Brominated flame retardants
BPA	Bisphenol A
Cd	Cadmium
CO_2	Carbon dioxide
DDT	Dichlorodiphenyltrichloroethane
DNA	Deoxyribonucleic acid
ECs	Emerging contaminants
EDCs	Endocrine-disrupting chemicals
EQS	Environmental Quality Standards
E-waste	Electronic waste
FTIR	Fourier transform infrared spectroscopy
GC	Gas chromatography
HPLC	High-performance liquid chromatography
MPs	Microplastics
MS	Mass spectrometry
NMs	Nanomaterials
NPs	Nanoparticles
NTS	Non-targeted screening
OPFRs	Organophosphate flame retardants
Pb	Lead
PBDEs	Polybrominated diphenyl ethers
PBFR	Polybrominated flame retardants
PCBs	Polychlorinated biphenyls
PCPs	Personal care products
PFASs	Per- and poly-fluoroalkyl substances
PFOA	Perfluorooctanoic acid
PFOS	Perfluorooctane sulfonate
POPs	Persistent organic pollutants
PPCPs	Pharmaceutical and personal care products
RO	Reverse osmosis
TiO_2-NPs	Titanium dioxide nanoparticles
UNEP	United Nations Environment Programme
VOCs	Volatile organic compounds
WFD	Water Framework Directive
WHO	World Health Organization

1.1 Introduction

New compounds are described as ECs that come from drugs, PCPs, domestic and industrial products, agriculture, and wastewater (Chaturvedi et al. 2021). They are generally categorized as artificial polymers and natural polymers. These compounds are commonly identified as micropollutants and are generated globally. They exist in low concentrations in aquatic systems and are considered major pollutants; however, they have not been thoroughly studied yet. These chemicals are widespread in the aquatic environment and enter the water system through several routes. These pollutants are detected in water bodies due to the increase in anthropogenic activities (Wang and Yang 2016; Khan et al. 2022). These chemicals can hinder hormone systems and are known as endocrine-disrupting chemicals (EDCs) (De Coster and Van Larebeke 2012). Chemicals that are used in PCPs, electronics, and medicines are uncontrolled and highly persistent. In recent years, these toxic pollutants have attracted attention because of their rapid use in daily life (Dhangar and Kumar 2020).

1.1.1 Classification/Types of Emerging Contaminants (ECs)

ECs can come from a variety of sources and are typically divided into several categories. They include pesticides, herbicides, antibiotics, nanomaterials (NMs), industrial waste chemicals, PCPs, pharmaceutical compounds, EDCs, per- and polyfluoroalkyl substances (PFAS), and microplastics (MPs) (Das et al. 2024). The main features of organometallic compounds are based on their chemical properties or molecular structures. The sources of these pollutants are quite complicated (Mofijur et al. 2024).

- **Personal Care Products (PCPs) and Synthetic Musks**

PCPs include cosmetics, sunscreens, fragrances, and other care products. The chemicals that are used in PCPs can leak into the atmosphere and may be detrimental to the health of human beings, and other organisms (Caliman and Gavrilescu 2009). Triclosan is a disinfectant and is used in toothpaste, soaps, sprays, etc. It moves into aquatic bodies through the drain and can bio-concentrate in aquatic creatures. In human beings, it links to hormonal differences and increased allergic reactions. Artificial musks are made by humans and are used in body spray, roll-ons, lotions, and soaps. They are designed to replicate the scent of natural musk. Nitro musks (musk ketone and xylene), polycyclic musks (tonalide and galaxolide), macrocyclic musks (ethylene brassylate and ambrettolide), and alicyclic musks (helvetolide and romandolide) are some common synthetic musks. These compounds are versatile because they are stable and have a good smell. Nonetheless, their use has significantly reduced because of environmental and health threat aspects.

- **Household Products**

Cleaning agents, laundry detergent, sanitizers, disinfectants, etc. are examples of domestic goods. Surfactants are the functional constituents of wiping causes and laundry detergent. Phosphates that are present in the detergent have water-softening properties. It may be introduced to the environment through domestic sewage and is known to be hazardous to water life (Gupta et al. 2023). It is present in water bodies in high quantities causing eutrophication and can create a dead zone because it may have negative effects on aquatic organisms and water oxygen content.

- **Microplastics (MPs)**

They are tiny particles of plastic that are formed from the breakdown of larger plastic particles through different mechanisms (Bhardwaj 2024; Bhardwaj et al. 2024a). The size of these particles is <5 mm. They are found in all habitats of Earth including oceans, rivers, groundwater, and soil as well as in polar regions. They are found in the form of granules, microbeads, fragments, and fibers. They persist for a long time in the environment. They may be more hazardous when they interact with heavy metals, dyes, and organic pollutants (Goh et al. 2022). They get ingested by aquatic species and have been reported to cause both physical as well as chemical harm during later stages. Pollutants like heavy metals, microfiber, plastic, etc. initiate water toxins and pollute the soil (Bashir et al. 2020; Mairaj et al. 2022; Bhardwaj et al. 2023b; Alam et al. 2023).

- **Pesticides, Herbicides, and Persistent Organic Pollutants (POPs)**

These chemicals are applied by agricultural manufacturing to supervise infestation by insects and wild plants in crop-growing spaces. Atrazine, chlordane, dichlorodiphenyltrichloroethane (DDT), glyphosate, methoprene, and neonicotinoids are commonly used chemicals (Tlak Gajger and Dar 2021). They interfere with microbial communities in the soil and change the capacity of the soil. Industrialization, the rapid pace of urbanization, and the boost in the application of agrochemicals are the primary determinants that affect the quality and quantity of water (Dong et al. 2023). Persistent organic pollutants (POPs) are moderately new compounds and are also being detected in the aquatic environment (Bhardwaj et al. 2019; Bhardwaj and Jindal 2020, 2022). They have long biological half-lives and bioaccumulate in living bodies. They are hazardous even at low concentrations and can easily pass through the food chain (El-Shahawi et al. 2010). They become a global concern due to their persistent nature and bioaccumulating characteristics (Fernández and Grimalt 2003).

- **Per- and Poly-Fluoroalkyl Substances (PFAS) and Flame Retardants**

PFAS are man-made chemicals and are produced from two routes, direct fluorination and oligomerization (Evich et al. 2022). They are known as forever chemicals. They have water and stain-resistance properties and are used in firefighting foams, waterproof apparel, non-stick cookware, industrial processes, and stain-resistant clothes.

These persistent pollutants have widespread applications. The brominated flame retardants (BFRs) and organophosphate flame retardants (OPFRs) are the types of flame retardants. They are used in electronics, textiles, furniture, and building materials to reduce fire dangers. Both chemicals are persistent in the environment and toxic to biota and the environment.

- **Pharmaceutical Compounds and Endocrine Disrupting Chemicals (EDCs)**

Recreational and agriculture chemicals utilized for human and veterinary uses can percolate into the environment because of inefficient disposal, accretion, and production phases and pollute water and soil resources. EDCs can disturb the hormonal secretion in organisms. Phthalates, pesticides, and bisphenol A (BPA) are well-known EDCs in the surroundings (Bhardwaj and Sharma 2021; Bhardwaj 2022). These chemicals can cause reproductive and developmental issues. They can also affect the immune systems of humans and animals. They are normally recognized as invertebrate and vertebrate hormones (Pal et al. 2014).

- **Nanomaterials (NMs)**

NMs are engineered particles of range 1–100 nm (Tokas et al. 2024). These particles have unique properties making them potential candidates in the fields of electronics, environment, and medicine (Cheriyamundath and Vavilala 2021). They are capable of entering the environment due to activities of manufacturing and disposal. They can be used in the remediation of pollutants in the treatment of wastewater (Singh and Batra 2018). Because of their relatively smaller size as well as peculiar properties of adsorption, they have become threats to the health of humans and ecosystems. Kunduru et al. (2017) and Bhardwaj et al. (2023a) described some nanoparticles (NPs) like titanium dioxide nanoparticles (TiO_2-NPs), carbon nanotubes, and silver nanoparticles (Ag-NPs) in detail.

Organizations such as the United Nations Environment Programme (UNEP) and the World Health Organization (WHO) have studied several ECs. The Water Framework Directive (WFD) is crucial for maintaining the favorable ecological status of different water bodies and for implementing effective management techniques to control pollution levels. Environmental Quality Standards (EQS) has proposed the highest acceptable level of ECs that could be allowed and tolerated. Improper utilization of ECs exposes society to numerous risks hence the need to study them so as to enhance health and safety, environmental conservation as well as orchestrate sustainable development. This chapter provides information on the presence, types, sources, effects, detection methods, challenges, and treatment of ECs. Different types of ECs with their common sources are shown in Table 1.1. A diagrammatic representation of the objectives of the study related to ECs is shown in Fig. 1.1.

Table 1.1 Different types of emerging contaminants (ECs) with their common sources

S. No	Type of Emerging Contaminants (ECs)	Common sources	Examples
1	Pharmaceutical and personal care products (PPCPs)	Human and veterinary drugs, personal care products (PCPs)	Antibiotics, anti-depressants, painkillers, cosmetics, sunscreens
2	Endocrine-disrupting chemicals (EDCs)	Plastics, industrial chemicals	Bisphenol A (BPA), phthalates, polychlorinated biphenyls (PCBs)
3	Per- and Poly-fluoroalkyl Substances (PFAS)	Non-stick cookware, water-repellent clothing, firefighting foams	Perfluorooctanoic acid (PFOA), perfluorooctane sulfonate (PFOS)
4	Nanomaterials (NMs)	Numerous industrial and consumer products	Carbon nanotubes, silver nanoparticles (Ag-NPs), titanium dioxide nanoparticles (TiO_2-NPs)
5	Industrial chemicals	Manufacturing processes, consumer products	Flame retardants, plasticizers, surfactants
6	Pesticides and herbicides	Agricultural and landscaping functions	Neonicotinoids, glyphosate, atrazine
7	Microplastics (MPs)	Fragmentation of larger plastic debris, personal care products (PCPs), textiles	Microbeads, synthetic fibers
8	Flame retardants	Consumer products, building materials	Polybrominated diphenyl ethers (PBDEs), organophosphate flame retardants (OPFRs)
9	Persistent organic pollutants (POPs)	Industrial processes, waste incineration	Dioxins, furans
10	Personal and household products	Cleaning agents, disinfectants	Triclosan, detergents
11	Artificial musks	Fragrances, personal care products (PCPs)	Artificial musk compounds
12	Veterinary drugs	Livestock manufacture	Antibiotics, antiparasitic, growth promoters

1.2 Sources of Emerging Contaminants

ECs are one of the major pollutants that originate from a variety of anthropogenic sources, for example, agricultural waste, hospital waste, industrial chemicals, laboratory waste, municipal waste, urban runoff, wastewater treatment plants, and improper disposal of products (Banik and Kaur 2024). The most common source of ECs is the industrial and domestic wastewater that pollutes the rivers, lakes, and oceans (Dimpe and Nomngongo 2016). According to Tong et al. (2017), ECs can be stored in sediments and soil for a long time. Volatile organic compounds (VOCs) are known as airborne contaminants that can settle on the surface of water bodies. Recently, it has

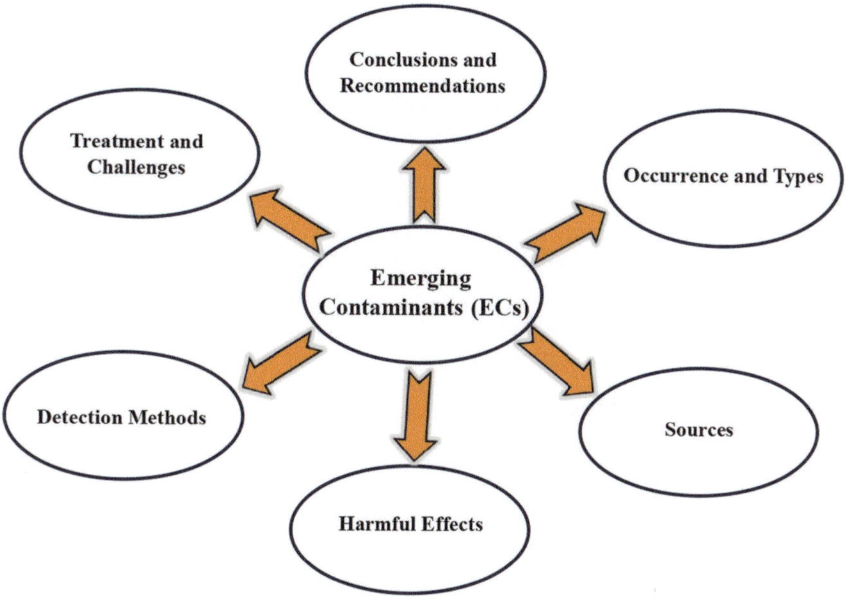

Fig. 1.1 Diagrammatic representation of the study

been observed that industrial and human activities have led to a rapid increase in EC levels found in the environment. The origins and pathways of ECs in the environment are shown in Fig. 1.2.

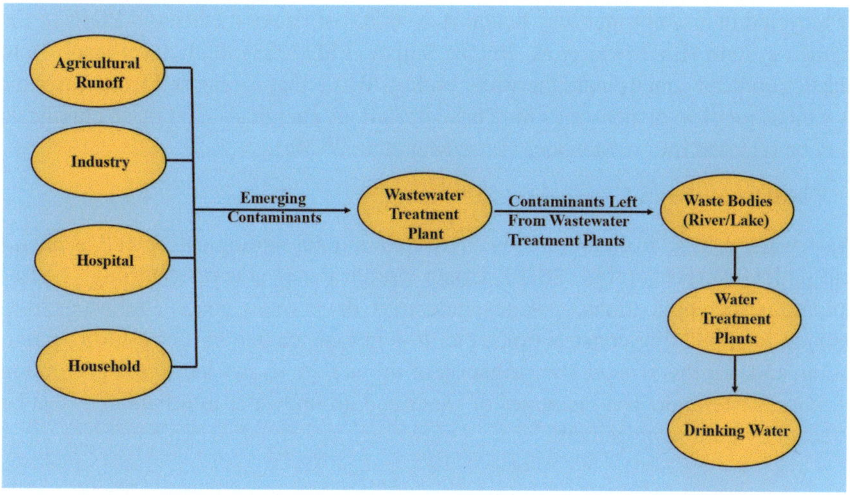

Fig. 1.2 Origins and pathways of emerging contaminants (ECs) in the environment

- **Agriculture Runoff**

Runoffs from agricultural sectors have large quantities of pollutants and then these pollutants enter water bodies (Vymazal and Brezinova 2015). They include animal medicines, metals, herbicides, and pesticides. It has been seen that farmers are using high concentrations of fertilizers and pesticides to increase the production of crops (Garcia-Galan et al. 2021). This over-usage of fertilizers and pesticides has been reported to affect agricultural activities. Nitrogen as well as phosphate-rich fertilizers have been observed to enhance nutrient pollution, thus initiating eutrophication in aquatic ecosystems (Bhardwaj et al. 2023c). Antibiotic agents, pain relievers, and anti-depressants are examples of pharmaceutical products that enter the environment mainly because of agricultural runoff.

- **Wastewater from Industries**

Industrial wastewater is the major trigger of pollution of the water and consists of paints, pesticides, dyes, heavy metals, etc. (Lokhande et al. 2011; Wang and Yang 2016). Textile industries produce wastewater containing dyes, detergents, and microfibers (Ramasamy et al. 2022). Petrochemical industries release oils, grease, and hydrocarbons in their wastewater (Jameel et al. 2011). The worldwide use of drugs and PCPs has increased over the last few years due to advances in research and development, and the rapid growth of the population (Van Boeckel et al. 2014). The first route of PCPs into nature is excreted by humans into the wastewater system. PCPs include chemicals such as parabens and triclosan which are said to find their way into water bodies. Polybrominated flame retardants (PBFR) are used in electrical and electronics industries, and textile and furniture industries, and can be discharged into wastewater and concentrated in living tissues (Bhardwaj et al. 2024b; Bhardwaj et al. 2025a). Electronic waste (e-waste) collected according to unauthorized methods or recycled in an improper way becomes a source of heavy metals and POPs in the aquatic system (Bhardwaj et al. 2025b; Bajpai et al. 2025). Phthalates leach from plastics and can contaminate the water bodies. Waste that is released from hospitals as well as medical dispensaries may have hazardous chemicals, and these chemicals may be released into wastewater (Bhardwaj et al. 2024c).

- **Wastewater from Urban Areas**

Wastewater release from urban areas may contain microbiological as well as chemical pollutants (Hsiao et al. 2020). Urban wastewater is generated through household/domestic applications, sewages, industries, drain water, coagulated water from parking lots, building constructions, etc. It is in essence due to the resulting population, and, for this reason, the irrigation of urban regions has led to the generation of more wastewater. The resources of water and groundwater in urban areas can be affected by poor management.

1.3 Harmful Effects of Emerging Contaminants (ECs)

ECs that are generated from agricultural sources may pose hazardous risks to the aquatic environment and human health (Ouda et al. 2021). The release of these pollutants into water sources poses a threat to water bodies (Arman et al. 2021). They are particularly concerning because conventional drinking water treatment technologies often fail to remove them, leading to their accumulation in water bodies and drinking water sources, which poses risks to both animals and humans. For these contaminants, since less or no knowledge and data exist, these contaminations may be more dangerous (Escher et al. 2020). These contaminants may be very hazardous to the environment and human health if not treated. Due to prolongation and bioaccumulation potential, these pollutants may be hazardous to aquatic organisms. ECs have been seen to affect internal organs, endocrine systems, reproductive issues, developmental problems, and behavioral patterns in living species. The long-term effects of ECs are yet to be fully discovered. They have been associated with increasing cancer risks, especially breast and prostate cancers, and may contribute to population decline in aquatic organisms (Richardson and Kimura 2017). Additionally, they can trigger algal blooms, oxygen depletion, loss of biodiversity, and harm pollinators (Goulson 2013). These contaminants can penetrate cellular membranes and cause cellular damage.

PFAS can cause immune system disorders, cancer, thyroid disease, and liver damage (Gebbink and van Leeuwen 2020). PCPs and their metabolites have been reported in high concentrations in water bodies (Chopra and Kumar 2018). Fetal exposure to these chemicals is through the placenta and breastfeeding (Rolfo et al. 2020), while adults are exposed via food, inhalation, and skin control (Di Nisio and Foresta et al. 2019). Some of the ECs have been associated with neurotoxic impacts on children. BPA and phthalates disrupt the normal functioning of hormones, causing reproductive and developmental problems, and can also induce cancers (Vandenberg et al. 2007). The effects of heavy metals and metalloids in drinking water, soil, and air include skin lesions and tumors, mood swings, carcinogenesis, and damage to the lungs, gastrointestinal tract, central nervous system, circulatory system, reproductive system, liver, and kidney (Silva-Gigante et al. 2023). Organisms known as pharmacophores can contribute to antimicrobial resistance and unwanted neurological impacts whether ECs are present in low concentrations in aquatic systems (Samal et al. 2022).

1.4 Detection Methods of Emerging Contaminants (ECs)

ECs are found in wastewater, municipal sewage, and natural water sources (EL-Kalliny et al. 2023). The concentration of ECs is not monitored systematically. Due to their low concentration and complex chemical structure, the detection of ECs becomes a difficult process. Techniques such as gas chromatography (GC) and high-performance liquid chromatography (HPLC) are used with mass spectrometry (MS) to know the structural information of unknown ECs. These techniques are very sensitive and accurate for the estimation of ECs. The Non-Targeted Screening (NTS) technique is a high-resolution technique for the identification of unknown ECs.

Through spectroscopic techniques like Fourier Transform Infrared Spectroscopy (FTIR) and Raman spectroscopy, different functional groups can be identified in the chemical entities, gases, and, in particular, those VOCs. These techniques are also used for the identification of small-sized plastic particles such as MPs. In both gaseous and liquid forms of the enzymes, electrochemical sensors enable sensation. Biosensors, unlike other sensors, possess some unique elements like protein molecules such as enzymes or antibodies dedicated to sensing biological molecules, such as glucose, toxic molecules, pathogenic organisms, and specific deoxyribonucleic acid (DNA) sequences. Immunoassays make use of specific antibodies that bind to specific EC. This is considered as one of the inexpensive and rapid techniques which has been utilized for initial screening procedures.

1.5 Treatment of Emerging Contaminants (ECs)

Most of the ECs could not be eliminated effectively by using conventional wastewater treatment techniques (Tran et al. 2018). Traditional methodologies include physical, chemical, and biological processes. Techniques such as filtration are considered physical methods, whereas methodologies involving coagulation, flocculation, or disinfection are included under chemical methods (Pandey et al. 2024). Treated wastewater can still contain traces of pharmaceutical compounds, PCPs, and industrial chemicals. Advanced treatment techniques like membrane filtration, reverse osmosis (RO), and oxidation process are better solutions to remove ECs from aquatic systems. RO is used to remove small molecules of ECs.

Advanced oxidation processes (AOPs) like ozonation, photocatalysis, and Fenton's reaction are used to remove/degrade ECs. In the AOP method, active oxygen species, like hydroxyl radicals, break pollutants into less toxic substances. Ozonation is used for water treatment, by removing PCPs and pharmaceutical compounds. ECs could be eliminated from aquatic ecosystems by using two main processes, i.e., adsorption and degradation. During the process of adsorption, the pollutants get attached to or bind to the NM surfaces. This traps and blocks the pollutants from further returning to the ecosystem. Adsorption has been proposed as a long-term remedy for this problem. Although various adsorbents are currently available and efficient in removing ECs, their cost might be prohibitively high. Degradation is the technique in which the transformation of pollutants into less harmful processes occurs. Some commonly used methodologies for degradation processes include advanced oxidation methodologies, biodegradation, bioremediation, chemical treatments, photodegradation, and redox reactions (Mishra et al. 2023).

Phytoremediation and phytoaccumulation are the biological processes to remove ECs from wastewater (Shah 2024a, b). In this technique, absorption and removal of ECs can be performed. Aquatic macrophytes like hyacinths and water lilies can eliminate arsenic (As), cadmium (Cd), and lead (Pb) (Ali et al. 2013). Bioremediation is eco-friendly, inexpensive, and sustainable, as well as a natural way of cleaning polluted sites. In this approach, microorganisms including bacteria and fungi are utilized in the breakdown of anthropogenic products such as hydrocarbons and pesticides, respectively (Vasilachi et al. 2021; Amran et al. 2022). Bioaugmentation is a biological treatment that has a special microorganism for the removal of ECs (Vasilachi et al. 2021).

Effective monitoring and analysis are important to understand the ecological impacts of ECs. Thorough monitoring programs of both chemical and biological processes are used to develop effective management programs (Duarte et al. 2023). Effective management of ECs in marine ecosystems needs advanced technologies (Wang et al. 2023). Environmentally sustainable techniques and green chemistry are required to understand the pollution due to ECs. Green practices are essential for reducing the negative impacts caused by industrial operations and the consumption of resources. Green chemistry suggests that renewable raw materials can be used for process development, energy efficiency can be practiced minimizing risk from hazardous chemicals, and many more such input applications can be included. For example, environmentally friendly, nonorganic solvents such as supercritical carbon dioxide (CO_2) and water samples replace toxic organic solvents. They have been seen reducing pollution levels and improving consumption levels. This is achieved without compromising the safety issues of workers or the environment.

Another sustainable method is the use of a catalyst which facilitates the reaction and gives a high turnover of reactants with less energy. For example, in enzyme catalysis the process of chemical reactions becomes easier under little energy, therefore little energy is used, and toxic products are formed (Rath et al. 2024). By applying these techniques, there will be a way of ensuring that industries have minimum effect on the environment thus over time, supporting sustainable development goals with little exploitation of resources in the future. To overcome the issue of ECs some wastewater management technologies are used, and these technologies include regulation of industrial waste, sustainable agriculture activities, advanced wastewater treatment, and proper disposal of plastic waste and solid waste (Jambeck et al. 2015).

1.6 Challenges in the Treatment of Emerging Contaminants (ECs)

The removal of ECs from aquatic ecosystems is a complex and challenging task. One reason for this difficulty is EC's presence in minute or trace quantities (ng/L or μg/L). Furthermore, ECs exhibit a variety of chemical structures and properties. Sedimentation, filtration, and coagulation are conventional wastewater treatment technologies, and these techniques are not effective for the removal of ECs. Some ECs are resistant to biodegradation and pass through treatment systems unchanged. The advanced treatment technologies such as ozonation, activated carbon adsorption, and AOPs are effective for the removal of ECs but these technologies are expensive and require energy for the operation. It has been reported that some ECs possess the ability to get converted into other by-products as they undergo the removal processes. These by-products might have more detrimental effects on the health of humans and the environment. Therefore, there is a need for standardized methodologies and guidelines for the elimination and monitoring of EC levels.

1.7 Recommendations

Research is used as evidence to make regulations and guidelines that can restrict human exposure to ECs. Knowing the effects of these contaminants assists in safeguarding threatened species and preserving biodiversity. Analyzing and controlling the sources of ECs is paramount in reducing their impact on ecosystems and human health. To cater to the aforementioned issues, there is a need to plan important strategies that would be beneficial in detecting, monitoring, and enhancing waste management and waste treatment systems. Proper waste management as well as pharmaceutical disposal has been reported to be used for source-reducing processes. Developing green chemistry principles could be helpful in reducing pollutants.

Researchers are studying various aspects of ECs to get detailed knowledge about their source, effects, and behaviors. An extensive global approach, including a standardized analytical program and harmonized regulation, is important for the proper management of ECs in aquatic ecosystems. The partnership of research, chemists, biologists, ecologists, engineers, policymakers, and stakeholders can develop an amazing solution. Public awareness and frameworks lead to sustainable practices. There is a requirement for effective communication between the general public and regulatory authorities regarding the risk of ECs. To combat the issue of ECs, continuous research is aimed at enhancing detection technologies, formulating cost-effective and sustainable treatment methods, and strengthening regulatory policies. There is a need to disseminate information regarding ECs and their harmful impact on organisms, to make people aware of the ECs used by them.

To control and degrade these contaminants, sufficient attention must be paid to advanced techniques. To control the source of ECs, strict regulation of industrial discharge, wastewater treatment, use of sustainable practices and clean technology, use of proper farming practices, use of organic fertilizers, create proper distance between agricultural land and water bodies effective and proper waste disposal, public awareness and sustainable coastal planning are required. Sources as well as different pathways of ECs have to be identified to reduce their release and exposure levels in the environment. Industries and consumers need to be encouraged to adopt practices that are sustainable. This would minimize the usage as well as release of different harmful chemicals. Communities have to be empowered to participate in different practices used for monitoring and advocacy. This would lead to increased effective local environmental protection initiatives. Robust frameworks need to be developed. This would be beneficial for assessing various risks that are associated with different ECs for guiding policymakers during framing informed decisions. A regulatory standard could be established by utilizing research findings on ECs. This would be helpful in ensuring different environments as well as public health protection.

1.8 Conclusions

ECs have become a global environmental concern due to their rapidly increasing harmful effects on humans and aquatic organisms. In this chapter, the authors have compiled and highlighted how regular habits, urbanization, industrialization, and other factors have led to the negative interference of ECs in different stages of biological and ecological processes. The main point is to inform society about the effect of ECs on the environment, which affects both organisms and biodiversity and indirectly on our own health. Anthropogenic activities are the primary source of ECs in aquatic systems. The presence of ECs is important for environmental risk assessment and policymaking to safeguard human health and ecosystems from potentially harmful exposures. New pollutants are being added to the aquatic ecosystem through industrial processes and a number of other anthropogenic activities, and it is the primary reason too challenging to design efficient removal strategies. Studying ECs drives innovation in analytical techniques and monitoring technologies, improving detection and quantification. Research collaborations among different disciplines such as biology, chemistry, environmental science studies, and public health would be helpful in promoting and fostering a detailed understanding of different aspects of ECs. Furthermore, identifying the pathways through which ECs enter human health such as through food, air, and water will assist in assessing the health risks associated with exposure to EC levels of these contaminations.

Acknowledgements The authors are thankful to their respective organizations for their support and cooperation in this study.

References

Alam S, Bhardwaj LK, Mallick R, Rai S (2023) Estimation of heavy metals and fluoride ion in vegetables grown nearby the stretch of River Yamuna, Delhi (NCR). India. Indian J Environ Protection 43(1):64–73

Ali H, Khan E, Sajad MA (2013) Phytoremediation of heavy metals—concepts and applications. Chemosphere 91(7):869–881

Amran RH, Jamal MT, Pugazhendi A, Al-Harbi M, Ghandourah M, Al-Otaibi A, Haque MF (2022) Biodegradation and bioremediation of petroleum hydrocarbons in marine ecosystems by microorganisms: a review. Nat Environ Pollution Technol 21(3)

Arman NZ, Salmiati S, Aris A, Salim MR, Nazifa TH, Muhamad MS, Marpongahtun M (2021) A review on emerging pollutants in the water environment Existences, health effects and treatment processes. Water 13(22):3258

Bajpai S, Bhardwaj LK, Rath P, Rahman F, Kumar A, Abdulraheem MI (2025) Electronic wastes and circular economy: a path towards sustainable management. Electronic waste: impact on health, animals, and the environment, pp 173–180. CRC Press

Banik N, Kaur N (2024) Industrial application of hydrotalcite-based materials. Hydrotalcite-based materials: synthesis, characterization and application, p 107

Bashir I, Lone FA, Bhat RA, Mir SA, Dar ZA, Dar SA (2020) Concerns and threats of contamination on aquatic ecosystems. Bioremediation and biotechnology: sustainable approaches to pollution degradation, pp 1–26

Bhardwaj L, Jindal T (2022) Polar ecotoxicology: sources and toxic effects of pollutants. New frontiers in environmental Toxicology, pp 9–14

Bhardwaj L, Sharma S, Ranjan A, Jindal T (2019) Persistent organic pollutants in lakes of Broknes peninsula at Larsemann Hills area, East Antarctica. Ecotoxicology 28:589–596

Bhardwaj LK, Jindal T (2020) Persistent organic pollutants in lakes of Grovnes Peninsula at Larsemann Hill area. East Antarctica. Earth Syst Environ 4(2):349–358

Bhardwaj LK, Sharma A (2021) Estimation of physico-chemical, trace metals, microbiological and phthalate in PET bottled water. Chemistry Africa 4(4):981–991

Bhardwaj LK (2022) Evaluation of bis (2-ethylhexyl) phthalate (DEHP) in the PET bottled mineral water of different brands and impact of heat by GC–MS/MS. Chemistry Africa 5(4):929–942

Bhardwaj LK (2024) Occurrence of microplastics (MPs) in Antarctica and its impact on the health of organisms. Maritime Technol Res 6(2):265418. https://doi.org/10.33175/mtr.2024.265418

Bhardwaj LK, Rath P, Choudhury M (2023a) A comprehensive review on the classification, uses, sources of nanoparticles (NPs) and their toxicity on health. Aerosol Sci Eng 7(1):69–86

Bhardwaj LK, Kumar D, Kumar A (2023b) Phytoremediation potential of Ocimum sanctum: a sustainable approach for remediation of heavy metals. In: Phytoremediation potential of medicinal and aromatic plants, pp 46–57. CRC Press

Bhardwaj LK, Sharma S, Jindal T (2023c) Estimation of physico-chemical and heavy metals in the lakes of Grovnes & Broknes Peninsula, Larsemann Hill. East Antarctica. Chem Africa 6(5):2677–2694

Bhardwaj LK, Rath P, Tokas R (2024b) E-waste management in developing countries: current practices, challenges, disposal, and impact on human health and environment. Sustainable solutions for E-waste and development, pp 90–102

Bhardwaj LK, Rath P, Bajpai S, Upadhyay D, Jain H, Kumar N, Sinha S (2024c) COVID-19 and the interplay with antibacterial drug resistance. In: Frontiers in combating antibacterial resistance: current perspectives and future horizons, pp 246–273. IGI Global

Bhardwaj LK, Rath P, Choudhury M, Metzdorff A, Jodder PK (2025a) Electronic wastes and contamination of water. Electronic waste: impact on health, animals, and the environment, p 48. CRC Press

Bhardwaj LK, Rath P, Jain H, Choudhury M (2025b) Exploring the effects of e-waste on soil, water quality and human health. Discover Civil Eng 2(1):1–16

Bhardwaj LK, Rath P, Yadav P, Gupta U (2024c) Microplastic contamination, an emerging threat to the freshwater environment: a systematic review. Environ Syst Res 13(1):8
Caliman FA, Gavrilescu M (2009) Pharmaceuticals, personal care products and endocrine disrupting agents in the environment–a review. CLEAN–Soil, Air, Water 37(4–5):277–303
Chaturvedi P, Shukla P, Giri BS, Chowdhary P, Chandra R, Gupta P, Pandey A (2021) Prevalence and hazardous impact of pharmaceutical and personal care products and antibiotics in environment: a review on emerging contaminants. Environ Res 194:110664
Cheriyamundath S, Vavilala SL (2021) Nanotechnology-based wastewater treatment. Water Environ J 35(1):123–132
Chopra S, Kumar D (2018) Pharmaceuticals and personal care products (PPCPs) as emerging environmental pollutants: toxicity and risk assessment. Advances in animal biotechnology and its applications, pp 337–353
Das S, Parida VK, Tiwary CS, Gupta AK, Chowdhury S (2024) Emerging Contaminants in the aquatic environment: fate, occurrence, impacts, and toxicity. In: Bioremediation of emerging contaminants in water. Volume 1 (pp 1–32). American Chemical Society. https://doi.org/10.1021/bk-2024-1475.ch001
De Coster S, Van Larebeke N (2012) Endocrine-disrupting chemicals: associated disorders and mechanisms of action. J Environ Public Health 2012(1):713696
Dhangar K, Kumar M (2020) Tricks and tracks in removal of emerging contaminants from the wastewater through hybrid treatment systems: a review. Sci Total Environ 738:140320
Di Nisio A, Foresta C (2019) Water and soil pollution as determinant of water and food quality/contamination and its impact on male fertility. Reprod Biol Endocrinol 17:1–13
Dimpe KM, Nomngongo PN (2016) Current sample preparation methodologies for analysis of emerging pollutants in different environmental matrices. TrAC, Trends Anal Chem 82:199–207
Dong M, He L, Jiang M, Zhu Y, Wang J, Gustave W, Wang S, Deng Y, Zhang X, Wang Z (2023) Biochar for the removal of emerging pollutants from aquatic systems: a review. Int J Environ Res Public Health 20(3):1679
Duarte JAP, Ribeiro AKN, De Carvalho P, Bortolini JC, Ostroski IC (2023) Emerging contaminants in the aquatic environment: phytoplankton structure in the presence of sulfamethoxazole and diclofenac. Environ Sci Pollut Res 30(16):46604–46617
El-Kalliny AS, Abdel-Wahed MS, El-Zahhar AA, Hamza IA, Gad-Allah TA (2023) Nanomaterials: a review of emerging contaminants with potential health or environmental impact. Discover Nano 18(1):68
El-Shahawi MS, Hamza A, Bashammakh AS, Al-Saggaf WT (2010) An overview on the accumulation, distribution, transformations, toxicity and analytical methods for the monitoring of persistent organic pollutants. Talanta 80(5):1587–1597
Escher BI, Stapleton HM, Schymanski EL (2020) Tracking complex mixtures of chemicals in our changing environment. Science 367(6476):388–392
Evich MG, Davis MJ, McCord JP, Acrey B, Awkerman JA, Knappe DR, Lindstrom AB, Speth TF, Tebes-Stevens C, Strynar MJ, Wang Z (2022) Per-and polyfluoroalkyl substances in the environment. Science 375(6580):eabg9065
Fernández P, Grimalt JO (2003) On the global distribution of persistent organic pollutants. Chimia 57(9):514–514
García-Galán MJ, Matamoros V, Uggetti E, Díez-Montero R, García J (2021) Removal and environmental risk assessment of contaminants of emerging concern from irrigation waters in a semi-closed microalgae photobioreactor. Environ Res 194:110278
Gebbink WA, van Leeuwen SP (2020) Environmental contamination and human exposure to PFASs near a fluorochemical production plant: review of historic and current PFOA and GenX contamination in the Netherlands. Environ Int 137:105583
Goh PS, Kang HS, Ismail AF, Khor WH, Quen LK, Higgins D (2022) Nanomaterials for microplastic remediation from aquatic environment: Why nano matters? Chemosphere 299:134418
Goulson D (2013) An overview of the environmental risks posed by neonicotinoid insecticides. J Appl Ecol 50(4):977–987

Gupta P, Sharma A, Bhardwaj LK (2023) Solid waste management (SWM) and its effect on environment & human health. Futuristic trends in agriculture engineering & food sciences 3, 7. IIP Proceedings

Hsiao TC, Lin AYC, Lien WC, Lin YC (2020) Size distribution, biological characteristics and emerging contaminants of aerosols emitted from an urban wastewater treatment plant. J Hazard Mater 388:121809

Jambeck JR, Geyer R, Wilcox C, Siegler TR, Perryman M, Andrady A, Narayan R, Law KL (2015) Plastic waste inputs from land into the ocean. Science 347(6223):768–771

Jameel AT, Muyubi SA, Karim MIA, Alam MZ (2011) Removal of oil and grease as emerging pollutants of concern (EPC) in wastewater stream. IIUM Eng J 12(4)

Khan S, Naushad M, Govarthanan M, Iqbal J, Alfadul SM (2022) Emerging contaminants of high concern for the environment: current trends and future research. Environ Res 207:112609

Kunduru KR, Nazarkovsky M, Farah S, Pawar RP, Basu A, Domb AJ (2017) Nanotechnology for water purification: applications of nanotechnology methods in wastewater treatment. Water purification, pp 33–74

Lokhande RS, Singare PU, Pimple DS (2011) Toxicity study of heavy metals pollutants in waste water effluent samples collected from Taloja industrial estate of Mumbai. India. Resour Environ 1(1):13–19

Mairaj S, Nagar RD, Bhardwaj LK, Rehman F, Punnakal A, Chauhan A, Jindal T (2022) Antioxidant responses of arsenite-induced oxidative stress in rice (Oryza sativa L.) and its modulation by eugenol (extracted from Ocimum sanctum). Indian J Agric Res 56(4):396–400

Mishra SR, Gadore V, Ahmaruzzaman M (2023) A critical review on In2S3-based nanomaterial for emerging contaminants elimination through integrated adsorption-degradation technique: effect of reaction parameters and co-existing species. J Hazardous Mater Lett, 100087

Mofijur M, Hasan MM, Ahmed SF, Djavanroodi F, Fattah IMR, Silitonga AS, Kalam MA, Zhou JL, Khan TY (2024) Advances in identifying and managing emerging contaminants in aquatic ecosystems: analytical approaches, toxicity assessment, transformation pathways, environmental fate, and remediation strategies. Environ Pollut 341:122889. https://doi.org/10.1016/j.envpol.2023.122889

Ouda M, Kadadou D, Swaidan B, Al-Othman A, Al-Asheh S, Banat F, Hasan SW (2021) Emerging contaminants in the water bodies of the Middle East and North Africa (MENA): a critical review. Sci Total Environ 754:142177

Pal A, He Y, Jekel M, Reinhard M, Gin KYH (2014) Emerging contaminants of public health significance as water quality indicator compounds in the urban water cycle. Environ Int 71:46–62

Pandey JK, Tauseef SM, Manna S, Patel RK, Singh VK, Dasgotra A (eds) (2024) Application of nanotechnology for resource recovery from wastewater. CRC Press

Ramasamy R, Aragaw TA, Balasaraswathi Subramanian R (2022) Wastewater treatment plant effluent and microfiber pollution: focus on industry-specific wastewater. Environ Sci Pollut Res 29(34):51211–51233

Rath P, Bhardwaj LK, Yadav P, Bhardwaj AK (2024) A synthesis of biogenic nanoparticles (NPs) for the treatment of wastewater and its application: a review. Biogenic wastes-enabled nanomaterial synthesis: applications in environmental sustainability, pp 127–148

Richardson SD, Kimura SY (2017) Emerging environmental contaminants: challenges facing our next generation and potential engineering solutions. Environ Technol Innov 8:40–56

Rolfo A, Nuzzo AM, De Amicis R, Moretti L, Bertoli S, Leone A (2020) Fetal–maternal exposure to endocrine disruptors: correlation with diet intake and pregnancy outcomes. Nutrients 12(6):1744

Samal K, Mahapatra S, Ali MH (2022) Pharmaceutical wastewater as Emerging Contaminants (EC): treatment technologies, impact on environment and human health. Energy Nexus 6:100076

Shah MP (ed) (2024b) Emerging innovative trends in the application of biological processes for industrial wastewater treatment. Elsevier

Shah MP (2024) Trends in biological processes in industrial wastewater treatment. IOP Publishing

Silva-Gigante M, Hinojosa-Reyes L, Rosas-Castor JM, Quero-Jiménez PC, Pino-Sandoval DA, Guzmán-Mar JL (2023) Heavy metals and metalloids accumulation in common beans (Phaseolus vulgaris L.): a review. Chemosphere 335:139010

Singh SP, Batra RD (2018) Nanotechnology in wastewater treatment: a review. Novel applications in polymers and waste management, pp 173–182

Tlak Gajger I, Dar SA (2021) Plant allelochemicals as sources of insecticides. Insects 12(3):189

Tokas R, Bhardwaj LK, Kumar N, Jindal T (2024) Nanotechnology for sustainable development and future: a review. Green and sustainable approaches using wastes for the production of multifunctional nanomaterials, pp 221–233

Tong L, Qin L, Xie C, Liu H, Wang Y, Guan C, Huang S (2017) Distribution of antibiotics in alluvial sediment near animal breeding areas at the Jianghan Plain, Central China. Chemosphere 186:100–107

Tran NH, Reinhard M, Gin KYH (2018) Occurrence and fate of emerging contaminants in municipal wastewater treatment plants from different geographical regions-a review. Water Res 133:182–207

Van Boeckel TP, Gandra S, Ashok A, Caudron Q, Grenfell BT, Levin SA, Laxminarayan R (2014) Global antibiotic consumption 2000 to 2010: an analysis of national pharmaceutical sales data. Lancet Infect Dis 14(8):742–750

Vandenberg LN, Hauser R, Marcus M, Olea N, Welshons WV (2007) Human exposure to bisphenol A (BPA). Reprod Toxicol 24(2):139–177

Vasilachi IC, Asiminicesei DM, Fertu DI, Gavrilescu M (2021) Occurrence and fate of emerging pollutants in water environment and options for their removal. Water 13(2):181

Vymazal J, Březinová T (2015) The use of constructed wetlands for removal of pesticides from agricultural runoff and drainage: a review. Environ Int 75:11–20

Wang H, Wang Y, Dionysiou DD (2023) Advanced oxidation processes for removal of emerging contaminants in water. Water 15(3):398

Wang Q, Yang Z (2016) Industrial water pollution, water environment treatment, and health risks in China. Environ Pollut 218:358–365

Chapter 2
Sources of Emerging Contaminants

Dolly Sharma, Abhiram Kumar, Amaresh Mishra, Jyoti Chaprana, Rashmi Mishra, and Bhaswati Banerjee

Abstract Emerging contaminants (ECs) are biological and chemical pollutants that are not often identified in the ecosystem but pose a significant hazard to both humans and the environment. The aim of this study is to investigate the distinct classes of ECs from both anthropogenic and natural sources that are infiltrating aquatic systems universally. Additionally, we classified the ECs in a broad range. This chapter deciphers several anthropogenic origins such as industrial wastes, agricultural runoff (pesticides, herbicides, and fertilizers), urban sewage (untreated wastewater contains pharmaceuticals, microplastics (MPs), and harmful microbes), plastic pollution, household chemicals (detergents, personal care products (PCPs), and cleaning agents), and vehicle emissions. Potential difficulties of EC monitoring and control within a variety of aquatic ecosystems are discussed, thereby entailing the call for better surveillance and the development of more advanced wastewater treatment systems in an effort to mitigate the risks of uncontrolled drainage from wastewater treatment and discharge of ECs to water features. In conclusion, this chapter provides better insight into controlling ECs and provides a pollutant-free environment. The targeted audience of this chapter is environmental scientists, policymakers, chemical fertilizers, and the pharmaceutical industry.

D. Sharma · R. Mishra
Department of Biotechnology, Noida Institute of Engineering and Technology, Greater Noida, Uttar Pradesh, India

A. Kumar
Birla Institute of Technology and Sciences Pilani, Hyderabad Campus, Telangana, India

A. Mishra
Department of Biotechnology, Parul Institute of Technology, Parul University, Vadodara, Gujarat, India

J. Chaprana · B. Banerjee (✉)
Department of Environmental Science, SoVSAS, Gautam Buddha University, Greater Noida, Uttar Pradesh, India
e-mail: bhaswati@gbu.ac.in

B. Banerjee
School of Biotechnology, Gautam Buddha University, Greater Noida, Uttar Pradesh, India

H. Jain et al. (eds.), *Emerging Contaminants in the Aquatic Environment*, Advances in Wastewater Research, https://doi.org/10.1007/978-981-96-8548-6_2

Keywords Emerging contaminants (ECs) · Anthropogenic sources · Natural sources · Aquatic environment · Wastewater treatment · Human health · Environment

Abbreviations

AOPs	Advanced oxidation processes
Ag	Silver
As	Arsenic
BPA	Bisphenol A
Cd	Cadmium
ECs	Emerging contaminants
EDCs	Endocrine-disrupting chemicals
ED	Endocrine disruption
F	Fluoride
GC-MS	Gas chromatography-mass spectrometry
HABs	Harmful algal blooms
HBCD	Hexabromocyclododecane
Hg	Mercury
IAS	Invasive alien species
LC-MS	Liquid chromatography-mass spectrometry
MF	Membrane filtration
MPs	Microplastics
PAHs	Polycyclic aromatic hydrocarbons
PBTs	Persistent, bioaccumulative, and toxic substances
Pb	Lead
PBDEs	Polybrominated diphenyl ethers
PCBs	Polychlorinated biphenyls
PE	Polyethylene
PFAS	Per- and poly-fluoroalkyl substances
PFOA	Perfluorooctanoic acid
PFOS	Perfluorooctane sulfonate
PCPs	Personal care products
PCR	Polymerase chain reaction
POPs	Persistent organic pollutants
PPCPs	Pharmaceuticals and personal care products
PS	Polystyrene
TiO_2	Titanium dioxide
WHO	World Health Organization
WWTPs	Wastewater treatment plants

2.1 Introduction

ECs are biological and chemical pollutants that do not naturally occur in the environment and present serious threats to human health and the environment (Khan et al. 2022). These chemicals are medications, cosmetics, industrial chemicals, and agricultural runoff, which can percolate into aquatic environments through multiple routes. The distribution of ECs in aquatic ecosystems is a global issue due to its potential impact on ecological balance and negative health outcomes for humans. EC presence in aqueous systems has been studied for the last two decades, a period of time fueled mainly by the technological progress in the methods of detection that enable the detection of these compounds at large numbers of very low concentrations. This new insight has revealed the possibility of ECs being ambient and entering aquatic organisms or being transported by biomagnification along the food chain. For instance, per- and poly-fluoroalkyl substances (PFAS), also known as "forever chemicals," have been detected worldwide in concentrations in all the relevant water bodies, and there is an ongoing debate regarding their stability and toxicity (Cousins et al. 2022).

The routes of ECs entering aquatic ecosystems are diverse (Hiranmai and Kamaraj 2023). Anthropogenic activities such as unreactive or inadequately treated wastewater disposal, pesticides and fertilizers runoff from agriculture, and industrial discharge are significant contributors. In addition, the inappropriate disposal of medications and cosmetics leads to pharmaceuticals and PCP loadings in water sources. However, natural sources, although less common, still play a role in the presence of ECs in water bodies. The environmental and health implications of ECs are profound. Specific EC exposure has also been associated with adverse health consequences, including endocrine dysfunction, reproductive/developmental, and possibly the risk of certain cancers (Wang et al. 2022). Aquatic organisms, in particular, have also been proven to be sensitive, and ECs influence growth, reproduction, and behavior and, as a consequence, exert a cascading effect on biodiversity and ecosystem services. This is a necessity that calls for the introduction of advanced monitoring and treatment technologies and powerful regulatory standards. Public awareness and education remain significant determinants of proper management and containment of ECs released into the environment. As a result, emerging infections are of great significance in both the health of the aquatic environment and human beings. Research and development, policy regulation, and public communications are all crucial to the mitigation and protection of surface water health. This chapter describes the occurrence, impacts, sources, challenges and future prospects, implications for environmental policy, and resource management for ECs (Fig. 2.1).

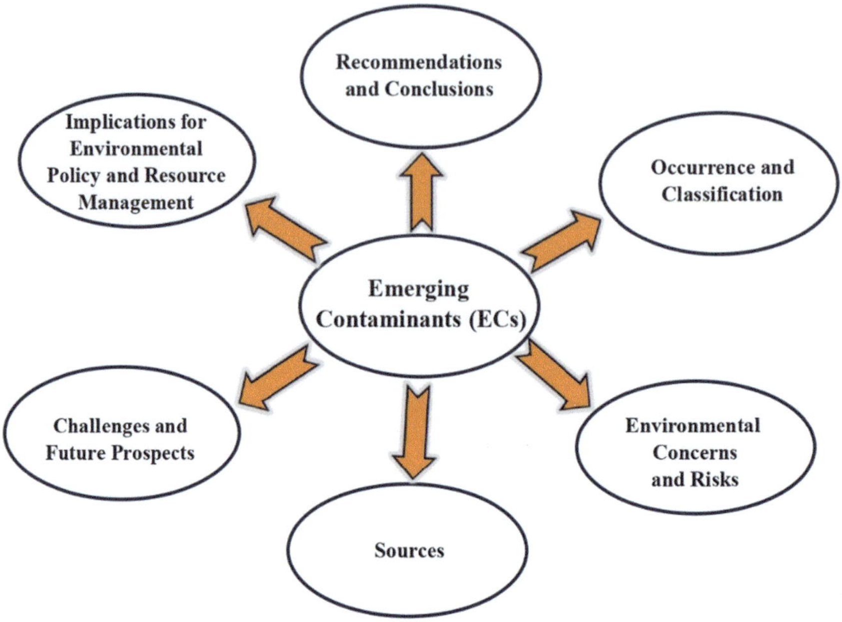

Fig. 2.1 Occurrence, impacts, sources, challenges and future prospects, implications for environmental policy, and resource management for ECs

2.2 Classification of Emerging Contaminants (ECs)

2.2.1 Biological Contaminants

The biological contaminants, as one of the most significant groups of ECs, refer to all microorganisms and biological agents that, when entering an aquatic environment, could lead to a great threat to ecosystem function and human health. The pollutants comprised bacteria, viruses, protozoa, algae, and invasive vegetation, all of which have a specific impact on ecologic impairment and human health concerns. Pathogenic microorganisms, fungi, bacteria, and viruses are the most significant biotic pollutants in water bodies (Prasad and Grobelak 2020). The presence of EC is commonly associated with uncontrolled or under-control wastewater, agricultural runoff, and stormwater discharges. For example, some pathogens (e.g., *Escherichia coli and Salmonella*) can consume contaminants in drinking water (e.g., pathogenic fecal contamination), become contaminants in drinking water, and also become waterborne diseases upon human contact with drinking water (Bhardwaj and Jindal 2022a).

The World Health Organization (WHO) has also stated that waterborne pathogens are the greatest cause of diseases globally, so water treatments and sanitation should be attempted as much as they can in order to reduce their effects (WHO 2022).

Cyanobacterial bloom, or other biological contaminants such as algal bloom, is another important group of biological contaminants (Sukenik and Kaplan 2021). Eutrophication triggered by nutrient-contaminated runoff from agricultural land and urban settlements upgrades the propagation of this algal bloom. Some cyanobacteria are toxigenic, issuing cyanobacteria poisons, which have harmful effects on aquatic ecosystems and human health (Zanchett and Oliveira-Filho 2013). Human exposure has been linked to neurological and hepatic diseases in human mortalities in the water bodies. The increasing frequency and severity of harmful algal blooms (HABs) have been linked with climate change and nutrient pollution, raising the urgency to adopt the full spectrum of management strategies for this problem (Gobler 2020).

Invasive alien species (IAS) are biological contaminants with which ecosystems' balance has substantial deleterious ecological implications. The introduction of alien species into aquatic ecosystems can be triggered by the loss of native biodiversity, habitat structure alteration, and modified ecosystem performance (Çinar et al. 2014). For example, the zebra mussel (*Dreissena polymorpha*), an invasive species that originated from Eastern Europe, has spread to the wild waters around the world, causing enormous ecological and economic damage. IAS can, at the same time, be a source of competition for native species to resources, a reservoir of pathogens, and a shaper of habitat structure, inducing a loss of ecological synergies and services, the sustainability of which is threatened by IAS. Detection and monitoring of biological pollutants are extremely difficult due to their divergency, in addition to their complicated ecological connections (Pourchet et al. 2020).

These pollutions are determined and differentiated both with conventional microbiological techniques and sensitive molecular methods (e.g., polymerase chain reaction (PCR) and metagenomics). However, biological impurities are also vibrant and short-lived, so their levels constantly need to be observed to accurately assess and supervise the hazards correlated with them. Biological contaminant removal tactics are a combination of preventive and removal/clean-up techniques. Powerful (effluent treatment, management of agricultural runoff with best management practices, and water quality monitoring) management of these actions is considered to be of significant importance. Also, public perception of clean waste disposal and the risk of waterborne pathogens are important pathways for source control of contamination. Invasion treatment requires a coordinated effort, including the implementation of broad, comprehensive biosecurity, early warning, and early response measures to prevent establishment and spread.

2.2.2 *Chemical Contaminants*

As one of the most common ECs, chemical contaminants are complex chemical features found in both synthetic and endogenous compounds in aquatic environments, resulting in adverse environmental and human consequences. Identified emerging pollutants include pharmaceuticals and personal care products (PPCPs), industrial chemicals, pesticides, and MPs, which are all biological challenges because of their unique chemical features (presence/forms of uptake and ecotoxicity). PPCPs are increasingly detected in water bodies due to the pervasiveness of use and release through domestic sewer and industrial discharges (Becker and Stefanakis 2017). The environmental stability of antibiotic compounds and analgesic or painkiller compounds is known to cause bacterial resistance against antibiotics and endocrine disruption (ED) in aquatic organisms. Individual exposure to pharmacologically active compounds has been linked to changes in reproductive behavior and physiology in fishes (Sloman and McNeil 2012).

Industrial substances, such as PFAS, are of instance importance since they are constant and bioaccumulative (Parolini et al. 2022). PFAS are utilized in a plethora of user supplies because of their aquatic and oil-repelling properties, which are appropriate (Cook et al. 2023). Ubiquitous presence has been documented globally in water supplies, and this has the potential to contribute to health problems such as high cholesterol, liver disease, and certain cancers (Panieri et al. 2022). Recent regulatory efforts have been directed toward reducing PFAS levels in drinking water to lower the associated risks (Abunada et al. 2020).

Pesticides that leach from agricultural runoff are one of the main routes of chemical stressors (Singh et al. 2023). Herbicides, insecticides, and fungicides can pollute the surface and ground waters and enter non-target individuals as they come from the soil and the surface and groundwater table. Neonicotinoid insecticides have been linked with pollinator decline and poisonousness to freshwater invertebrates (Morrissey et al. 2015). Due to the persistence and mobility of such chemicals, containment, and management of such chemicals, restrictions on monitoring and management of water quality are essential. MPs are plastic particles that are less than 5 mm in size, and they are pervasive contaminants of aquatic environments (Bhardwaj 2024; Bhardwaj et al. 2024, 2025). They are generated from the physical breakdown of large plastic particles. They are able to adsorb and concentrate a greater variety of contaminants and can enter the food chain.

Ingestion of MP by aquatic animals can cause mechanical damage, compromised feeding success, and exposure to toxins (Franzellitti et al. 2019). As MPs are pervasive, integrated waste management strategies and initiatives to increase public awareness are urgently needed to address plastic contamination. Detection and quantification of organic pollutants in aquatic surroundings are made simpler by the use of analytical systems, including liquid chromatography-mass spectrometry (LC–MS) and gas chromatography-mass spectrometry (GC–MS) (Meher and Zarouri 2025).

These systems enable the recognition of pollutants at minimal concentrations that are otherwise unnoticed and provide further insights and data for hazard assessment and regulatory activities. Nevertheless, the continuous release of novel chemicals into the market creates and sustains a challenge for monitoring and assessment mechanisms. Regulatory, technological, and public engagement tools are control strategies for chemical pollutants. Implementing stricter discharge limits, promoting the development of biodegradable and less toxic alternatives, and enhancing wastewater treatment processes are essential steps. In addition, public awareness efforts against improper disposal practices and the environmental hazards associated with chemical agents might reduce contamination sources.

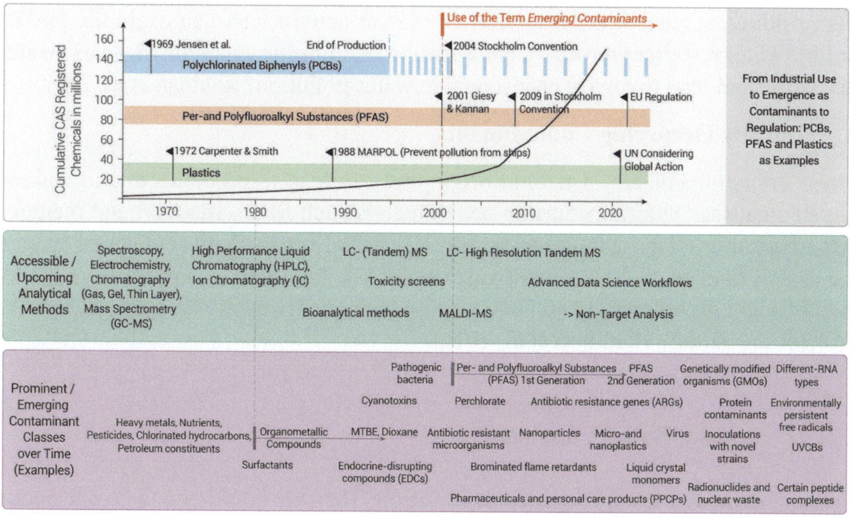

Fig. 2.2 The evolution of emerging contaminants (ECs) about advancements in detection and monitoring of trace-level toxic chemicals: Since the early 2000s, the term "emerging contaminants" (ECs) has been used to describe newly identified classes of pollutants. Substances such as polychlorinated biphenyls (PCBs), per- and poly-fluoroalkyl substances (PFAS), and plastics serve as notable examples of chemicals that have been widely used for decades (represented by gray bars). Over time, these substances were recognized as contaminants (marked by pins) and subsequently regulated or phased out (indicated by the faded-out shadow), though the timelines for restriction varied. The lower panel illustrates how certain ECs emerged as substitutes for previously banned or restricted pollutants, as indicated by the arrows (Wang et al. 2024). (https://doi.org/10.1016/j.xinn.2024.100612) CC by 4.0. e (http://creativecommons.org/licenses/by-nc-nd/4.0/)

The evolution of ECs in relation to advancements in the detection and monitoring of trace-level toxic chemicals is shown in Fig. 2.2.

2.2.3 *Other Classifications Based on Origin and Impact*

ECs are capable of being carefully classified by both the origin and the effect, providing a more focused understanding of their origin and any likely environmental and human health consequences.

2.2.3.1 Origin-Based Classification

- **Anthropogenic Contaminants**

These pollutants are direct resultants of human activities. As an example, PPCPs pollute water resources through a combination of domestic and industrial wastewater discharge and lead to highly unsustainable water pollution (Rahman et al. 2009).

- **Naturally Occurring Contaminants**

These contaminants are of natural origin and can lead to problems due to anthropogenic causes. Natural occurring heavy metals such as arsenic (As) and mercury (Hg) that can enter the aquatic stream as a result of drilling or land application change are cases (Alam et al. 2023; Bhardwaj et al. 2023). Additionally, cyanotoxins generated by algal blooms are natural but can replicate greatly due to nutrient filling from agricultural runoff (Hamilton et al. 2016).

2.2.3.2 Impact-Based Classification

These chemicals are stable against degradation. They are confirmed to be poisonous (e.g., polybutylene terephthalates (PBTs)) and can move extended ranges, stay in the natural environment, and be biomagnified through the food sequence, to become a substantial danger to the ecosystem and human health (Pereira et al. 2015).

- **Endocrine-Disrupting Chemicals (EDCs)**

EDCs reduce the hormonal axis, leading to continuing irregularities, reproductive dysfunction, and neurological and immune interests in animals and humans. Regular EDCs are various pesticides and plasticizers in common produce, such as phthalates (Bhardwaj and Sharma 2021; Bhardwaj 2022).

- **Carcinogenic and Mutagenic Contaminants**

These toxins may be risky and can begin cancer and genetic mutations. These chemicals comprise manufacturing by-products, polycyclic aromatic hydrocarbons (PAHs), and heavy metals (Bhardwaj et al. 2021, 2023; Alam et al. 2023). Introduction to pollutants has been related to long-term health-related effects, including other classes of cancer (Feng et al. 2023).

Knowing ECs from the perspective of cause and consequence is necessary for the improvement of identifiable mitigation measures and policy controls. Stakeholders can converge on preserving the environment and public health by characterizing the beginnings and potential impacts that these pollutants may have.

2.3 Environmental Concerns and Risks

2.3.1 *Impact on Human Health*

The rising presence of ECs in aquatic ecosystems corresponds to a dangerous risk to human well-being as various exposure channels are used, such as drinking water, utilization of poisoned aquatic organisms, and recreational events. These contaminants involve a varied range of compounds such as medicines, cosmetic preparations, industrial chemicals, pesticides, and MPs, which each add to the danger they appear to be. PPCPs are ubiquitous and are present in water bodies due to their extensive use and subsequent discharge through domestic sewage and industrial discharges (Dey et al. 2019). Environmental movement of compounds like drugs/antibiotics/hormones and analgesics can result in accidental human exposure. For example, the existence of antibiotics in the water supplies may lead to the emergence of antibiotic-resistant bacteria and create challenges for public health by diminishing, in a way, the effectiveness of antibiotic treatments (Serwecińska 2020).

Industrial chemicals, e.g., PFAS, are used in many consumer products and are reported in drinking sources worldwide with several health-associated risks (Panieri et al. 2022). A recent study found contamination of PFAS in drinking water reservoirs, putting pressure on the lifelong health effects (Levin et al. 2024). Pesticides from agricultural wastewater are one of the major chemical polluting agents (Saleh et al. 2020). Herbicides, insecticides, and fungicides could contaminate surface and ground waters and, subsequently, non-target organisms, as well as ultimately humans, through drinking water or food. The influence of pesticides has been associated with ED, reproductive and developmental, and carcinogenesis (Birnbaum and Fenton 2003).

2.3.2 *Ecosystem Disruptions*

ECs are of great concern to aquatic ecosystems and pose a serious threat in the form of negative impacts on biodiversity, food webs, and ecosystem performance as a whole. PPCPs (surfactants and other additives with possible cytotoxicity) pollute the water mainly by domestic and crude wastewater discharges. Compounds such as antibiotics, hormones, and analgesics can stay in the surroundings, leading to unintentional introduction to aquatic animals. For example, antibiotics brought into aquatic bodies have been known to speed up the rise of antibiotic-resistant bacteria, which can adversely affect aquatic animals by altering microbial groups and nutrient cycling (Singh et al. 2024). Their presence has been confirmed in all freshwater and marine aquatic environments around the world, with potential risks to the ecosystem, including bioaccumulation in aquatic organisms, with subsequent biomagnification through the food chain impacting trophic levels (Saidon et al. 2024). Certain pesticides have been associated with ED, reproduction and developmental toxicity, and also aquatic mortality.

2.3.3 Global Distribution Patterns

Presence distributions of ECs are complicated and extended over the world, and they are controlled by production rates, application methods, and the stability/movement abilities of ECs within the environment. Their occurrence has been reported in a wide range of aquatic ecosystems around the globe, from highly populated urban settlements to very remote areas, demonstrating their abundance. Exposure of aquatic organisms to an unintended substance is possible through persistence in the environment, antibiotics, hormones, and analgesics. Their global distribution has been reported in water sources the world over with possible ecological risks, including bioaccumulation from aquatic organisms, biomagnification through trophic levels, and impacts on predator species (Saidon et al. 2024). Pesticides spilled in agricultural runoff are also one of the primary causes of chemical pollution.

2.4 Sources of Emerging Contaminants (ECs)

Several ECs with their sources and harmful effects are shown in Table 2.1. The diverse pathways of EC generation, use, and environmental release are represented in Fig. 2.3.

2.4.1 Anthropogenic Sources

Human activities are the primary drivers by which ECs contaminate the aquatic system. These anthropogenic sources add a wide range of contaminants—pharmaceutical compounds, PCPs, industrial chemicals, pesticides, and MPs—to water environments, resulting in serious environmental and public health issues. Human defecation and unlawful removal of fresh medications lead to pharmaceutical compounds entering wastewater treatment plants (WWTPs), which often lack the capability to fully eliminate these complexes. Thus, compounds such as antibiotics, hormones, and analgesics have been observed in samples of surface waters. This leads to interruptions in environmental systems as well as antibiotic conflicts. Industrial effluents possess a variety of substances, for example, solvents, heavy metals, and artificial chemicals. These get emitted into aquatic bodies (Sathya et al. 2022).

Even due to regulatory oversight, spills and non-compliant event points can trigger massive contamination events. Persistent organic pollutants (POPs) have been

Table 2.1 Contribution of various sources to emerging contaminant (EC) pollution

S. no	Emerging contaminants	Sources	Examples	Harmful effects	References
1	Pharmaceuticals	Wastewater treatment plants (WWTPs), hospital effluents, improper disposal	Antibiotics, painkillers, antidepressants	Development of antibiotic-resistant bacteria, endocrine disruption in aquatic organisms	Kümmerer (2009)
2	Personal Care Products (PCPs)	Domestic wastewater, stormwater runoff	Sunscreens, fragrances, preservatives	Toxicity to aquatic species, disruption of endocrine systems, bioaccumulation	Montes-Grajales et al. (2017)
3	Pesticides	Agricultural runoff, urban landscapes	Herbicides, insecticides	Bioaccumulation, neurotoxicity, reproductive and developmental issues in aquatic organisms	Åkerblom (2004)
4	Industrial Chemicals	Effluents from manufacturing plants, spills	Flame retardants, plasticizers (e.g., bisphenol A (BPA), phthalates)	Endocrine disruption, impaired reproduction, bioaccumulation	Wilkinson et al. (2016)
5	Heavy Metals	Industrial discharges, mining activities	Mercury (Hg), lead (Pb), cadmium (Cd)	Neurological damage, bioaccumulation, reduced growth and reproduction in aquatic organisms	Wang et al. (2022)
6	Microplastics (MPs)	Breakdown of larger plastics, wastewater discharges	Polyethylene (PE), polystyrene (PS)	Physical ingestion by aquatic life, chemical toxicity due to associated pollutants	Gallo et al. (2020)
7	Per- and Poly-fluoroalkyl Substances (PFAS)	Industrial and domestic use, fire-fighting foams	Perfluorooctanoic acid (PFOA), perfluorooctane sulfonate (PFOS)	Persistent in the environment, bioaccumulation, liver and thyroid toxicity, potential carcinogenic effects	Mazumder et al. (2023)
8	Nanoparticles	Industrial and consumer products, medical applications	Silver (Ag), titanium dioxide (TiO_2)	Cellular toxicity, oxidative stress, bioaccumulation	Handy et al. (2008)

(continued)

Table 2.1 (continued)

S. no	Emerging contaminants	Sources	Examples	Harmful effects	References
9	Hormones	Human and animal excretion, pharmaceutical disposal	Estrogens, androgens	Endocrine disruption, altered reproductive behaviors in fish and amphibians	da Silva et al. (2018)
10	Flame Retardants	Electronics, textiles, building materials	Polybrominated diphenyl ethers (PBDEs), hexabromocyclododecane (HBCD)	Bioaccumulation, endocrine disruption, developmental neurotoxicity	Hsu et al. (2018)

Fig. 2.3 Schematic representation of the diverse pathways of EC generation, use, and environmental release: Various sectors, including industries, agriculture, households, hospitals, and wastewater treatment plants, contribute to the distribution of emerging contaminants (ECs). These contaminants enter the environment through multiple pathways, such as industrial emissions, agricultural runoff, household waste, medical discharges, and effluent from treatment facilities. Collectively, these sources play a significant role in the widespread dissemination of ECs, impacting ecosystems and public health (Wang et al. 2024). (https://doi.org/10.1016/j.xinn.2024.100612) CC by 4.0. e (http://creativecommons.org/licenses/by-nc-nd/4.0/)

noticed in aquatic surroundings and serve as long-term dangers due to their continuity and bioaccumulation (Bhardwaj et al. 2018, 2019; Bhardwaj and Jindal 2019; Bhardwaj and Jindal 2020, 2022b). Usage of pesticides and fertilizer organic in food manufacturers is directed to runoff that transfers pesticides and fertilizer into immediate streams and rivers, most clearly during precipitation. Pesticides can have bad effects on non-target water species, affecting food webs and environmental events. MPs arise from an array of sources, including the deprivation of macroscopic plastic trash and the application of microbeads in user supplies.

These particles are omnipresent in aquatic systems and can adsorb and transport other pollutants in order to "acquire" the food web. Intake of MP in aquatic organisms can result in physical damage and toxicant exposure, which, in turn, affect growth rate, reproductive success, and survival (Franzellitti et al. 2019). Urban populations make ECs through stormwater discharge that collects pollutants from streets, rooftops, and other impervious surfaces that prevent infiltration to the ground. This runoff can comprise hydrocarbons, heavy metals, and nutrients that can contaminate water bodies and be toxic to aquatic fauna (Sonone et al. 2020). WWTPs are the central objects of ECs. Advanced treatment technologies are required to effectively reduce the release of these contaminants (Morin-Crini et al. 2022; Shah 2023). Mitigation of anthropogenic ECs requires multipronged management solutions such as effective waste management policies, improved wastewater treatment capabilities, and robust regulatory controls to reduce their entry into aquatic environments.

2.4.2 *Natural Sources*

ECs in aquatic environments are mainly released by anthropogenic activity but can also be released naturally. Understanding of such natural sources is the basis of comprehensive environmental risk assessment as well as the initiation of appropriate strategies of mitigation. Some organic compounds, for example, PAHs, have been observed to be products of natural events, such as forest fires or volcanic eruptions. These events release PAHs into the air. They further enter and contaminate different water systems through rainfalls, thereby leading to enhanced pollution levels (Hussain et al. 2018).

Weathering of minerals and geothermal activity are able to provide, naturally, elements and compounds of water. Metals such as As and fluoride (F), for instance, can seep from rock formations into the groundwater, which produces contaminated groundwater with high As and F concentrations that are a health threat to humans and aquatic organisms (Ghosh and Mukhopadhyay 2019). Natural biological activity explains the presence of certain contaminants. In particular, cyanobacterial blooms are toxic to both aquatic organisms as well as humans. This blooming often arises as a result of natural nutrient fluxes but can be boosted by human activities (Singh and Babele 2020). Natural leaching of soils/rocks can leach out in the form of humic acid/metals to groundwater sources. Further, under natural rainfall conditions, runoff can transport organic and inorganic contaminants from the terrestrial to the aquatic environment, and, as a result, a range of associated contaminants may be present (Weldeslassie et al. 2018). Natural sources are a factor in the presence of certain contaminants, but very often, they are amplified by human causes. Identification of and distinguishing between endogenous and anthropogenic sources are prerequisites for reliable environmental surveillance and the implementation of appropriate pollution control strategies.

2.5 Challenges and Future Prospects

EC remediation in aquatic ecosystems has become a complex problem. This requires solutions that are innovative and collaborative. Such challenges require detection complexities, regulatory gaps, and technological limitations, as well as including credible risk assessments. Because of the high concentration of ECs, it becomes difficult to detect and determine their presence in aquatic systems. Traditional analytical methods are less efficient in identifying these contaminants at low concentrations, which leads to an underestimation of the phenomenon's incidence and potential consequences. Enhanced analytical techniques, particularly ultra-high resolution-mass spectrometry, are critical in order to improve the sensitivity of detection and to provide the necessary accurate data for environmental monitoring (Hernandez et al. 2012). There are several ECs that are not sufficiently regulated, and, consequently, monitoring and management differ across regions.

The absence of a uniform guide makes it challenging to coordinate measures to contain EC poisoning. The establishment of comprehensive regulatory frameworks that include a broad spectrum of ECs is essential to achieve effective control strategies that safeguard aquatic environments and human health (Li et al. 2024). Unfortunately, conventional wastewater treatment facilities are not designed to effectively eliminate ECs, and, therefore, discharge remains widespread. Further, the technologies, i.e., membrane filtration (MF), advanced oxidation processes (AOPs), and bioremediation, that can be combined to optimize removal efficiencies are needed. As a result, the high expense and technical complexity of these technologies present a major challenge to adoption on a grand scale. Due to their wide chemical diversity and heterogeneous toxicity profiles, risk assessment of ECs is challenging. Uncertainty is also partially due to the insufficient understanding of the biological chronic effects resulting from the low-level exposure and the possible additive effects of the number of contaminants. Sophisticated risk assessment models combining these risk variables are vitally needed to inform managerial practices and regulatory response.

With regard to future prospects, many of the most challenging issues require an extreme, multidisciplinary solution. The development of sensitive and selective analytical approaches plays an equally important role in the accurate determination of ECs. Advances in sensor technologies and bioanalytical methods hold the promise of real-time detection and monitoring (Deroco et al. 2022). It is essential to formulate comprehensive regulations across the whole range of ECs. In an international collaboration, it is possible to standardize procedures while promoting sound management practices. Increasing awareness not only of where and how but also of the consequences of ECs may promote adaptive behavior with the potential to minimize ECs in the environment.

Educational intervention and interaction with stakeholders are at the heart of this exercise (Melnick et al. 2016). Above all, a study of ECs' environmental behavior and effects is essential. However, emerging treatments will be enhanced by developing sustainable, affordable treatment technologies that will improve our ability to counter their effects (Pal 2018). As such, a concerted action among scientific, regulatory, and public spheres is needed to offset the tendency that ECs pose in aquatic media. By creative problem-solving and group work together, we are able to establish effective approaches for safeguarding water quality while preserving both the environment and human health.

2.6 Implications for Environmental Policy and Resource Management

Environmental policy and resource management are challenged by the pervasive nature of aquatic ECs. These contaminants are still not in full regulatory control and may impact the ecosystem as well as human health. The identification and tracking of ECs are a challenging task due to their widely varying chemical structure and the constraints that exist in present strategies. Traditional water quality monitoring programs may not trace all of these pollutants, leading to their underestimation in terms of their presence and impact. Measurements obtained by such general analytical methods, e.g., high-resolution mass spectrometry, are decisive for the enhancement of detection capacity so that correct information can be provided in environmental monitoring (Hernandez et al. 2012).

Regulatory frameworks, however, are often reactive to new contaminants, with corresponding discrepancies in the monitoring and control protocols in jurisdictions. Considering that there are no standard guidelines, it is hard to define and set up standardized strategies for the mitigation of EC pollution. Nevertheless, the development of general regulatory frameworks for a large number of ECs is critical to guarantee an efficient control mechanism, as well as the protection of the aquatic environment and human health (Puri et al. 2023). Conventional methods of WWTPs have been observed to be less effective against complete decontamination of ECs. Thus, it becomes a significant pollutant in water bodies. There is also a need to increase removal efficiency. This necessitates the requirement of introducing sophisticated treatment systems. These include MF, AOPs, and bioremediation.

However, the elevated price and the practical encounters in the use of these tools still suffer from a substantial obstacle in their widespread use. Chemical structure heterogeneity and varying potencies also pose challenges in EC risk assessment. Lack of data on the long-term effects of low-dose exposure and the potential toxic effects of different toxicants pose uncertainties. There is a demand for the development of holistic risk assessment models on the grounds of these parameters to inform management practices and reinforce control measures.

2.7 Recommendation

Curbing the removal of ECs in aquatic environments requires a multidisciplinary solution. Advanced treatment technologies such as MF, AOPs, and bioremediation should be prioritized in WWTPs to enhance removal efficiencies. The integration of green chemistry principles in designing less toxic and biodegradable alternatives can minimize (i.e., reduce the initial release of ECs). Public health campaigns aimed at the correct disposal of PPCPs are needed in order to reduce contamination at their source. Introducing strict regulatory environments, along with near-real-time monitoring tools, like utilizing biosensors, could help in the process of early detection of ECs and also help in its preventive measures. Furthermore, interdisciplinary interactions among policymakers, environmental scientists, and industrial partners have been seen to be important in generating sustainable approaches for the management of ECs, the health of the ecosystem, and protection of public goods.

2.8 Conclusions

Growing ECs are a significant aspect of the environmental threat posed by their multi-pathway, persistence, and broader environmental footprint from aquatic environments to human health. EC contaminants are now increasingly present in water resources globally and are in urgent need of efficient control measures. The chemical complexity of ECs has been observed to arise due to chemical variability. This leads to different toxicity levels, as well as gives rise to multiple influx pathways into aquatic ecosystems. Anthropogenic activities such as discharge from industries, runoffs from agricultural fields, and improper waste disposal are some of the major sources of EC pollution. Despite its rarity, natural sources also play a role in contaminant behavior. Thus, although the challenge associated with ECs is quite critical, the rapidly growing body of knowledge and technical advances open some promising approaches to deal with the impact. By actively promoting and mutually supporting collaborative research among scientists, policymakers, various industry sectors, and communities, we can protect aquatic ecosystems and public health and provide sustainable access to water for future decades.

Acknowledgements The authors express the most profound gratitude to Dr. Laxmi Kant Bhardwaj and the entire team of editors and reviewers for their invaluable insights and rigorous reviews that significantly enhanced the quality of this chapter. Special thanks to the environmental research community for their ongoing efforts in studying emerging contaminants, which inspired and informed our discussions throughout this work.

References

Abunada Z, Alazaiza MY, Bashir MJ (2020) An overview of per-and polyfluoroalkyl substances (PFAS) in the environment: source, fate, risk and regulations. Water 12(12):3590

Åkerblom N (2004) Agricultural pesticide toxicity to aquatic organisms: a literature review (p 31). Sveriges lantbruksuniv

Alam S, Bhardwaj LK, Mallick R, Rai S (2023) Estimation of heavy metals and fluoride ion in vegetables grown nearby the stretch of River Yamuna, Delhi (NCR). India. Indian J Environ Protection 43(1):64–73

Becker JA, Stefanakis AI (2017) Pharmaceuticals and personal care products as emerging water contaminants. In: Pharmaceutical sciences: breakthroughs in research and practice, pp 1457–1475. IGI Global

Bhardwaj LK, Chauhan A, Ranjan A, Jindal T (2018) Persistent organic pollutants in biotic and abiotic components of Antarctic pristine environment. Earth Syst Environ 2:35–54

Bhardwaj L, Jindal T (2019) AJ Csian OURNALOF HEMISTRY AJ Csian OURNALOF HEMISTRY. Asian J Chem 31(7):1574–1580

Bhardwaj LK, Sharma S, Ranjan A, Jindal T (2019) Persistent organic pollutants in lakes of Broknes peninsula at Larsemann Hills area, East Antarctica. Ecotoxicology 28:589–596

Bhardwaj LK, Jindal T (2020) Persistent organic pollutants in lakes of Grovnes Peninsula at Larsemann Hill area. East Antarctica. Earth Syst Environ 4(2):349–358

Bhardwaj LK, Sharma A (2021) Estimation of physico-chemical, trace metals, microbiological and phthalate in PET bottled water. Chem Africa 4(4):981–991

Bhardwaj LK, Sharma S, Jindal T (2021) Occurrence of polycyclic aromatic hydrocarbons (PAHs) in the lake water at Grovnes Peninsula Over East Antarctica. Chem Africa 4:965–980
Bhardwaj LK (2022) Evaluation of bis (2-ethylhexyl) phthalate (DEHP) in the PET bottled mineral water of different brands and impact of heat by GC–MS/MS. Chem Africa 5(4):929–942
Bhardwaj LK, Jindal T (2022a) Evaluation of coliform and Faecal coliform bacteria in the lakes of Broknes and Grovnes Peninsula, East Antarctica. Nat Environ Pollution Technol, 21
Bhardwaj LK, Jindal T (2022b) Polar ecotoxicology: Sources and toxic effects of pollutants. New Frontiers in environmental Toxicology, pp 9–14
Bhardwaj LK, Sharma S, Jindal T (2023) Estimation of physico-chemical and heavy metals in the lakes of Grovnes & Broknes Peninsula, Larsemann Hill. East Antarctica. Chem Africa 6(5):2677–2694
Bhardwaj LK (2024) Occurrence of microplastics (MPs) in Antarctica and its impact on the health of organisms. Maritime Technol Res 6(2):265418
Bhardwaj LK, Rath P, Yadav P, Gupta U (2024) Microplastic contamination, an emerging threat to the freshwater environment: a systematic review. Environ Syst Res 13(1):8
Bhardwaj LK, Rath P, Rehman F, Yadav P, Massahi T, Mittal R, Jain H (2025) Status of Micro/Nano-Plastic (MNP) pollution in the aquatic system. In: Societal and environmental ramifications of plastic pollution, pp 43–62. IGI Global
Birnbaum LS, Fenton SE (2003) Cancer and developmental exposure to endocrine disruptors. Environ Health Perspect 111(4):389–394
Çinar ME, Arianoutsou M, Zenetos A, Golani D (2014) Impacts of invasive alien marine species on ecosystem services and biodiversity: a pan-European review. Aquat Invasions 9(4):391–423
Cook EK, Brown-Augustine M, Creel E, Raj S (2023) Greener solutions to achieve durable water repellency without using PFASs
Cousins IT, Johansson JH, Salter ME, Sha B, Scheringer M (2022) Outside the safe operating space of a new planetary boundary for per-and polyfluoroalkyl substances (PFAS). Environ Sci Technol 56(16):11172–11179
da Silva APA, De Oliveira CDL, Quirino AMS, Da Silva FDM, de Aquino Saraiva R, Silva-Cavalcanti JS (2018) Endocrine disruptors in aquatic environment: effects and consequences on the biodiversity of fish and amphibian species. Aquat Sci Technol 6:35–51
Deroco PB, Wachholz Junior D, Kubota LT (2022) Recent advances and future trends in bioanalytical chemistry. Tools and trends in bioanalytical chemistry, pp 543–558
Dey S, Bano F, Malik A (2019) Pharmaceuticals and personal care product (PPCP) contamination—a global discharge inventory. In: Pharmaceuticals and personal care products: waste management and treatment technology, pp 1–26. Butterworth-Heinemann
Feng W, Deng Y, Yang F, Miao Q, Ngien SK (2023) Systematic review of contaminants of emerging concern (CECs): distribution, risks, and implications for water quality and health. Water 15(22):3922
Franzellitti S, Canesi L, Auguste M, Wathsala RH, Fabbri E (2019) Microplastic exposure and effects in aquatic organisms: a physiological perspective. Environ Toxicol Pharmacol 68:37–51
Gallo F, Fossi C, Weber R, Santillo D, Sousa J, Ingram I, Nadal A, Romano D (2020) Marine litter plastics and microplastics and their toxic chemicals components: the need for urgent preventive measures. In Analysis of Nanoplastics and Microplastics in Food, pp 159–179. CRC Press
Ghosh G, Mukhopadhyay DK (2019) Human health hazards due to arsenic and fluoride contamination in drinking water and food chain. Groundwater development and management: issues and challenges in South Asia, pp 351–369
Gobler CJ (2020) Climate change and harmful algal blooms: insights and perspective. Harmful Algae 91:101731
Hamilton DP, Salmaso N, Paerl HW (2016) Mitigating harmful cyanobacterial blooms: strategies for control of nitrogen and phosphorus loads. Aquat Ecol 50:351–366
Handy RD, Von der Kammer F, Lead JR, Hassellöv M, Owen R, Crane M (2008) The ecotoxicology and chemistry of manufactured nanoparticles. Ecotoxicology 17:287–314

Hernandez F, Sancho JV, Ibáñez M, Abad E, Portolés T, Mattioli L (2012) Current use of high-resolution mass spectrometry in the environmental sciences. Anal Bioanal Chem 403:1251–1264

Hiranmai RY, Kamaraj M (2023) Occurrence, fate, and toxicity of emerging contaminants<? index value=" emerging contaminants"?> in a diverse ecosystem<? index value=" contaminants"?>. Phys Sci Rev 8(9):2219–2242

Hsu YC, Arcega RAD, Gou YY, Tayo LL, Lin YH, Lin SL, Chao HR (2018) Levels of non-PBDE halogenated fire retardants and brominated dioxins and their toxicological effects in indoor environments-a review. Aerosol Air Qual Res 18(8):2047–2063

Hussain K, Hoque RR, Balachandran S, Medhi S, Idris MG, Rahman M, Hussain FL (2018) Monitoring and risk analysis of PAHs in the environment. Handbook of environmental materials management, pp 1–35

Khan S, Naushad M, Govarthanan M, Iqbal J, Alfadul SM (2022) Emerging contaminants of high concern for the environment: current trends and future research. Environ Res 207:112609

Kümmerer K (2009) The presence of pharmaceuticals in the environment due to human use–present knowledge and future challenges. J Environ Manage 90(8):2354–2366

Levin R, Villanueva CM, Beene D, Cradock AL, Donat-Vargas C, Lewis J, Martinez-Morata I, Minovi D, Nigra AE, Olson ED, Schaider LA (2024) US drinking water quality: exposure risk profiles for seven legacy and emerging contaminants. J Eposure Sci Environ Epidemiol 34(1):3–22

Li X, Shen X, Jiang W, Xi Y, Li S (2024) Comprehensive review of emerging contaminants: detection technologies, environmental impact, and management strategies. Ecotoxicol Environ Saf 278:116420

Mazumder NUS, Hossain MT, Jahura FT, Girase A, Hall AS, Lu J, Ormond RB (2023) Firefighters' exposure to per-and polyfluoroalkyl substances (PFAS) as an occupational hazard: a review. Frontiers Mater 10:1143411

Meher AK, Zarouri A (2025) Environmental applications of mass spectrometry for emerging contaminants. Molecules 30(2):364

Melnick ER, Probst MA, Schoenfeld E, Collins SP, Breslin M, Walsh C, Kuppermann N, Dunn P, Abella BS, Boatright D, Hess EP (2016) Development and testing of shared decision making interventions for use in emergency care: a research agenda. Acad Emerg Med 23(12):1346–1353

Montes-Grajales D, Fennix-Agudelo M, Miranda-Castro W (2017) Occurrence of personal care products as emerging chemicals of concern in water resources: a review. Sci Total Environ 595:601–614

Morin-Crini N, Lichtfouse E, Fourmentin M, Ribeiro ARL, Noutsopoulos C, Mapelli F, Fenyvesi É, Vieira MGA, Picos-Corrales LA, Moreno-Piraján JC, Giraldo L (2022) Removal of emerging contaminants from wastewater using advanced treatments. a review. Environ Chem Lett 20(2):1333–1375

Morrissey CA, Mineau P, Devries JH, Sanchez-Bayo F, Liess M, Cavallaro MC, Liber K (2015) Neonicotinoid contamination of global surface waters and associated risk to aquatic invertebrates: a review. Environ Int 74:291–303

Pal P (2018) Treatment and disposal of pharmaceutical wastewater: toward the sustainable strategy. Sep Purif Rev 47(3):179–198

Panieri E, Baralic K, Djukic-Cosic D, Buha Djordjevic A, Saso L (2022) PFAS molecules: a major concern for the human health and the environment. Toxics 10(2):44

Parolini M, De Felice B, Rusconi M, Morganti M, Polesello S, Valsecchi S (2022) A review of the bioaccumulation and adverse effects of PFAS in free-living organisms from contaminated sites nearby fluorochemical production plants. Water Emerg Contaminants Nanoplastics 1(4):1–13

Pereira LC, de Souza AO, Bernardes MFF, Pazin M, Tasso MJ, Pereira PH, Dorta DJ (2015) A perspective on the potential risks of emerging contaminants to human and environmental health. Environ Sci Pollut Res 22:13800–13823

Pourchet M, Debrauwer L, Klanova J, Price EJ, Covaci A, Caballero-Casero N, Oberacher H, Lamoree M, Damont A, Fenaille F, Vlaanderen J (2020) Suspect and non-targeted screening of chemicals of emerging concern for human biomonitoring, environmental health studies and

support to risk assessment: from promises to challenges and harmonisation issues. Environ Int 139:105545
Prasad MNV, Grobelak A (eds) (2020) Waterborne pathogens: detection and treatment. butterworth-heinemann
Puri M, Gandhi K, Kumar MS (2023) Emerging environmental contaminants: a global perspective on policies and regulations. J Environ Manage 332:117344
Rahman MF, Yanful EK, Jasim SY (2009) Endocrine disrupting compounds (EDCs) and pharmaceuticals and personal care products (PPCPs) in the aquatic environment: implications for the drinking water industry and global environmental health. J Water Health 7(2):224–243
Saidon NB, Szabó R, Budai P, Lehel J (2024) Trophic transfer and biomagnification potential of environmental contaminants (heavy metals) in aquatic ecosystems. Environ Pollut 340:122815
Saleh IA, Zouari N, Al-Ghouti MA (2020) Removal of pesticides from water and wastewater: chemical, physical and biological treatment approaches. Environ Technol Innov 19:101026
Sathya K, Nagarajan K, Carlin Geor Malar G, Rajalakshmi S, Raja Lakshmi P (2022) A comprehensive review on comparison among effluent treatment methods and modern methods of treatment of industrial wastewater effluent from different sources. Appl Water Sci 12(4):70
Serwecińska L (2020) Antimicrobials and antibiotic-resistant bacteria: a risk to the environment and to public health. Water 12(12):3313
Shah MP (2023) Correction to: advanced and innovative approaches of environmental biotechnology in industrial wastewater treatment. In: Advanced and innovative approaches of environmental biotechnology in industrial wastewater treatment, pp C1–C1. Singapore: Springer Nature Singapore
Singh A, Babele PK (2020). Dynamics of harmful cyanobacterial blooms and their toxins: environmental and human health perspectives and management strategies. In: Advances in cyanobacterial biology, pp 301–317. Academic Press
Singh A, Pratap SG, Raj A (2024) Occurrence and dissemination of antibiotics and antibiotic resistance in aquatic environment and its ecological implications: a review. Environ Sci Pollut Res 31(35):47505–47529
Singh NK, Sanghvi G, Yadav M, Padhiyar H, Christian J, Singh V (2023) Fate of pesticides in agricultural runoff treatment systems: occurrence, impacts and technological progress. Environ Res, 117100
Sloman KA, McNeil PL (2012) Using physiology and behaviour to understand the responses of fish early life stages to toxicants. J Fish Biol 81(7):2175–2198
Sonone SS, Jadhav S, Sankhla MS, Kumar R (2020) Water contamination by heavy metals and their toxic effect on aquaculture and human health through food Chain. Lett Appl Nanobioscience 10(2):2148–2166
Sukenik A, Kaplan A (2021) Cyanobacterial harmful algal blooms in aquatic ecosystems: a comprehensive outlook on current and emerging mitigation and control approaches. Microorganisms 9(7):1472
Wang Y, Li Q, Yun X, Zhou J, Wang J (2022) A review on the ecotoxicological effects of heavy metals on aquatic organisms. J Environ Earth Sci 6:148–161
Wang F, Xiang L, Leung KSY, Elsner M, Zhang Y, Guo Y, Pan B, Sun H, An T, Ying G, Brooks BW (2024) Emerging contaminants: a One Health perspective. The Innovation
Weldeslassie T, Naz H, Singh B, Oves M (2018) Chemical contaminants for soil, air and aquatic ecosystem. Modern age environmental problems and their remediation, pp 1–22
Wilkinson JL, Hooda PS, Barker J, Barton S, Swinden J (2016) Ecotoxic pharmaceuticals, personal care products, and other emerging contaminants: a review of environmental, receptor-mediated, developmental, and epigenetic toxicity with discussion of proposed toxicity to humans. Crit Rev Environ Sci Technol 46(4):336–381
World Health Organization (2022). Guidelines for drinking-water quality: Fourth edition incorporating the first and second addenda (4th ed + 1st add + 2nd add). World Health Organization. https://apps.who.int/iris/handle/10665/352532

Zanchett G, Oliveira-Filho EC (2013) Cyanobacteria and cyanotoxins: from impacts on aquatic ecosystems and human health to anticarcinogenic effects. Toxins 5(10):1896–1917

Chapter 3
Transport and Fate of Emerging Contaminants in Aquatic Systems

Laxmi Kant Bhardwaj, Rupam Priya, Moharana Choudhury, Amit Kumar, and Prangya Rath

Abstract Aquatic environments experience continuous increases in the concentration of emerging contaminants (ECs) that drive escalating risks toward human health and ecological systems. Studies of material movement patterns across extensive distances will help create effective programs that decrease their propagation. Water bodies are protected against contaminants through physicochemical transformation operations, biological mechanisms, adsorption processes, and degradation and accumulation activities. These processes manage wastes from wastewater discharges, agricultural runoff, and airborne pollutants. Factors like hydrodynamics, sediment interactions, and microbial activity influence the persistence and distribution of these contaminants. Environmental contaminant behavior strongly depends on temperature-controlled parameters, which modify their natural modifications according to recent research findings. Scientists applied high-resolution mass spectrometry and spectroscopic methods to develop detection systems for analyzing transformation processes. Emphasizing computer models' essential role in predicting pollutant movements throughout the water basins while defining extended operation

L. K. Bhardwaj (✉)
Amity Institute of Environmental Toxicology, Safety and Management, Amity University, Noida, Uttar Pradesh, India
e-mail: bhardwaj.laxmikant@gmail.com

R. Priya
Department of Bioscience and Biotechnology, Banasthali University, Tonk, Rajasthan, India

M. Choudhury
Environmental Research and Management Division, Voice of Environment (VoE), Guwahati, Assam, India

A. Kumar
Molecular Biology and Biotechnology Division, CSIR-National Botanical Research Institute, Lucknow, Uttar Pradesh, India

P. Rath
Amity Institute of Environmental Sciences, Amity University, Noida, Uttar Pradesh, India

H. Jain et al. (eds.), *Emerging Contaminants in the Aquatic Environment*, Advances in Wastewater Research, https://doi.org/10.1007/978-981-96-8548-6_3

damage models. Successful resolution of water pollution requires scientists to integrate various research fields. This integration will enable the development of monitoring systems and risk evaluation tools, supported by next-generation water cleanup methods, to protect aquatic environments and human health.

Keywords Emerging contaminants (ECs) · Aquatic systems · Contaminants Transport & Fate · Human health · Wastewater

Abbreviations

μm	Micrometer
AI	Artificial intelligence
AOPs	Advanced oxidation processes
BMPs	Best Management Practices
CWA	Clean Water Act
ECs	Emerging contaminants
EDCs	Endocrine-disrupting chemicals
EMS	Environmental Management System
FTIR	Fourier Transform Infrared Spectroscopy
GC	Gas chromatography
GC-MS	Gas-chromatography-mass spectrometry
HPLC	High-performance liquid chromatography
HRMS	High-Resolution Mass Spectrometry
IWRM	Integrated Water Resources Management
LC	Liquid chromatography
LC-MS	Liquid chromatography-mass spectrometry
MF	Membrane filtration
ML	Machine learning
MPs	Microplastics
MS	Mass spectrometry
NMs	Nanomaterials
NPs	Nanoparticles
NT	Nanotechnology
PCPs	Personal care products
PFAS	Per- and polyfluoroalkyl substances
PM	Particulate matter
POPs	Persistent organic pollutants
PPCPs	Pharmaceutical and personal care products
ROS	Reactive oxygen species
USA	United States of America
USEPA	United States Environmental Protection Agency
VFAS	Volatile fatty acids
WWTPs	Wastewater treatment plants

3.1 Introduction

Recent environmental pollutants called ECs appear in the environment after detection, yet scientists lack sufficient understanding of their effects, behavior, and regulatory standards (Gavrilescu et al. 2015). The contaminants emerge because of human activities and technological developments. Different industrial activities, agricultural drainage, and domestic waste create emerging environmental contaminants. Most wastewater treatment plants (WWTPs) act as main channels because they do not eliminate contaminants from water before releasing them into the environment. Forest landfill leachates are along with atmospheric chemical depositions and improper chemical disposal patterns, resulting in their release into the environment's water and soil systems (Kumar and Patyal 2025). Pesticides, microplastics (MPs), and endocrine-disrupting chemicals (EDCs) are among the most concerning ECs. They persist in the environment and harm natural systems and human health (Gavrilescu et al. 2015; Gwenzi et al. 2018; Choudhury and Roy 2025).

The various water contaminants group themselves according to chemical makeup and environmental impact patterns. The entrance of antibiotics, hormones, and cosmetics into aquatic reservoirs occurs through sewage systems comprising pharmaceutical and personal care products (PPCPs) (Dey et al. 2019; Chaturvedi et al. 2021). Artificial industrial substances known as per- and polyfluoroalkyl (PFAS) persist in the ecosystem. Over time, plastic debris from synthetic fiber manufacturing or degradation ends up on land and water. Different contaminant groups pose varying environmental and toxicological hazards, including industrial chemicals, nanomaterials (NMs), and insecticides. We must develop innovative monitoring and treatment methods, as many pollutants present serious environmental challenges.

As they linger in the environment and build up across food-chain systems, ECs, which include drugs, pesticides, personal care products (PCPs), and MPs, can generate environmental problems. Wastewater discharges, agricultural drainage, and industrial operations produce pollution that distributes toxins in freshwater and marine habitats. Wastewater management at its basic tier fails to clear multiple pollutants, which permits such contaminants to survive in ecosystems, endangering aquatic environments. EDCs transform fish and amphibian hormonal activities, which causes reproductive disruption in these species. MP-caused environmental destruction escalates since these compounds move up in food chains by registering toxic matter on their surfaces during their journey (Bhardwaj 2024; Bhardwaj et al. 2024).

The rising levels of pollution damage natural environments, both oceanic and ground-based, and pose health risks to humans. Persistent organic pollutants (POPs) and heavy metals in soil harm soil biology and plant health (Bhardwaj et al. 2018; Alam et al. 2023). Food web systems allow biological pollutants to enter different levels where they accumulate for living organisms. Exposure to environmental pollutants for extended periods requires medical attention because it leads to immune dysfunctions, neuron damage, and cancer development risks (Thompson et al. 2015). To protect the environment, we need better tools for evaluation and strict laws that back up sound protection systems. Modern engineering is also needed to devise

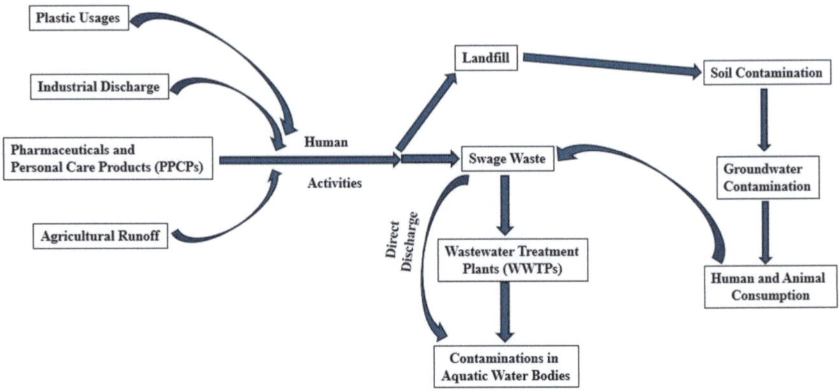

Fig. 3.1 Transport of emerging contaminants (ECs)

ways to stop new pollution sources before they happen. The transport of ECs is well presented in Fig. 3.1.

3.2 Transport Mechanisms of Emerging Contaminants (ECs)

3.2.1 Hydrodynamic Transport (Advection and Dispersion)

ECs paired with MPs move through water systems through advection and dispersion procedures. Bulk movement of contaminants passes through streaming water due to gravity and pressure characteristics that we define as advection. Surface water contaminants move by advection at the same speed as the flowing water, while groundwater transmission depends on the permeability of subsurface strata. The dispersion of environmental pollutants relies on advection, as it influences the spatial distribution patterns of pollutants and can cause significant damage to aquatic ecosystems and water quality. ECs with poor degradability spread over extensive ranges from their original points to various locations, creating secondary risks of biological poisoning and environmental destruction.

Dispersion, conversely, describes the spreading of contaminants caused by velocity variations within the flow. It occurs when molecular diffusion combines with turbulence-driven effects, shear forces, and interactions with porous medium. Plume dispersion stretches along the flow path because of longitudinal diffusion but develops a perpendicular spread from transverse dispersion that decreases concentration over time. Different destinies are applied to aquatic contaminants, which result from advection interacting with dispersion. Exposure risks and remedy strategies depend on the extent of dispersion through the combination of flow velocity with medium heterogeneity and contaminant properties. A detailed understanding of these transport processes is required to conduct predictive analyses and implement efficient mitigation approaches.

3.2.2 Sorption and Desorption Processes

The destiny and movement of ECs within the environment depend primarily on sorption as a controlling process. ECs adhere to solid surfaces, including soil, sediments, and organic matter, by adsorption and partitioning processes, during which contaminants become fixed. Sorption efficiency for ECs depends on their physicochemical properties, temperature values, pH levels, and the nature of compounds present in the system. Organic carbon is a partitioned site for hydrophobic contaminants, yet polar contaminants form bonds through electrostatic and hydrogen interactions (Ahmed et al. 2015). The robust sorption process confines both the movement and the accessibility of ECs, initially preventing near-term environmental consequences resulting in chronic sediment and soil accumulation.

The ability of ECs to return to water or soil solution primarily depends on desorption processes, which control their solution duration and transport ability. Changes in environmental conditions, such as pH variations, ionic strength, and natural organic matter, enhance the release of ECs from soil and sediment. This increases the threat of water pollution and the continued accumulation of ECs in biological systems. Additionally, the environmental retention of ECs is prolonged because some ECs exhibit hysteresis characteristics, causing desorption to occur more slowly than sorption. Accurate prediction of EC behavior depends heavily on a thorough understanding of sorption–desorption processes, which is crucial for risk assessment and site cleanup strategies.

3.2.3 Bioaccumulation and Biomagnification

Bioaccumulation and biomagnification permit an ecosystem impact from ECs that consist of MPs, pharmaceuticals, and pesticides. The prolonged retention of organic substances in an organism's tissue causes their build-up in organisms until they reach their capacity, which results in elevated concentrations. The permanent organic pollutant accumulation in organisms occurs due to the accumulation process inside their biological tissues and organs. Water pollutants have the most detrimental effects on fish and shellfish species because these organisms stay in direct contact with water contaminants (Qadri and Faiq 2020; Malik et al. 2020). A combination of different body tissue contaminants induces endocrine disruption effects while prompting reactive oxygen species (ROS) generation and reproductive interferences.

The concentration of pollutants natural ecological ecosystems receive escalates as food-chain systems advance toward the highest levels. High-level predators track pollutants in prey, and this accumulation results in increased pollutants in their substantial body tissues. Tissue accumulation of toxic substances presents considerable risks to natural predators as well as human consumers of seafood because toxic substances amplify within their body tissues (Bansal 2021). Environmental pollutants, including POPs, heavy metals, and MPs, gradually accumulate as they move upward through different species toward ecological sustainability and human health risks (Bhardwaj and Jindal 2020). The natural accumulation of pollutants helps create protection systems for biodiversity alongside human health measures and regulatory frameworks for pollutant control and pollution reduction.

3.2.4 *Role of Colloids and Particulates*

ECs move through aquatic and terrestrial environments because of their transport by colloids and particulates. Fusing small sizes with a large surface area that promotes high reactivity enables colloidal particles to bind pollutants such as pesticides, pharmaceuticals, and MPs (Ogbeh et al. 2024; Hublikar et al. 2024). The chemical binding between contaminants and colloids increases the movement of substances over longer distances as compared to conditions in which the pollutants exist as dissolved molecules. The stability of colloid-contaminant compounds determines the environmental hazard levels because they maintain slow contaminant release throughout extended periods.

Two types of particulate matter (PM) that exceed colloid dimensions strongly influence the fate of ECs: suspended sediments and MPs. MPs act as concentrated areas for contaminants because they absorb hydrophobic organic pollutants, heavy metals, and POPs, facilitating the long-distance migration of these substances throughout marine and freshwater ecosystems. Environmental requirements determine whether deposited contaminants remain once suspended sediments bind to them or if the sediments later become suspended through environmental changes. The movement patterns of these particulates depend on three key factors, including pH, ionic strength, and organic matter content, because these components manage how much contamination sticks to and unfastens from the matter. By looking at these processes, we can determine how to reduce the risks of new contaminants and evaluate their hazards. The transport process of ECs and its description and key factors are presented well in Table 3.1.

Table 3.1 Transport process of emerging contaminants (ECs) with description and key factors

S. no	Process	Description	Key factors	References
1	Sorption	Adsorption or absorption of contaminants onto sediments or organic matter	Organic carbon content, sediment characteristics, contaminant polarity	Zhang et al. (2018)
2	Degradation	Breakdown of contaminants by chemical, biological, or photolytic processes	pH, temperature, microbial activity, sunlight	Boxall et al. (2012), Luo et al. (2014)
3	Bioaccumulation	Uptake of contaminants by aquatic organisms	Lipophilicity, trophic level, metabolic transformation	Wilkinson et al. (2017), Miller et al. (2020)
4	Photolysis	Decomposition of contaminants under sunlight exposure	Light intensity, wavelength, presence of catalysts	Fenner et al. (2013), Richardson and Kimura (2019)
5	Hydrolysis	Chemical breakdown of contaminants due to reaction with water	pH, temperature, chemical structure	Yamamoto et al. (2009)
6	Volatilization	Transfer of contaminants from water to air	Henry's law constant, temperature, wind speed	Limmer and Burken (2016)
7	Sedimentation	Settling of particulate-bound contaminants to the bottom	Particle size, density, water flow rate	Koelmans et al. (2015), Zhang et al. (2021)
8	Leaching	Movement of contaminants from sediment to groundwater	Permeability, water flow, and contaminant solubility	Jurado et al. (2012), Lapworth et al. (2018)

3.3 Fate of Emerging Contaminants (ECs) in Aquatic Systems

3.3.1 Chemical and Biological Degradation

The future path of ECs in water bodies mainly depends on chemical breakdown methods and biological conversion pathways, which you must fully comprehend (Wilkinson et al. 2017). Chemical decomposition operations, which include hydrolysis, oxidation, and photodegradation, cause alterations to chemical structures through which various toxic substances are formed. Due to light-assisted ROS formation and hydroxyl radical production, surface water contaminants substantially degrade through photodegradation. Hydrolysis breakdown mechanisms transform pollutants, including pharmaceuticals and pesticides, and these processes mainly depend on environmental pH levels and temperature fluctuations. The degradation of products depends heavily on environmental variables since resulting chemical byproducts may exhibit more extended existence or produce more severe detrimental outcomes than their initial compositions.

Microbial metabolic activities break down contaminants into safer breakdown products through enzymatic reactions, leading to biological degradation. Biodegradation occurs when microorganisms use contaminants to obtain energy or carbon sources while producing stabilized breakdown end products (Doukani et al. 2022). The frequency of biodegradation progresses at different rates based on three principal variables: microbial communities, oxygen distribution points, and nutritional needs. Under anoxic circumstances, biodegradation processes sometimes slow down considerably, which causes intermediate compounds to accumulate. Though modern pollutants, including drugs and MPs, show resistance to microbial breakdown, leading to extended environmental hazards, the capacity of biological processes to lower contaminant levels remains constant. Scientists must thoroughly understand the breakdown of chemicals and biological agents to predict how EC survives in aquatic ecosystems.

3.3.2 Photodegradation and Hydrolysis

The breakdown process of ECs in aquatic systems occurs mainly through photodegradation and hydrolysis mechanisms (Benitez et al. 2013). Photodegradation occurs in sunlight and breaks down contaminants through fundamental and peripheral photochemical reactions (Bolton and Cater 2018; Sánchez et al. 2022). Pollutants break down through photochemical processes either by direct photodegradation, where solar radiative energy is absorbed by contaminants, or by indirect photodegradation, in which the hydroxyl radicals or singlet oxygen are generated by photosensitizers and dissolved organic matter. Processes using photodegradation methods at various rates and extents occur based on light intensities and depth of water, along with structural characteristics of contaminants. Highly photoreactive substances transform less threatening chemical compounds or additional toxic substances through degradation (Saravanan et al. 2022).

Water molecules serve as reactants in hydrolysis, a process through which contaminants break down into more minor, more biodegradable compounds. The hydrolysis reaction rate depends on pH levels, temperature, and the specific chemical structure of the pollutants, as some functional groups, such as esters, amides, and carbamates, tend to hydrolyze quickly. Affecting their environmental survival time and biological availability, the hydrolytic process in water bodies serves as a primary mechanism to break down pesticides, medicines, and industrial chemicals. Hydrolysis and photodegradation reduce the concentration of ECs through a joint mechanism, which affects their environmental impact, aquatic ecosystem risk, and potential human health threats (Vasilachi et al. 2021).

3.3.3 *Persistence and Transformation Products*

Multiple environmental processes determine how ECs from pharmaceuticals, pesticides, and PCPs survive and which structures they modify. The endurance of such contaminants against breakdown mechanisms creates long-lasting residence times in water bodies, which defines their persistence. Chemical structures, environmental conditions, and microbial activities influence their longevity. Dangerous environmental problems occur from extremely long-lasting chemicals because they accumulate poisonous contaminants in aquatic organisms. These environmental pollutants persist in the food chains due to their strong ability to stay in the ecosystem. The adverse effects on human health occur because people consume water and seafood (Choudhury et al. 2022).

Different transformation products that form during the degradation and metabolism of ECs often result in comparable or upgraded levels of harm in contrast to their original substances. The hydrolysis, photolysis, and microbial degradation processes form different products through transformation paths. Evaluating risks related to transport pathway formation proves challenging due to complex paths of substance transformation, making substances outlast their original forms while showing unfamiliar environmental behavior patterns (Hemond and Fechner 2022). Improved aquatic ecosystem protection in the long term requires scientists to investigate how ECs persist and transform because this information delivers better water treatment solutions.

3.4 Key Contaminants and Their Behavior

PPCPs represent new contaminants, which consist of medicinal drugs, cosmetic items, and personal hygiene items. Various substances from PPCPs find their way into the environment using wastewater outflows as well as inadequate disposal methods and agricultural drainage (Wang et al. 2022; Nguyen et al. 2023; Choudhury et al. 2025a). The primary environmental pollutants consist of antibiotics, hormones, painkillers, synthetic musks that remain in water areas, and underlying sediments and drinking supplies. The environmental behavior of substances depends on their chemical composition, capability to dissolve, and the rate at which they decompose. PPCPs exist at varying decomposition rates, producing environmental and human health dangers (Ziylan-Yavas et al. 2022). The growing need to manage PPCPs effectively in water requires immediate attention because these substances accumulate in organisms and disrupt hormone systems while developing antibiotic resistance.

Pesticides and herbicides exist as the primary contaminants found in agricultural lands and urban sites because their various properties determine how they remain in the environment and spread through ecosystems (Meftaul et al. 2020). Organophosphates and pyrethroids, together with glyphosate, invade different water systems before causing groundwater and surface water contamination. The life cycle of pesticides depends on their chemical makeup and properties, including soil content, acidity, and microbial activity. While organochlorines persist in the environment due to their hydrophobic nature and tendency to accumulate in biological systems, neonicotinoids exhibit high water solubility, facilitating leaching. Their disintegration also depends much on volatilization, photodegradation, and microbial degradation. The presence of particular pesticide compounds still exists in the environment as harmful substances. They affect unintended species and disturb aquatic ecosystems as they build up along the food chain (Bhardwaj and Jindal 2019).

The different behavioral patterns of MPs and nanoparticles (NPs) establish them as the primary environmental contaminants, leading to the decline of biodiversity and harm to human health (Bhardwaj et al. 2023). MPs result from plastic wear, industrial production, and product usage to develop particles smaller than one micrometer (μm). MPs survive in soil combined with water and air to give toxic chemicals a route as they create habitats for microbes. NPs that measure smaller than 100 nm exhibit quick movements because of their physicochemical reactions that increase their membrane penetration efficiency and accumulation of organisms (Bhardwaj et al. 2023). A wide range of complex reactions occurs between pollutants and compounds, resulting in different levels of pollutant availability and risk for organisms. Their enduring nature and capacity to build up generate substantial harmful ecological consequences and health issues, thus demanding advanced methods of detection and prevention procedures.

The industrial chemical group, which consists of PFAS, acts as a persistent environmental pollutant since it resists decomposition processes effectively. PFAS can be found in firefighting foams, nonstick materials, and water-repellent products because of their carbon–fluorine bonds that deliver superb heat and water and oil defensive properties (Glüge et al. 2020). These compounds display three fundamental natural patterns, including their increased living creatures, capacity to travel large distances through water and air, and continued presence in soil and sediments. Drinking water pollution provoked by volatile fatty acids (VFAS) creates health risks because these chemicals disturb hormone regulation, damage immune systems, and possibly lead to cancer development (Dahiya et al. 2024). The extensive use of PFAS substances and their long-lasting nature creates a significant environmental protection issue since available regulations and cleanup procedures are challenging.

3.5 Environmental and Human Health Implications

Multiple dangerous effects arise from environmental health and human health conditions due to pollutant substances such as pesticides and MPs. Eco-pollutants accumulating in ecosystems create bioaccumulation effects until they occupy more prominent roles in the food chain, influencing human populations and animal groups. Along with other medical problems, pesticides affect endocrine systems in pollinators and aquatic life, which results in reproductive health and neurological issues. Inadequate pollution control produces an oxidative strain that produces an inflammatory response before potentially inducing human carcinogenesis. This problem's resolution needs strong environmental regulations, sustainable replacement options, and improved wildlife monitoring to minimize ecological damage and health consequences (Choudhury and Roy 2025).

The evaluation process to determine pollution risks affecting human and environmental health requires a thorough bioremediation analysis of pollutant availability and accumulation rates. Bioavailability describes the degree to which contaminants become accessible to organisms since this determines their toxic potential based on the absorbability of different substances. The ability of substances to enter organism systems emerges from their chemical makeup and the influence of natural environments and biological adaptive responses. Bioaccumulation is a method that allows pollutants to accumulate in animal tissues at levels exceeding the environment (Jakimska-Nagórska et al. 2011). Specific pesticide applications and MPs pose significant risks to environmental health due to their ability to accumulate in diminishing environmental concentrations, thus enhancing problems for wildlife and human health. Human bodies that receive ongoing toxic exposure develop harmful damage that strikes their neurological system, their endocrine system, and their immune system. Environmental regulations need strict implementation to address ecological and health deterioration caused by these risks, while new materials and improved cleanup methods must be developed and established (Liu et al. 2024).

Environmental conditions caused by simultaneous pesticides and MP occurrence severely affect human health through aquatic food networks. These contaminants guide habitat breakdown in aquatic systems by targeting important food-chain producers such as plankton and algae. Pesticide practices and their detrimental impact on important species produce poisonous accumulation, damaging fish and sea life throughout different eating levels (Sabra and Mehana 2015). The consumption of MPs by tiny aquatic organisms causes mechanical and chemical destruction, while substance pollutants that absorb these plastics lead to the worst type of damage. Consumers who eat seafood face health dangers caused by toxic accumulation and the ongoing adverse outcomes from pollutant transfer between different food-chain tiers, biodiversity reduction, and decreased fish supplies. Protecting public health and aquatic ecosystems depends on law enforcement that combines waste management, sustainable farming practices, and proper environmental standards (Boyd 2003).

3.6 Analytical Approaches for Detection and Monitoring of Emerging Contaminants (ECs)

Advanced spectroscopic and chromatographic tools have become necessary for detecting ECs in complex mixtures. Multiple spectroscopic approaches that combine Raman with Fourier Transform Infrared Spectroscopy (FTIR) and Mass spectrometry (MS) enable researchers to obtain detailed knowledge of molecular structures, recognize functional groups, and precisely identify trace amounts of contaminants. MS operated through both liquid chromatography (LC) and gas chromatography (GC) provides high sensitivity, selectivity, and efficient separation for various samples when performing chemical pollutant detection. Complete assessment analysis occurs through a qualified analytical method combined with quantitative evaluation tools that monitor food safety alongside environmental conditions precisely.

Modern analytical chemistry utilizes united spectroscopic and chromatographic techniques that enable real-time precise contaminant detection with better accurate results and heightened efficiency levels. The scientific analysis uses high-performance liquid chromatography (HPLC) and gas-chromatography-mass spectrometry (GC–MS) instruments in combination with FTIR and Raman spectroscopy to detect ECs precisely in environmental specimens. Trace analysis becomes more effective through hyphenated techniques involving liquid chromatography-mass spectrometry (LC–MS) and GC–MS since these methods combine improved specificity with heightened sensitivity when analyzing complex mixtures. Food safety and water quality protection, regulatory purposes, and risk evaluations depend heavily on these techniques because they identify dangerous pollutants to keep ecological systems secure (Meher and Zarouri 2025).

MS is an analytical method that develops efficient strategies for detecting and tracking pesticides alongside MPs and environmental and biological contaminants (Bhardwaj and Jindal 2020). The detection approaches offer suitable sensitivity levels and specific chemical identification features that allow the detection of chemical compounds at low concentrations. MS assessment depends mainly on GC–MS and LC–MS systems. The analytical techniques mentioned enable researchers to break apart complex blends. High-Resolution Mass Spectrometry (HRMS) incorporates two chief analytical characteristics to distinguish molecules based on weight while revealing precise structural characteristics. MS-based techniques operate crucially for regulatory oversight and forensic work, while research investigations identify contaminants, track environmental degradation, and determine health threats. Enhancements in instrumentation and data processing lead to improved performance of MS-based methods, making detection operations more efficient and accurate (Donato et al. 2012).

Governments have difficulty taking static measurements of pesticide and MP pollution in environmental samples because the chemicals are complicated, and the amounts are tiny. The analytical methods GC–MS and LC–MS give accurate and selective results, but they need expensive equipment and specialized staff to process samples in a complicated way. Various spectroscopic methodologies should be combined with microscopic methods for MP characterization due to their diverse dimensions and constituting polymers. Detection methods become complicated by matrix factors originating from soil, water, and biological tissues, which require specific cleaning procedures. When artificial intelligence (AI) sensors are used, they can connect to high throughput screening platforms, allowing for improved sensitivity, shorter testing times, and real-time observations. Acquiring dependable and comparable study data depends on established standard methods and method harmonization practices.

3.7 Remediation and Mitigation Strategies

Using both treatment and removal methods for wastewater helps protect the environment by removing all the contaminants in industrial, agricultural, and municipal wastewater before it is reused or thrown away. The wastewater treatment sequence begins with physical approaches. However, biological techniques follow these initial steps before the process concludes with chemical and advanced methods to reduce organic content pathogens and dangerous pollutants. Advanced oxidation processes (AOPs), membrane technology, and biological remediation methods are now used to deal with MPs, heavy metals, and POPs. When built wetlands are combined with electrocoagulation technology and nanotechnology (NT) solutions, long-lasting ways to deal with pollution are found. Such protective measures safeguard the aquatic infrastructure and the health of humans and environmental systems.

New monitoring systems based on AI technologies are now essential tools for improving how wastewater is handled by making decisions in real time about the best ways to clean it up (Alprol et al. 2024). The industry and scientific researchers adopt photocatalysis and biochar adsorption with enzymatic degradation because these systems target pesticides and pharmaceutical residues in contaminated areas. Remote areas lacking water face success in wastewater management through decentralized systems that deliver improved water management abilities, power conservation, and reduced operational expenses. The circular economy combines collecting nutrients from wastewater with recycling water to turn polluted water resources into valuable, long-lasting assets for managing water resources (Gogoi and Choudhury 2022).

National management of environmental pollutants depends on two core methods: natural attenuation and biodegradation. The bizarre natural attenuation mechanism requires biological systems to unite with chemical and physical elements that activate automatic pollutant degradation. The successive microbial processes and dilution, sorption, and volatilization functions enable hazardous substance breakdown. Dangerous substances transform into less toxic end products through natural environmental microorganisms with self-regulating decay cycles. Organic pollutants from petroleum hydrocarbons and pesticides undergo their most significant biodegradation process since microbes transform them into carbon dioxide, water, and biomass by utilizing contaminants as an energy source. Natural attenuation functions at low costs, while active environmental monitoring enables microbial breakdown surveillance to ensure no harm exists to humans or natural systems (Surampalli et al. 2004).

Operational strategies for site cleanup receive planning from combined environmental policies working with regulatory mechanisms. Institutions generate rules for controlling pollutant sources by establishing sustainable practices that enforce required procedures during cleanup activities. The REACH EU regulatory system and the United States Environmental Protection Agency (USEPA) pesticide regulations create maximum safe substance boundaries that promote environmentally friendly substitute development (Barceló 1993). Bioremediation policies, cutting-edge filtration methods, and circular economic ideas must all be implemented as a necessary environmental protection system. Sustainable contamination control systems require several vital elements, including stronger enforcement measures, environmentally beneficial technology incentives, and international collaborations.

3.8 Environmental Regulations and Management Strategies

Environmental standards have been studied to contribute an important role in studies associated with the enhancement of water quality as well as the protection of water ecosystems. In countries like the United States of America (USA), the Clean Water Act (CWA) (1972) has been one of the fundamental statutes that regulates pollutant discharge into surface waters. It also sets some effluent standards for each type of industry. It maintains standards for preserving water bodies as they are used for sources of potable water and ground recreation and simultaneously supports aquatic ecosystems. In India, the Environment Protection Act of 1986 prevents and controls pollution in the country; the government authorities are authorized to fix pollution standards and undertake remedial action if necessary. Sustainability and adequate water conservation standards are also associated with such development regulations.

Best Management Practices (BMPs) are efficient, low-cost measures to minimize the emission of pollutants and enhance water quality. Some agricultural BMPs are contouring, the application of buffer strips, and cover crops, which reduce soil erosion and nutrient-polluting effects. Similarly, urban BMPs concern stormwater through green roofs, rain gardens, and permeable pavements. Such practices are increasingly being used to reduce the negative impacts of human activities on water supply. Water quality management should, therefore, embrace an ecological, economic, and social system. There is a tool called Integrated Water Resources Management (IWRM), which encourages the optimal utilization of water and land resources and related sources while simultaneously considering people's needs and the health of the natural environment. Moreover, habitat conservation measures such as wetlands and riparian forest corridors rely on natural means of physical and chemical water treatment and biological diversification. The stated strategies are good examples of how new and environmentally conscious solutions can effectively combat water quality problems in an ever-complicated environment (Choudhury et al. 2025b).

3.9 Case Studies and Practical Implications

(i) ***Examples of Pollutant Transport and Mitigation in Different Aquatic Environments***

A new threat to freshwater ecosystems is MP particles, which enter from the disintegration of more substantial plastic debris and microbeads. They have been reported in rivers, lakes, estuaries, and marine environments, where pollutants and other organisms accumulate. Indian and global researchers have evaluated the emission and removal of these pollutants by field investigations and controlled experiments. Some general control measures that have revealed signs of success include high-efficiency filters in wastewater treatment facilities and strict cardinal policular policies. Moreover, MPs depend on public awareness to support campaigns and industries to adapt and develop environmentally friendly products.

(ii) ***Lessons Learned and Applications in Environmental Management Systems***

Environmental Management System (EMS) has been adopted by many organizations that cut across industries for ecological issues and sustainable development. Successful EMS implementation highlights several key lessons: In the first place, commitment and active management support, role clarification, and resource sufficiency positively predict long-term improvement of available management practices. **In addition, early involvement of stakeholders and interdepartmental collaboration guarantee** the smooth implementation of the EMS practices. These lessons feed into broader environmental programs and seek the development of strong frameworks that tackle the transport of pollutants and enhance the ecosystem health of the water bodies (Dixit et al. 2025).

3.10 Future Perspectives and Research Directions

Rapid progress in pollution measurement techniques allows for better evaluations of environmental and health effects, especially regarding pesticides and MPs. Both detection precision and sensitivity will improve because future research will combine high-resolution spectroscopy with sensor systems and machine learning (ML). AI predictive simulations collaborating with advanced computational models will strengthen pollution risk prediction abilities and ecological pollutant tracking predictions. Using analytical methods in their environment without destructive effects will decrease the time needed for laboratory testing. Innovative contaminant monitoring and mitigation solutions need interdisciplinary collaborations between professionals of environmental science, data analytics, and material science.

Predictions about climate change modifications will affect pollutant movement patterns in ecosystems such as temperatures, rainfall patterns, and extreme weather occurrence patterns. Higher temperatures cause surface pollutants to escape into the air, while floodwater distributes pesticides and MPs through runoff that eventually seeps into water bodies through soil leaching. Pollutant concentrations increase within different soil and sediment layers because droughts force them to alter their chemical makeup, which affects their time spent in the environment. Predictive models must combine weather prediction information with contaminant movement standards to study human avoidance measures and weak points in the environmental system. Scientific studies about how climate changes affect ECs during behavioral phases will contribute to building sustainable environmental management regulations.

The complete sustainability governance framework demands legal systems that follow pollutants throughout their manufacturing process until they reach their waste disposal stages. Future research should prioritize the creation of green materials, better tracking systems for pollution dispersion, and enhanced waste management structure development. Research-based environmental risk reduction requires policymakers to work with industry representatives to create a multidisciplinary approach to locating and resolving regulatory gaps. Sustainable practice incentives combined with evidence-based global policies will decrease environmental and human health risks as they work toward long-term sustainability objectives.

3.11 Recommendations

To manage ECs in aquatic systems, a combination of activated carbon adsorption, AOP, and membrane filtration (MF) technology should be used. Sequential wastewater treatment eliminates pharmaceuticals, pesticides, and MPs, protecting natural water bodies. Governments need enhanced monitoring programs to determine maximum levels of ECs from industrial, agricultural, and municipal sources. Environmental impacts decrease as individuals choose low-toxicity and easily decomposable products. A monitoring system is essential for tracking change in cross-environmental pollution. Long-term research on the EC effects on aquatic ecosystems will provide vital information for informed decision-making. Proper disposal practices will become standard when citizens participate in awareness programs to prevent contamination sources. Success in controlling water contaminants relies on integrating technological advancements with policy guidelines and community participation strategies.

3.12 Conclusions

Aquatic ecosystems trigger physical and chemical transformations of ECs through environmental elements combined with biological and physical operational mechanisms. Different transformation processes, including adsorption, degradation, and bioaccumulation, affect the enduring nature of PPCPs and MPs in ecological environments. Hydrodynamic processes and biogeochemical modifications of sediments influence the situation and dispersion rate of contaminants in the environment. Accurately predicting contaminant behavior requires understanding relevant mechanisms to mitigate risks to aquatic ecosystems and human health.

Developing innovative detection systems requires parallel research alongside investigation of pollution effects on the environment and removal processes to reduce environmental pollution levels. Advanced models and interdisciplinary research practices allow improved pollutant environmental behavior prediction determination. Operational improvements in regulatory standards and effective, sustainable wastewater strategies will reduce the number of ECs released into aquatic bodies. Collaborative actions between policy development processes and science-based approaches will better protect long-term aquatic ecosystem health.

Acknowledgements The authors are thankful to their respective organizations for their support and cooperation in this study.

References

Ahmed AA, Thiele-Bruhn S, Aziz SG, Hilal RH, Elroby SA, Al-Youbi AO, Leinweber P, Kühn O (2015) Interaction of polar and nonpolar organic pollutants with soil organic matter: sorption experiments and molecular dynamics simulation. Sci Total Environ 508:276–287

Alam S, Bhardwaj LK, Mallick R, Rai S (2023) Estimation of heavy metals and fluoride ion in vegetables grown nearby the stretch of River Yamuna, Delhi (NCR). India. Indian J Environ Protection 43(1):64–73

Alprol AE, Mansour AT, Ibrahim MEED, Ashour M (2024) Artificial intelligence technologies revolutionizing wastewater treatment: current trends and future prospective. Water 16(2):314

Bansal OP (2021) Accumulation of potentially toxic metals and pesticides in fish other seafood and their impact on human. Int J Recent Scientific Res 12(8):42848–42868

Barceló D (1993) Environmental Protection Agency and other methods for the determination of priority pesticides and their transformation products in water. J Chromatogr A 643(1–2):117–143

Benitez FJ, Acero JL, Real FJ, Roldan G, Rodriguez E (2013) Photolysis of model emerging contaminants in ultra-pure water: kinetics, by-products formation and degradation pathways. Water Res 47(2):870–880

Bhardwaj L, Chauhan A, Ranjan A, Jindal T (2018) Persistent organic pollutants in biotic and abiotic components of Antarctic pristine environment. Earth Syst Environ 2:35–54

Bhardwaj LK, Jindal T (2019) Contamination of lakes in Broknes peninsula, East Antarctica through the pesticides and PAHs. Asian-J Chem 31(7):1574–1580

Bhardwaj LK, Jindal T (2020) Persistent organic pollutants in lakes of Grovnes Peninsula at Larsemann Hill area. East Antarctica. Earth Syst Environ 4(2):349–358

Bhardwaj LK (2024) Occurrence of microplastics (MPs) in Antarctica and its impact on the health of organisms. Maritime Technol Res 6(2):265418–265418

Bhardwaj LK, Rath P, Choudhury M (2023) A comprehensive review on the classification, uses, sources of nanoparticles (NPs) and their toxicity on health. Aerosol Sci Eng 7(1):69–86

Bhardwaj LK, Rath P, Yadav P, Gupta U (2024) Microplastic contamination, an emerging threat to the freshwater environment: a systematic review. Environ Syst Res 13(1):8

Bolton JR, Cater SR (2018) Homogeneous photodegradation of pollutants in contaminated water: an introduction. In: Aquatic and surface photochemistry, pp 467–490. CRC Press

Boxall AB, Rudd MA, Brooks BW, Caldwell DJ, Choi K, Hickmann S, Innes E, Ostapyk K, Staveley JP, Verslycke T, Ankley GT (2012) Pharmaceuticals and personal care products in the environment: what are the big questions? Environ Health Perspect 120(9):1221–1229

Boyd CE (2003) Guidelines for aquaculture effluent management at the farm-level. Aquaculture 226(1–4):101–112

Chaturvedi P, Shukla P, Giri BS, Chowdhary P, Chandra R, Gupta P, Pandey A (2021) Prevalence and hazardous impact of pharmaceutical and personal care products and antibiotics in environment: a review on emerging contaminants. Environ Res 194:110664

Choudhury M, Roy P (2025) Challenges with microplastic pollution in the regime of UN sustainable development goals. World development sustainability, p 100216

Choudhury M, Adhikari MD, Agarwal S, Samanta P, Sharma A, Kundu D, Kumar S (2025a) Pharmaceuticals and personal care products in soil: sources, impacts and Myco-remediation strategies. Emerging contaminants, p 100488

Choudhury M, Majumdar S, Goswami S, Sillanpää M (eds) (2025b) Smart wastewater systems and climate change: innovations through spatial intelligence (Vol 15). Royal Society of Chemistry

Choudhury M, Sharma A, Pervez A, Upadhyay P, Dutta J (2022) Growing menace of microplastics in and around the coastal ecosystem. Coastal ecosystems: environmental importance, current challenges and conservation measures, pp 117–137

Dahiya P, Kumari S, Behl M, Kashyap A, Kumari D, Thakur K, Devi M, Kumari N, Kaushik N, Walia A, Bhatt AK (2024) Guardians of the Gut: harnessing the power of probiotic microbiota and their exopolysaccharides to mitigate heavy metal toxicity in human for better health. Probiotics and antimicrobial proteins, pp 1–17

Dey S, Bano F, Malik A (2019) Pharmaceuticals and personal care product (PPCP) contamination—a global discharge inventory. In: Pharmaceuticals and personal care products: waste management and treatment technology, pp 1–26. Butterworth-Heinemann

Dixit G, Majumdar S, Choudhury M (2025) Introduction to ecosystem services for climate resilience: a theoretical framework. In: Sustainable synergy: harnessing ecosystems for climate resilience, pp 3–16. Cham: Springer Nature Switzerland

Donato P, Cacciola F, Tranchida PQ, Dugo P, Mondello L (2012) Mass spectrometry detection in comprehensive liquid chromatography: basic concepts, instrumental aspects, applications and trends. Mass Spectrom Rev 31(5):523–559

Doukani K, Boukirat D, Boumezrag A, Bouhenni H, Bounouira Y (2022) Biodegradation of pollutants. Handbook of Biodegradable Materials. Springer International Publishing, Cham, pp 1–27

Fenner K, Canonica S, Wackett LP, Elsner M (2013) Evaluating pesticide degradation in the environment: blind spots and emerging opportunities. Science 341(6147):752–758

Gavrilescu M, Demnerová K, Aamand J, Agathos S, Fava F (2015) Emerging pollutants in the environment: present and future challenges in biomonitoring, ecological risks and bioremediation. New Biotechnol 32(1):147–156

Glüge J, Scheringer M, Cousins IT, DeWitt JC, Goldenman G, Herzke D, Lohmann R, Ng CA, Trier X, Wang Z (2020) An overview of the uses of per-and polyfluoroalkyl substances (PFAS). Environ Sci Process Impacts 22(12):2345–2373

Gogoi N, Choudhury M (2022) Green materials synthesis for wastewater treatment. In: Encyclopedia of green materials, pp 1–14. Singapore: Springer Nature Singapore

Gwenzi W, Mangori L, Danha C, Chaukura N, Dunjana N, Sanganyado E (2018) Sources, behaviour, and environmental and human health risks of high-technology rare earth elements as emerging contaminants. Sci Total Environ 636:299–313

Hemond HF, Fechner EJ (2022) Chemical fate and transport in the environment. Academic Press

Hublikar LV, Shilar FA, Mathada BS, Ganachari SV (2024) Comprehensive investigation of green solutions for sustainable wastewater remediation: a review. J Molecular Liquids, p 124532

Jakimska-Nagórska A, Konieczka P, Skóra K, Namieśnik J (2011) Bioaccumulation of metals in tissues of marine animals, part I: the role and impact of heavy metals on organisms. Pol J Environ Stud 20:1117–1125

Jurado A, Vàzquez-Suñé E, Carrera J, de Alda ML, Pujades E, Barceló D (2012) Emerging organic contaminants in groundwater in Spain: a review of sources, recent occurrence and fate in a European context. Sci Total Environ 440:82–94

Koelmans AA, Besseling E, Shim WJ (2015) Nanoplastics in the aquatic environment. Critical review. Marine anthropogenic litter, pp 325–340

Kumar A, Patyal A (2025) Sources and hazardous effects of environmental contaminants. Epidemiology and environmental hygiene in veterinary public health, p 309

Lapworth DJ, Das P, Shaw A, Mukherjee A, Civil W, Petersen JO, Gooddy DC, Wakefield O, Finlayson A, Krishan G, Sengupta P (2018) Deep urban groundwater vulnerability in India revealed through the use of emerging organic contaminants and residence time tracers. Environ Pollut 240:938–949

Limmer M, Burken J (2016) Phytovolatilization of organic contaminants. Environ Sci Technol 50(13):6632–6643

Liu X, Sathishkumar K, Zhang H, Saxena KK, Zhang F, Naraginiti S, Rajendiran R, Aruliah R, Guo X (2024) Frontiers in environmental cleanup: recent advances in remediation of emerging pollutants from soil and water. J Hazardous Mater Adva, 100461

Luo Y, Guo W, Ngo HH, Nghiem LD, Hai FI, Zhang J, Liang S, Wang XC (2014) A review on the occurrence of micropollutants in the aquatic environment and their fate and removal during wastewater treatment. Sci Total Environ 473:619–641

Malik DS, Sharma AK, Sharma AK, Thakur R, Sharma M (2020) A review on impact of water pollution on freshwater fish species and their aquatic environment. Adv Environ Pollution Manag Wastewater Impacts Treatment Technol 1:10–28

Meftaul IM, Venkateswarlu K, Dharmarajan R, Annamalai P, Megharaj M (2020) Pesticides in the urban environment: a potential threat that knocks at the door. Sci Total Environ 711:134612

Meher AK, Zarouri A (2025) Environmental applications of mass spectrometry for emerging contaminants. Molecules 30(2):364

Miller ME, Hamann M, Kroon FJ (2020) Bioaccumulation and biomagnification of microplastics in marine organisms: a review and meta-analysis of current data. PLoS ONE 15(10):e0240792

Nguyen MK, Lin C, Nguyen HL, Hung NTQ, La DD, Nguyen XH, Chang SW, Chung WJ, Nguyen DD (2023) Occurrence, fate, and potential risk of pharmaceutical pollutants in agriculture: challenges and environmentally friendly solutions. Sci Total Environ 899:165323

Ogbeh GO, Ogunlela AO, Akinbile CO, Iwar RT (2024) Adsorption of organic micropollutants in water: a review of advances in modelling, mechanisms, adsorbents, and their characteristics. Environ Eng Res 30:230733

Qadri R, Faiq MA (2020) Freshwater pollution: effects on aquatic life and human health. Fresh water pollution dynamics and remediation, pp 15–26

Richardson SD, Kimura SY (2019) Water analysis: emerging contaminants and current issues. Anal Chem 92(1):473–505

Sabra FS, Mehana ESED (2015) Pesticides toxicity in fish with particular reference to insecticides. Asian J Agric Food Sci 3(1)

Sánchez M, Ruiz I, Soto M (2022) The potential of constructed wetland systems and photodegradation processes for the removal of emerging contaminants—a review. Environments 9(9):116

Saravanan A, Kumar PS, Jeevanantham S, Anubha M, Jayashree SJEP (2022) Degradation of toxic agrochemicals and pharmaceutical pollutants: effective and alternative approaches toward photocatalysis. Environ Pollut 298:118844

Surampalli RY, Ong SK, Seagren E, Nuno J, Banerji S (eds) (2004) Natural attenuation of hazardous wastes. ASCE Publications

Thompson PA, Khatami M, Baglole CJ, Sun J, Harris SA, Moon EY, Al-Mulla F, Al-Temaimi R, Brown DG, Colacci AM, Mondello C (2015) Environmental immune disruptors, inflammation and cancer risk. Carcinogenesis 36(Suppl_1):S232–S253

Vasilachi IC, Asiminicesei DM, Fertu DI, Gavrilescu M (2021) Occurrence and fate of emerging pollutants in water environment and options for their removal. Water 13(2):181

Wang C, Lu Y, Sun B, Zhang M, Wang R, Li X, Mao R, Cao Z, Song S (2022) Contamination, transport, and ecological risks of pharmaceuticals and personal care products in a large irrigation region. Sci Total Environ 851:158179

Wilkinson J, Hooda PS, Barker J, Barton S, Swinden J (2017) Occurrence, fate and transformation of emerging contaminants in water: an overarching review of the field. Environ Pollut 231:954–970

Yamamoto H, Nakamura Y, Moriguchi S, Nakamura Y, Honda Y, Tamura I, Hirata Y, Hayashi A, Sekizawa J (2009) Persistence and partitioning of eight selected pharmaceuticals in the aquatic environment: laboratory photolysis, biodegradation, and sorption experiments. Water Res 43(2):351–362

Zhang H, Wang J, Zhou B, Zhou Y, Dai Z, Zhou Q, Chriestie P, Luo Y (2018) Enhanced adsorption of oxytetracycline to weathered microplastic polystyrene: Kinetics, isotherms and influencing factors. Environ Pollut 243:1550–1557

Zhang Q, Liu T, Liu L, Fan Y, Rao W, Zheng J, Qian X (2021) Distribution and sedimentation of microplastics in Taihu Lake. Sci Total Environ 795:148745

Ziylan-Yavas A, Santos D, Flores EMM, Ince NH (2022) Pharmaceuticals and personal care products (PPCPs): environmental and public health risks. Environ Prog Sustain Energy 41(4):e13821

Chapter 4
Health and Environmental Hazards

Erukulla Priyanka, Abhiram Kumar, and Kumar Pranav Narayan

Abstract The chapter explores the intricate relationship between health and environmental hazards, emphasizing how ecological factors significantly influence public health outcomes. Environmental hazards encompass a broad range of elements, including pollution, climate change, and natural disasters. Chemicals such as pharmaceutical wastes, personal care products (PCPs), microplastics (MPs), and nanomaterials (NMs) are detected in groundwater, soil, and air, which leads to an imbalance in the ecological system and directly or indirectly impacts human health. Traditional environmental monitoring methods often fail to adequately identify these chemicals, which can link exposure to serious health conditions such as cancer, neurological disorders, reproductive issues, and endocrine disruptions, all of which pose risks to human health. However, we also elaborate on some new detection techniques to easily monitor water, soil, and air pollution. Currently, pollutants and environmental hazards are the critical challenges for their management in metropolitan cities. However, emphasizes the urgent need for advanced detection technologies, rigorous regulatory frameworks, and sustainable waste management practices to address the challenges posed by emerging pollutants. Proactive and interdisciplinary approaches are essential for minimizing the associated risks and protecting both public health and environmental quality. This chapter discusses various pollutants, such as particulate matter (PM), toxic chemicals, pharmaceutical wastes, emerging contaminants (ECs), and their direct associations with chronic illnesses like respiratory diseases, cardiovascular disorders, and cancer. In conclusion, it asserts that addressing health and environmental hazards is essential for achieving a sustainable and equitable future.

Keywords Environmental hazards · Emerging contaminants (ECs) · Air pollution · Pharmaceutical wastes · Human health

---

Erukulla Priyanka and Abhiram Kumar are contributed equally.

E. Priyanka · A. Kumar · K. P. Narayan (✉)
Birla Institute of Technology and Sciences Pilani, Hyderabad Campus, Hyderabad, India
e-mail: pranav@hyderabad.bits-pilani.ac.in

H. Jain et al. (eds.), *Emerging Contaminants in the Aquatic Environment*, Advances in Wastewater Research, https://doi.org/10.1007/978-981-96-8548-6_4

Abbreviations

ATSDR	Agency for Toxic Substances and Disease Registry
AMR	Antimicrobial Resistance
BFRs	Brominated flame retardants
BPA	Bisphenol A
CPs	Chlorinated paraffins
DDT	Dichlorodiphenyltrichloroethane
DNA	Deoxyribonucleic acid
ECs	Emerging contaminants
EDCs	Endocrine-disrupting chemicals
EDs	Endocrine disruptors
EU	European Union
GC-MS	Gas chromatography-mass spectrometry
GHGs	Greenhouse gases
GIS	Geographic Information Systems
HBCDD	Hexabromocyclododecane
Hg	Mercury
HSPs	Heat shock proteins
Ig	Immunoglobulin
IPM	Integrated pest management
IoT	Internet of Things
LC-MS	Liquid chromatography-mass spectrometry
Li	Lithium
MIC	Methyl isocyanate
ML	Machine learning
Mn	Manganese
MPs	Microplastics
NCDs	Non-Communicable diseases
NMs	Nanomaterials
NPs	Nanoparticles
NT	Nanotechnology
OPEs	Organophosphate esters
OPs	Organophosphates
PAHs	Polycyclic Aromatic Hydrocarbons
Pb	Lead
PBDEs	Polybrominated diphenyl ethers
PCAs	Polychlorinated alkanes
PCBs	Polychlorinated biphenyls
PCCs	Personal care compounds
PCNs	Polychlorinated naphthalenes
PCPs	Personal care products
PFAS	Perfluoroalkyl substances
PFCs	Perfluorinated chemicals

PFOA	Perfluorooctanoate
PFOS	Perfluoro octane sulfonate
PhACs	Pharmaceutically active compounds
PM	Particulate matter
POPs	Persistent organic pollutants
ppt	Parts-per-trillion
PVC	Polyvinyl chloride
ROS	Reactive oxygen species
Sr	Strontium
Th	T helper
TiO_2	Titanium dioxide
TSLP	Thymic stromal lymphopoietin
UNEP	United Nations Environment Programme
USA	United States of America
USEPA	United States Environmental Protection Agency
W	Tungsten
WHO	World Health Organization
WWTPs	Wastewater treatment plants

4.1 Introduction

The urban population and rapid growth put city resources under considerable strain and continue to limit the equitable distribution of resources. Alongside climate change, impacts include escalating temperatures, storms, extreme droughts, and other hazards. This raises pressure on cities to sustain human health and well-being (Bhardwaj et al. 2024a). Urban lifestyles also fuel a surge in non-communicable diseases (NCDs), including obesity, diabetes, cancer, and cardiovascular disorders, which dominate the global disease burden and increase in low and middle-income nations (Vos et al. 2015). Numerous "emerging pollutants," ranging from xenobiotics to their metabolites, have been detected widely in aquatic environments. Even in complex matrices like sludge, these pollutants are found at parts-per-trillion (ppt) levels with advanced methods. The most comprehensive data for such studies is prioritized on consumption, environmental concentrations, and ecotoxicological data; however, the long-term effects of these substances on human and animal health have yet to be fully elucidated (Oulton et al. 2010). Emerging pollutants refer to a wide variety of substances, such as medicines, PCPs, illegal drugs, hormones, steroids, and industrial chemicals like benzothiazoles and benzotriazoles, as well as polychlorinated naphthalenes (PCNs) and polychlorinated alkanes (PCAs). They also include perfluorinated chemicals (PFCs), synthetic fragrances, quaternary ammonium compounds, common additives like bisphenol A (BPA), antimicrobial agents like triclosan and triclocarban, polar pesticides, veterinary drugs, industrial by-products, food additives, and even engineered NMs (Lapworth et al. 2012).

ECs are a different group of chemicals that have been detected in the natural environment due to everyday human actions, such as household, agricultural, and industrial practices. Although these substances might not be novel, recent advancements in highly sensitive analytical methods have made it possible to detect them in water, soil, and air, even at extremely low concentrations, sometimes as little as the ppt level. ECs include more than 3,000 different mixtures, and their end products, such as pesticides, heavy metals, fertilizers, MPs, pharmaceutically active compounds (PhACs), PCPs, and both natural and synthetic hormones, often make their way into the environment (Barboza and Gimenez 2015). ECs are polar substances with short half-lives, but their continuous introduction into the environment keeps them present at trace concentrations, earning them the classification of pseudo-persistent pollutants (Puri et al. 2023). Traditional urban wastewater treatment plants (WWTPs), which use processes like activated sludge, filtration, and disinfection, are not effective at removing ECs. As a result, these plants are considered a major source of ECs entering the environment. These contaminants pose significant risks to human health and wildlife, including reproductive and endocrine system disorders. Among the various chemical and biological pollutants, antibiotics and the spread of antibiotic-resistant bacteria and their acquired resistance genes stand out as a dangerous threat to ecosystems and public health. Recognizing the severity of this issue, the European Union (EU) and other international organizations have identified antibiotic-resistant bacteria and acquired resistance genes as critical public health challenges, given their widespread presence in the environment, the food chain, and even drinking water (Serwecińska 2020).

4.1.1 Heat Stress

Climate change is a factor in regulating high global temperatures, upregulating the difficulty and incidence of heatwaves, and is related to high death rates (García-Lledó et al. 2020; Bhardwaj 2023). During heat stress, excessive amounts of reactive oxygen species (ROS) are accumulated (Fig. 4.1). ROS induces oxidative stress and deprivation of biomolecules (e.g., proteins, lipids, carbohydrates, and deoxyribonucleic acid (DNA)) leading to cellular death, while also increasing the production of heat shock proteins (HSPs), antioxidants, other metabolites, anti-inflammatory cytokines, and stress granules that mitigate the effects of heat stress (Medina et al. 2021). There are studies showing that HSPs can impact the immune system, facilitating anti-inflammatory and pro-inflammatory routes. For example, HSP70-B29, a bacterial HSP70-derived peptide, has been reported to show HSP-specific Tregs that decrease arthritis. However, HSPs have also been proven to upregulate the secretion of pro-inflammatory cytokines such as IL-6 and IL-1 through the NF-kB passageway. In mice and human keratinocytes, HSP90 suppression vigorously blocked 12-O tetradecanoylphorbol-13-acetate-induced inflammation by regulating key signaling pathways and pro-inflammatory cytokines, making it a potent option for psoriasis

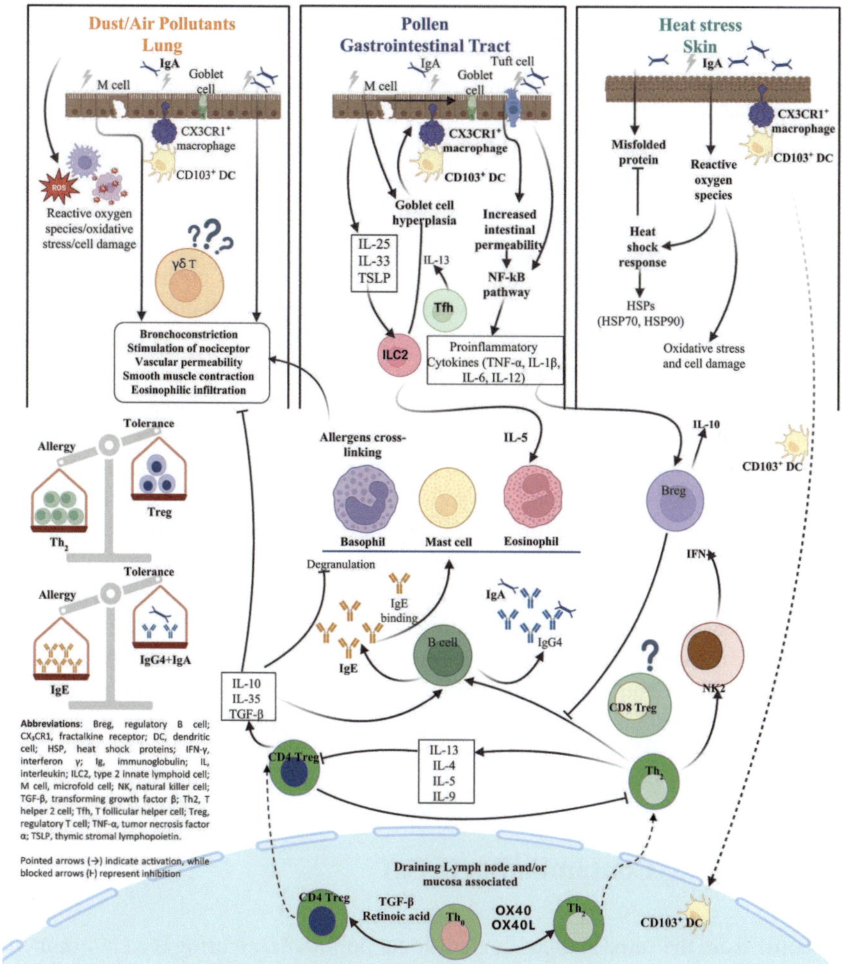

Fig. 4.1 Climate change-related factors like pollen, air pollution, and heat stress set off complex immune responses that can either promote inflammation or tolerance

(Schieber and Chandel 2014; van Eden 2018; Martine and Rébé 2019; Ben Abdallah et al. 2023).

4.1.2 *The Impact of Air Pollution on Health*

The 2019 Global Burden of Disease study revealed that airborne pollution, both inside and outside the environment, is responsible for approximately 10% of all deaths caused by NCDs (Dhimal et al. 2021). This makes it one of the top three threat factors contributing to mortality, with the majority of these deaths occurring in Africa and Asia. Additionally, the Lancet Commission on Pollution and Health has identified several serious health issues in children that are linked to air pollution. These include low birth weight and various NCDs. Furthermore, exposure to poor air quality is associated with neurodevelopmental disorders in young children (Landrigan et al. 2018; Fuller et al. 2022; Bhardwaj et al. 2023a).

When pollutants, allergens, and other ecological stressors breach a weakened epithelial barrier, the body releases pro-inflammatory molecules such as IL-33, IL-25, and thymic stromal lymphopoietin (TSLP). These signals push naïve T helper (Th) cells to develop into Th2 cells, which then trigger an inflammatory response by releasing cytokines like IL-4, IL-9, IL-5, and IL-13. This process encourages B cells to switch to produce immunoglobulin (Ig)E, a key player in allergic reactions. Upon re-exposure to allergens, IgE binds to mast cells and basophils, causing their activation and the release of histamine and prostaglandins. This leads to increased blood vessel permeability, smooth muscle contractions, and the recruitment of eosinophils—ultimately resulting in symptoms like bronchoconstriction. The effects of heat stress and air pollution further amplify these inflammatory pathways, making symptoms worse. However, the immune system also has built-in mechanisms to promote tolerance to everyday environmental exposures. Instead of turning into Th2 cells, some naïve Th cells differentiate into regulatory T cells, which help balance the immune response. These cells support B cells in producing anti-inflammatory antibodies like IgA and IgG4, reducing excessive immune reactions and promoting tolerance. (The figure is coming under CC 4.0 license; credit to Kumar et al. 2022; Agache et al. 2024.)

4.2 Emerging Contaminants (ECs)

ECs come from the same sources as traditional pollutants, entering the environment through industrial activities, emissions, wastewater, and effluent disposal. These are typically divided into two types: non-point and point sources. Point sources are specific, identifiable locations where pollutants are released into the environment. These sources often involve concentrated pollutant levels and include industrial discharges, sewage treatment plant effluents, and activities like mineral extractions. Non-point sources, also called diffuse sources, involve pollution that originates from broad, scattered areas rather than a single, identifiable location. Examples include runoff from fields treated with biosolids or manure and stormwater excess in urban or industrial areas. Unlike point sources, non-point sources tend to release pollutants at lower concentrations and are more likely to be naturally broken down. However, they are harder to trace back to their exact origin, making it challenging to recognize the specific source of contamination (Lapworth et al. 2012). Different ECs with their toxic effects on the environment are well described in Table 4.1.

Table 4.1 Categories of emerging contaminants (ECs) and their effects on the environment

Types of emerging contaminants (ECs)	Sub-categories of emerging contaminants (ECs)	Examples	Effects on environment	References
Organic emerging contaminants	Endocrine-disrupting chemicals (EDCs)	17-α-ethinylestradiol (EE2), 17-β-estradiol (E2), Estrone (E1), Phthalate acid esters, Bisphenols	Disturbance with hormones (mammals including humans)	Decision (2015)
	Pharmaceuticals, personal care products (PCPs), and hormones	Pentachlorophenol, Diclofenac, 2-Ethylhexyl-4-methoxycinnamate, Antibiotics, Illicit drugs	Mollusks can suffer from genetic damage, nervous system toxicity, and oxidative stress. Meanwhile, algae and fish may experience slower growth, and entire algae communities can struggle to develop, affecting the balance of the ecosystem	China (2022)
	Pesticides and herbicides	Methiocarb, Neonicotinoids, Oxadiazon, Tri-allate, Perchlorate, Dicofol	Harmful impacts on fish gills, the feminization of aquatic species, and serious disruptions to human reproductive and sexual health	Decision (2015)
	Persistent organic pollutants (POPs)	Polychlorinated biphenyls (PCBs), Brominated flame retardants (BFRs), Polycyclic aromatic hydrocarbons (PAHs), Perfluorinated chemicals, Per- and poly-fluoroalkyl substances (PFAS), Polybrominated diphenyl ethers (PBDEs)	Slowed growth, reduced reproduction, and lower survival rates in crustaceans and zebrafish	Rathi et al. (2021)
	Organic solvents/plastic additives	Hexachlorobutadiene, Dichloromethane, Chloroform, Trichloroethane	Obesity, male fertility, and puberty in humans	China (2022)
Inorganic emerging contaminants (ECs)	Metals	Strontium (Sr), Manganese (Mn), Tungsten (W), Lithium (Li)	High toxicity to aquatic organisms	Funding (2023)
	Nanoparticles (NPs)	Titanium nanoparticles (NPs)	Effects on GI, liver, and gills tract in Cyprinus carpio	Kanangire et al. (2016)

(continued)

Table 4.1 (continued)

Types of emerging contaminants (ECs)	Sub-categories of emerging contaminants (ECs)	Examples	Effects on environment	References
Biological emerging contaminants (ECs)	Pathogenic bacteria, Virus	*Legionella pneumophila*, *Pseudomonas aeruginosa*, and *Mycobacterium*	It causes several toxic effects on the environment	Funding (2023)
Other emerging contaminants (ECs)	Microplastics (MPs) and nanoplastics	Persistent organic chemicals and heavy metals	Oxidative stress in both adult and juvenile sea cucumbers, as well as humans, can cause cell toxicity and reproductive harm. In fish, whales, crabs, oysters, mussels, and plankton, it can also lead to blockages in the digestive system, further impacting their health	Thompson et al. (2004)

4.2.1 *Pesticides*

Pesticides have long been evaluated by regulatory bodies and research organizations to ensure their safety. However, the focus has recently shifted to their metabolites and by-products that form from pesticides breaking down. These metabolites can be biologically active and highly toxic and often make their way into groundwater. Because of this, some pesticides once considered safe are now under scrutiny, with growing concerns labeling them as ECs (Naidu et al. 2016).

4.2.2 *Pharmaceuticals*

Pharmaceutical chemicals have been known to end up in the environment for quite some time. Their sources are well understood and include human excretion, improper disposal of unused medications, and their use in veterinary practices (Poynton and Vulpe 2009). Stuart et al. (2012) highlighted that various pharmaceuticals have been found in both surface water and groundwater, with discarded wastewater identified as the primary source. These pharmaceuticals fall into broad categories, including antibiotics, prescription and over-the-counter medications, and iodinated X-ray contrast agents (Stuart et al. 2012).

4.2.3 *Lifestyle and Personal Care Compounds (PCCs)*

Caffeine and nicotine are the two most common "lifestyle compounds" found in groundwater as one of the sewage effluents. Meanwhile, PCPs often originate from WWTPs, as these compounds are not fully removed during the treatment process. A well-known example is triclosan, an active ingredient widely used in household items like toothpaste, soaps, and antimicrobial products (Naidu et al. 2016).

4.2.4 *Industrial Additives and By-Products*

Stuart et al. (2012) stated that most industrial chemicals released into the environment are regulated, either as controlled pollutants or with set limits for drinking water safety. Because of this, they're not usually classified as ECs. However, their by-products, like benzotriazole derivatives and dioxins, do not always fall under these regulations, meaning they can slip through the cracks and pose unrecognized environmental risks (Stuart et al. 2012).

4.2.5 Food Additives

Research has shown that sweeteners like acesulfame, saccharin, and sucralose often make their way into domestic wastewater through human excretion. Additionally, food preservatives, which include parabens, have been spotted in groundwater (Scheurer et al. 2009). A study by Tollefsen et al. (2012) estimated the ratio of the projected environmental concentration to the "no-effect" concentration for aquatic organisms, specifically for sucralose, to be 0.08—well below 1. This suggests that sucralose poses only a minimal risk to marine life, including plants, algae, crustaceans, and fish (Tollefsen et al. 2012).

4.2.6 Flame Retardants

Flame retardants are commonly added to household items to reduce fire risks and comply with safety regulations. Among the most researched are brominated flame retardants (BFRs), such as hexabromocyclododecane (HBCDD), and organophosphate esters (OPEs). HBCDD is widely used in thermal insulation and textile coatings and is now one of the most significant polymer additives after plasticizers. However, despite its benefits, HBCDD is extremely constant in the environment and, over time, tends to accumulate in the food chain (Das et al. 2013).

4.2.7 Plasticizers

Plasticizers are additives used to improve the flexibility of materials and make them easier to process. Common plasticizers are compounds like phthalates, BPA, and non-chlorinated OPEs. These substances reduce the material's yield strength and viscosity, enabling its manufacture at lower temperatures. As a result, plasticizers are found in a wide range of manufactured goods, including polyvinyl chloride (PVC) items, building materials, toys, perfumes, clothing, and food packaging (Annamalai and Namasivayam 2015). Among these, BPA is extensively studied in endocrine disruptors (EDs) globally (Beronius et al. 2010).

4.2.8 Perfluoroalkyl Substances (PFAS)

PFAS, or per- and poly-fluoroalkyl substances, have been used for around 70 years, primarily as waterproofing agents and surface protectors, including fire resistance (Harrad et al. 2019). In recent years, many PFAS that are being identified in the environment have been observed to be potentially bioaccumulative and toxic

(Conder et al. 2008). These substances encompass a broad group of chemicals with varying fluorocarbon chain structures (CF3[CF2] n-). Among them, perfluoro octane sulfonate (PFOS) and perfluorooctanoate (PFOA) are the most well-known, largely due to their widespread presence in the environment and their harmful effects (Wang et al. 2014).

4.2.9 Surfactants

Surfactants are a group of emerging organic compounds known for their emulsifying properties. They are primarily used as cleaning agents in detergents and sanitation products, helping to break down grease and dirt. Additionally, they play a role in other applications, such as pesticides and resins, by allowing water to spread more effectively and moisten different surfaces (Uğuz et al. 2009). Polybrominated diphenyl ethers (PBDEs) are a type of BFR that is normally utilized in plastics, textiles, electronic circuits, and other materials (Bhardwaj et al. 2025a, b). There are 209 different PBDE variations, and their extensive presence in the surroundings has raised concerns. They have been detected in humans, wildlife, and even in remote locations far from where they were originally produced.

PCAs, also known as chlorinated paraffins (CPs), are industrial compounds made up of chlorinated straight-chain hydrocarbons. Trace amounts of these compounds may be found in treated wastewater. PCAs are categorized based on their carbon chain length: short-chain PCAs have 10 to 13 carbon atoms, medium-chain PCAs contain 14 to 17 carbon atoms, and long-chain PCAs have 18 or more carbon atoms. There are more than 10,000 possible PCA variations. Their presence in sewage sludge, proof of accumulation in the environment and humans, and toxicity concerns highlight the need for further research to assess their potential risks. PCNs have been in use since the early 1900s for a variety of industrial applications, including dielectric fluids, lubricants, electroplating masking complexes, wood additives, engine oil additives, and dye creation. Structurally, they are similar to other persistent organic pollutants (POPs), sharing similar physical and chemical properties. There are 75 known PCN variations, each containing 1 to 8 chlorine atoms per each naphthalene molecule. These compounds are widespread ECs, and some PCNs have been found to exhibit dioxin-like toxicity, raising further concerns about their influence on the environment and human health (Clarke and Smith 2011).

4.2.10 Quaternary Ammonium Compounds

Quaternary ammonium compounds are a type of cationic surfactant, i.e., they carry a positive charge. Their molecular structure includes at least one water-repelling (hydrophobic) alkyl chain attached to a positively charged nitrogen atom, while the remaining alkyl groups are usually smaller, such as benzyl or methyl groups. Quaternary ammonium compounds primarily enter WWTPs through household products. In wastewater, they can break down through biodegradation, often within hours or days. However, WWTP discharges can still lead to environmental contaminations, particularly in marine sediments, where traces of quaternary ammonium compounds may accumulate (Lara-Martin et al. 2010; Clarke and Smith 2011).

4.2.11 Nanomaterials (NMs)

NMs are extremely small, ranging from 1 to 100 nm in size, and they possess unique properties such as high strength, low permeability, thermal stability, and excellent conductivity. Shortly, they are expected to play a major role in fields like chemotherapy, targeted drug delivery, and detecting foodborne pathogens. Their chemical structures vary widely and include nanotubes, fullerenes, metal oxides, quantum dots, titanium dioxide nanoparticles (TiO_2-NPs), nanoparticles (NPs), nanogold, nanosilver, and zerovalent iron NPs (Bhardwaj et al. 2023b). While much of the research on NMs focuses on developing innovative applications and products with enhanced properties, there is also growing apprehension about their potential impression on the environment. Scientists are actively studying how these materials travel through ecosystems, their long-term effects on health, and their overall environmental impact (Richardson 2012).

4.2.12 Microplastics (MPs)

MPs are minute plastic fragments that result from the break of larger plastic materials (Bhardwaj 2024; Bhardwaj et al. 2024c). They are widespread in the surroundings, affecting both humans and wildlife. These particles have been found in a variety of marine species, as well as in drinking water and everyday foods like salt, honey, and seafood. People can also be exposed to MPs by inhaling airborne particles. Studies on animals suggest that once MPs and nanoplastics enter the body, they can travel to vital organs such as the liver, thymus, spleen, lungs, reproductive organs, heart, kidneys, and even the brain by crossing the blood–brain barrier (Bhardwaj et al. 2025b). Additionally, MPs act as carriers for harmful pollutants like persistent organic chemicals and heavy metals, transferring them through the food chain from lower organisms to higher ones. When ingested, the additives and monomers in these plastics can interfere with essential biological functions, potentially disrupting the endocrine and immune systems. They have also been linked to issues with movement, reproduction, development, and even the risk of cancers (Ziani et al. 2023).

4.2.13 Endocrine-Disrupting Chemicals (EDCs)

Endocrine-disrupting chemicals (EDCs) are a group of substances that can inhibit the body's endocrine system. These compounds can affect the body's hormones by blocking, mimicking, or disrupting their normal action, which can have harmful effects on both people and wildlife. Some well-known EDCs comprise perchlorate, BPA, alkylphenols, dioxins, and polychlorinated biphenyls (PCBs)—many of which are commonly found in resins and plastics. Phthalates, commonly used in PCPs, and pesticides like atrazine and dichlorodiphenyltrichloroethane (DDT) are also recognized as EDCs. People can be exposed to EDCs through food, water, consumer products, and environmental contamination (Aydemir and Ulusu 2023). Research has linked EDCs to various health problems, including reproductive and developmental disorders.

Recent studies suggest that they may also impact key biological functions, such as the activity of the G-6-P dehydrogenase enzyme, which is important for many metabolic processes. Additionally, EDCs have been found to influence adipokines (hormones involved in metabolism) and reproductive health. As these chemicals build up in aquatic ecosystems, they pose an increasing threat to both wildlife and humans. A major concern is the widespread use of parabens, common EDs found in PCPs. Their potential to interfere with hormones has pushed scientists to explore safer alternatives (Kurowska et al. 2022).

4.3 Sources and Transport Pathways

ECs find their way into the environment through both natural processes and human activities. Key sources include wastewater discharge, agricultural runoff, industrial operations, improper waste disposal, and accidental spills, all of which contribute to the presence of these newly recognized pollutants. During manufacturing, a variety of hazardous chemicals and materials are often introduced into the environment, sometimes unintentionally. Poor waste management in industries further worsens pollution, making it a significant environmental concern. Additionally, landfills and waste treatment facilities act as major entry points for ECs, allowing these substances to spread into the surrounding ecosystem.

Antibiotics and growth hormones, commonly used in livestock farming, can enter the environment through animal wastes. According to the World Health Organization (WHO) (2020), these substances can pollute water and soil, contributing to the spread of antibiotic resistance and other environmental risks. Polluted water used for irrigation can introduce these contaminants into agricultural soils and crops, further impacting food systems (Bartelt-Hunt et al. 2011). Household activities also play a role in environmental contamination. Everyday tasks such as cleaning, doing laundry, and using PCPs can release harmful chemicals into the environment. A research study by Gore et al. (2015) found that home cleaning products, detergents, shampoos, lotions, and cosmetics contain hazardous substances that can enter wastewater systems, eventually making their way into natural ecosystems.

While human activities are the primary source of ECs, natural processes also contribute. Volcanic eruptions, forest fires, and emissions from biological sources can release pollutants into the air, water, and soil. During volcanic activity, heavy metals and other detrimental materials are discharged into the surrounding environment. Natural disasters can also introduce a variety of pollutants, including PAHs and other organic compounds. Pharmaceutical waste is another concern, particularly when medications are improperly disposed of, such as by flushing them down the toilet or pouring them into drains. Additionally, certain microbes and plants can release toxic chemicals that contribute to environmental pollution (Nel et al. 2006).

ECs can travel long distances through the air before settling into different parts of the environment (Galloway 2015). Once they enter ecosystems, these chemicals can impact aquatic life and may even contaminate drinking water sources, posing potential risks to human health (Daughton 2016). Environmental incidents like chemical spills and leaks can have immediate and severe effects on both ecosystems and public health in affected areas. Additionally, the buildup of ECs in food products raises further concerns, as it could pose serious health risks to consumers.

4.4 Impact on Human Health and Environment

Natural and man-made disasters contribute to the threat to human health, as they affect air, water, and soil. Extensive research has shown that ECs can have harmful effects on both humans and animals. Most studies on ECs have focused on surface water for several reasons. First, surface waters tend to contain higher concentrations and a wider variety of contaminants, especially when they are directly impacted by industrial discharges. Second, monitoring surface water is generally easier and more practical compared to tracking contaminants in groundwater (Stuart et al. 2012).

Environmental pollution remains a major challenge, as many contaminants are difficult to control and have become a part of everyday life. Steinemann (2004) emphasizes the importance of understanding how humans are exposed to these pollutants to help reduce their harmful effects. Persistent pollutants, such as pesticides, heavy metals, and industrial chemicals, are found in the air, water, and soil. Over time, these substances accumulate in living organisms, leading to long-term health risks. The threat is particularly severe in developing countries, where weaker environmental regulations and limited healthcare resources make populations more vulnerable. To combat pollution effectively, strong policies and international cooperation are essential. By working together, nations can reduce sources of contamination and ensure healthier living conditions for both current and future generations (Steinemann et al. 2004).

Identifying hazards involves recognizing substances or activities that could potentially harm human health or the environment. For instance, extensive research has shown that certain pesticides, like organophosphates (OPs), can have neurotoxic effects on humans (Costa 2006). The Agency for Toxic Substances and Disease Registry (ATSDR) has classified various industrial pollutants, such as PAHs and heavy metals, as carcinogens and EDs, meaning they can contribute to cancer and hormonal imbalances (ATSDR 2020). This process also includes assessing the amount, duration, and frequency of exposure to harmful substances. For example, children who have been exposed to lead-contaminated drinking water over a long period have been found to experience cognitive impairments (Lanphear et al. 2019).

4.5 Importance of Addressing Health and Environmental Hazards

(i) **Protecting Public Health**: Exposure to ECs can have serious health consequences, ranging from short-term illnesses to long-term chronic diseases. Understanding and reducing these risks is essential for preventing health issues and improving overall well-being. Since many emerging pollutants have unknown or potentially harmful effects, it is crucial to address them proactively to create safer and healthier living environments (WHO 2020).

(ii) **Environmental Conservation**: ECs pose a danger not only to human health but also to the environment. Safeguarding ecosystems is essential for preserving biodiversity, sustaining agriculture, and protecting natural resources. These pollutants can disrupt ecosystems, endanger wildlife, and degrade vital resources. Therefore, effective management is crucial to maintaining biodiversity and ensuring the long-term health of our environment.

(iii) **Economic Stability**: Health and environmental risks come with significant economic consequences, including higher healthcare costs, reduced productivity, and expensive cleanup efforts. Managing these hazards effectively can help ease these financial burdens. For example, ECs can lead to costly environmental restoration, biodiversity loss, and negative effects on key industries like fishing and tourism. Proactive mitigation measures can help minimize these economic impacts and protect both public health and the environment.

(iv) **Social Equity**: Certain groups, such as children, the elderly, and low-income communities, are more vulnerable to health and environmental risks. Addressing these hazards is essential for promoting social justice and equity. Managing ECs not only protects public health but also drives scientific progress and the creation of more effective policies and regulations.

(v) **Heavy Metals**: Toxic metals like lead (Pb) and mercury (Hg) can have serious health effects, including neurological damage and developmental disorders, particularly in children (Bhardwaj et al. 2023; Alam et al. 2023).

(vi) **Endocrine-Disrupting Chemicals (EDCs)**: EDCs are compounds that inhibit the body's hormonal system, potentially causing issues with development, reproduction, nervous system, and immune function. These substances can mimic or block natural hormones, leading to significant health concerns for both humans and wildlife (Gore et al. 2015).

(vii) **Antimicrobial Resistance (AMR)**: The overuse of antibiotics in medicine and agriculture has influenced the increase of drug-resistant bacteria. These bacteria can spread through the environment, making common infections more difficult to treat and posing a serious public health threat (WHO 2020).

(viii) **Nanomaterials (NMs)**: While NMs have unique properties that make them useful in various industries, they can persist in the environment and may pose dangers to both ecosystems and human health (Nel et al. 2006; Bhardwaj et al. 2023b).

(ix) **Persistent Organic Pollutants (POPs)**: POPs are long-lasting lethal compounds that can accumulate in the environment and living organisms (Bhardwaj et al. 2018). They have been related to serious health difficulties, such as immune system disorders, cancer, and reproductive problems.

4.6 Detection and Monitoring

The new development of threat detection relating to construction and the health of people around the area has aided in the advancement of instrumentation and monitoring systems. As highlighted by Pamukcu and Cheng (2017), the use of Geophysical Ground Penetrating Radar and other electromagnetic techniques such as sensors serve as indicators of perils that could damage buried structures. Such technology has proven to be useful when there is fear of damaging environmental conditions, the possibility of infrastructure collapse, and other threats buried in the ground. Their research indicates the focus on the work that makes usage of advanced sensing techniques for the protection and sustainability of tools intended to manage the environment.

Similarly, the chemical hazard identification and analysis capabilities have also improved tremendously with new surveillance instruments and systems. Gawlik-Kobylińska et al. (2021) propose the EU Sensor System that combines chemical surveying for the detection and tracking of dangerous substances in the environment. The combination of the powerful nodes needed for chemical detection and the chemical detection algorithms exemplifies how modern technologies can be used to increase awareness about the environment and increase surveillance of its health. Likewise, Wilson (2012) examined electronic-nose technologies, especially capabilities to monitor, in real-time, toxic compounds and emissions. All these developments underscore the pervasive value of accurate detection tools to deal with environmental and health challenges resulting from toxic chemicals. The comparison between conventional and new methods for detecting environmental hazards with their advantages and disadvantages is shown in Table 4.2.

Table 4.2 Comparison between conventional methods and new methods for detecting environmental hazards, along with their advantages and disadvantages

Aspect	Conventional Methods	New Methods
Detection techniques	Laboratory-based analysis (e.g., gas chromatography-mass spectrometry (GC–MS))	Real-time sensors (e.g., Internet of Things (IoT)-based air quality monitors)
	Manual sampling and analysis	Remote sensing (e.g., satellite imaging, drones)
	Chemical assays (e.g., colorimetric tests)	Biosensors (e.g., enzyme-based or DNA-based sensors)
Advantages	High accuracy and reliability	Real-time monitoring and rapid data collection
	Well-established protocols and standards	High sensitivity and specificity
	Suitable for detecting a wide range of pollutants	Portable and cost-effective for large-scale deployment
		Non-invasive and capable of remote operation
Disadvantages	Time-consuming and labor-intensive	May require calibration and validation against conventional methods
	Expensive equipment and maintenance	Limited by environmental conditions (e.g., weather for drones)
	Limited to point-source sampling, not suitable for large-scale monitoring	Potential for false positives or negatives due to sensor limitations
	Delayed results, not suitable for real-time decision-making	Higher initial investment for advanced technologies
References	Taylor et al. (2001), An et al. (2019), WHO (2021)	Kumar et al. (2013), Cynthia et al. (2019), Gavrilaş et al. (2022)

4.6.1 Conventional Methods

These are traditional techniques that have been used for decades, such as laboratory-based analyses and manual sampling. They are highly accurate but often slow, expensive, and limited in scope. For example, GC–MS and liquid chromatography-mass spectrometry (LC–MS) are extensively used for detecting air and water pollutants but require specialized equipment and expertise (Bhardwaj et al. 2019; Bhardwaj and Jindal 2020; Kumar et al. 2024, 2025).

4.6.2 New Methods

These include advanced technologies like IoT-based sensors, biosensors, and remote sensing. They offer real-time data, portability, and scalability but may require validation and are sometimes limited by environmental factors. For example, IoT-based air quality monitors provide continuous data but may need calibration to ensure accuracy.

4.7 Mitigation and Remediation Strategies

Some of the recent environmental problems include POPs and greenhouse gases (GHGs). Sources and distribution of the POPs, along with their negative impacts on the health and longevity of the environment, were thoroughly discussed and summarized by Mishra et al. (2022). Such pollutants upset ecological organization and are related to adverse health effects, including carcinogenicity and endocrine disruption, because of their toxic and accumulating character. The study highlighted the need to embrace newer remediation techniques, such as bioremediation and the use of chemicals in dealing with the effects of POPs. Some of these methods minimize human exposure to toxic material and, at the same time, help attain environmental standards.

Alleviating environmental risks is a complex process that involves risk evaluation, the application of treatment methods, and appropriate environmental management. Babaniyi et al. (2023) support the process of eliminating GHGs and other damaging compounds from industrial waste before they get discharged into the environment. Such practices, when used in conjunction with other technologies for catalytic fluid reduction and other novel techniques of cleaning up contaminated environments, go a long way in reducing the risks to the ecological systems and humans. The authors recommend the development of advanced preventive measures to be introduced and implemented in the long-term management of industries to enhance environmental conservation and population health.

4.8 Future Directions

To mitigate environmental risks and their adverse effects, expanding novel treatment methods is the way forward, as is the intersectoral approach. Optimizing the existing techniques or inventing new methods for the removal of pollutants, for instance, through nanotechnology (NT), bioengineering, and microbial applications, are conceivable, and hence cost-efficient yet environmentally friendly methods exist. Perhaps most exciting is the potential use of machine learning (ML) methodologies paired with real-time data to enhance other predictive risk assessment tools as a way to revolutionize how risk is assessed and preventions begin to be developed for potential hazards.

Equally important is the recapitalization of global policies that address sustainable industrial development and the stabilization of policies that seek to control pollution emissions. The governance strategies include cleaner production technologies for industries that involve carbon capture technologies and the circular economy. Stakeholder awareness will also be fundamental in ensuring that people living around sources of pollution develop an implementation democracy of environmental tasks. In addition, combining adaptation to climate change and the development of climate change measures linked to the climate-proofing of waste disposal systems and renewable energy frameworks will reduce the worst effects of the two scourges. Stakeholders can then strive to develop a healthier and more sustainable future along the paths identified above.

4.9 Case Studies

Examples of environmental dangers and the success of mitigation techniques are provided via case studies, which are based on real-world situations. Some instances to illustrate this point are as follows:

The Bhopal Gas Tragedy occurred in India in 1984. It was caused by the discharge of methyl isocyanate (MIC) from a pesticide facility that led to the deaths of more than 15,000 people and had long-term consequences on their health. According to Dhara and Dhara (2002), this debacle brought to light the need to implement strict workplace safety rules and emergency response capabilities. Atrazine, a herbicide that is commonly used, was discovered in groundwater, which has the potential to affect aquatic ecosystems as well as human health. According to Hayes and Gagnon (2024), the contamination led to the implementation of more stringent application rules and monitoring methods. The use of integrated pest management (IPM) strategies was encouraged to lessen the reliance on chemical pesticides throughout the mitigation process.

In the middle of the twentieth century, industrial wastewater from a chemical plant in Minamata, Japan, contaminated the environment and eventually affected people with Hg, resulting in the neurological illness known as Minamata disease. Methylmercury from the wastewater bioaccumulated in Minamata Bay's fish and shellfish. The local community suffered from severe Hg poisoning, which led to serious diseases and even death. The terrible effects of industrial pollution on people and the environment are brought home by the Minamata disaster (Vishwanathan et al. 2023). It emphasizes how crucial environmental stewardship and appropriate waste management are to avert such catastrophes.

4.10 Strategies for Efficient Mitigation and Prevention

Strategies for mitigation and prevention are essential components in the process of lowering environmental risks and the effects those risks have on human health and ecosystems. The legislative, technical, and social components are all included in these methods because of their multidimensional nature. Efficacious options for the management of pollution are made available by technological advancements. The cleaning of oil spills has been effective with the use of techniques such as bioremediation, which involves the utilization of microbial communities to degrade toxins (Das and Chandran 2011). According to Chen et al. (2020), emerging technologies, such as photocatalysis, have the potential to break down persistent environmental contaminants. The engagement of the general public is necessary for the effective implementation of mitigation techniques. For example, community-based environmental monitoring programs provide local communities with the ability to recognize and report environmental dangers (Conrad and Hilchey 2011).

In the process of creating legislation and determining the limits of exposure that are acceptable, governments play a crucial role. Through the implementation of severe effluent regulations, for instance, the Clean Water Act of 1972 in the United States of America (USA) seeks to limit the amount of water pollution (USEPA 2023). Hazardous chemicals are the focus of international accords such as the Stockholm Convention on POPs, which was established in 2001 by the United Nations Environment Programme (UNEP). Following the occurrence, India enacted the Environmental Protection Act (1986) to govern the handling of hazardous waste and industrial pollutants.

4.11 Recommendations

For successful risk management, it is essential to have an in-depth understanding of the routes, which include airborne, waterborne, and soil contamination. To identify and quantify the threats that are caused by these hazards, risk assessment methodologies are critical instruments that must be utilized. A better knowledge of the severity and probability of hazards may be achieved via the use of techniques such as hazard identification, exposure assessment, dose–response assessment, and risk characterization processes. To improve the quality and efficiency of these evaluations, contemporary methods such as Geographic Information Systems (GIS), computer modeling, and bio-monitoring have been used. Strategies for mitigating and preventing the effects of pollution are examined, with an emphasis placed on the significance of technological advancements such as enhanced pollution control technologies and renewable energy options. We can establish more reasonable rules and regulations to safeguard public health and the environment if we have a better understanding of how these hazards interact with one another and the consequences they have. This will ensure a future that is both sustainable and healthier.

To develop effective methods to reduce the consequences of these risks and preserve public health and the environment, it is vital to have a good understanding of these hazards. The implementation of policy interventions, such as regulatory frameworks, emission reduction objectives, and sustainable urban design, is an essential component in the process of mitigating these particular dangers. Ensuring proper workplace design and practices can help mitigate these risks (Centers for Disease Control and Prevention 2021). Having a thorough understanding of these channels and sources is very necessary to develop mitigation strategies that are capable of efficiently protecting both the environment and public health. A large reduction in exposure hazards may be achieved by the education of persons about the use of protective equipment, the employment of appropriate hygiene procedures, and the understanding of contamination sources. When working with insecticides, for instance, farmers are strongly encouraged to use protective gear like gloves and masks.

4.12 Conclusions

ECs are a set of chemicals and materials that, even though not routinely monitored or controlled, have the potential to cause considerable damage to both the environment and human health. It is essential to have an understanding of these dangers and take action to mitigate them to preserve public health and maintain the integrity of the ecosystem. It is essential to have a comprehensive understanding of the many sorts of dangers that are connected with ECs to devise effective methods for safeguarding both the environment and human health. We can effectively control the dangers that are provided by these pollutants and create a world that is safer and healthier if we address the hazards that are found in chemical, biological, physical, and ergonomic environments. There are many different sources of ECs, and they reach the environment through a diversity of different channels. The presence of these contaminants is caused by a variety of sources, including industrial, agricultural, residential, and natural sources. Each of these sources has its own set of obstacles for monitoring and mitigating the problem. A complete knowledge of the origins and routes of ECs is required to create effective strategies for safeguarding human health and the environment. This is necessary to address the hazards that are associated with emerging pollutants.

Acknowledgements The authors would like to acknowledge Frontiers in Sciences and the associate authors (DOI: https://doi.org/10.3389/fsci.2024.1279192) for providing open-access figure sources. We also extend our gratitude to the library and the Director of BITS Pilani Hyderabad for offering world-class infrastructure and research support.

References

Alam S, Bhardwaj LK, Mallick R, Rai S (2023) Estimation of heavy metals and fluoride ion in vegetables grown nearby the stretch of River Yamuna, Delhi (NCR). India. Indian J Environ Protection 43(1):64–73

Agache I, Akdis C, Akdis M, Al-Hemoud A, Annesi-Maesano I, Balmes J, Cecchi L, Damialis A, Haahtela T, Haber AL, Hart JE (2024) Immune-mediated disease caused by climate change-associated environmental hazards: mitigation and adaptation. Frontiers Sci 2:1279192

An Z, Huang RJ, Zhang R, Tie X, Li G, Cao J, Zhou W, Shi Z, Han Y, Gu Z, Ji Y (2019) Severe haze in northern China: a synergy of anthropogenic emissions and atmospheric processes. Proc Natl Acad Sci 116(18):8657–8666

Annamalai J, Namasivayam V (2015) Endocrine disrupting chemicals in the atmosphere: their effects on humans and wildlife. Environ Int 76:78–97

ATSDR (2020) Toxicological profile for polycyclic aromatic hydrocarbons (PAHs). Agency for toxic substances and disease registry

Aydemir D, Ulusu NN (2023) The impact of the endocrine-disrupting chemicals on the glucose-6-phosphate dehydrogenase enzyme activity. Front Pharmacol 14:1133741

Babaniyi BR, Ogundele OD, Bisi-Omotosho A, Babaniyi EE, Aransiola SA (2023) Remediation approaches in environmental sustainability. In: Microbiology for cleaner production and environmental sustainability, pp 321–346. CRC Press

Barboza LGA, Gimenez BCG (2015) Microplastics in the marine environment: current trends and future perspectives. Mar Pollut Bull 97(1–2):5–12

Bartelt-Hunt S, Snow DD, Damon-Powell T, Miesbach D (2011) Occurrence of steroid hormones and antibiotics in shallow groundwater impacted by livestock waste control facilities. J Contam Hydrol 123(3–4):94–103

Ben Abdallah H, Seeler S, Bregnhøj A, Ghatnekar G, Kristensen LS, Iversen L, Johansen C (2023) Heat shock protein 90 inhibitor RGRN-305 potently attenuates skin inflammation. Front Immunol 14:1128897

Beronius A, Rudén C, Håkansson H, Hanberg A (2010) Risk to all or none?: A comparative analysis of controversies in the health risk assessment of Bisphenol A. Reprod Toxicol 29(2):132–146

Bhardwaj L, Chauhan A, Ranjan A, Jindal T (2018) Persistent organic pollutants in biotic and abiotic components of Antarctic pristine environment. Earth Syst Environ 2:35–54

Bhardwaj L, Sharma S, Ranjan A, Jindal T (2019) Persistent organic pollutants in lakes of Broknes peninsula at Larsemann Hills area, East Antarctica. Ecotoxicology 28:589–596

Bhardwaj LK, Jindal T (2020) Persistent organic pollutants in lakes of Grovnes Peninsula at Larsemann Hill area. East Antarctica. Earth Syst Environ 4(2):349–358

Bhardwaj LK (2023) A comprehensive review on the climate change and its impact on health. https://doi.org/10.20944/preprints202305.0159.v1

Bhardwaj LK (2024) Occurrence of microplastics (MPs) in Antarctica and its impact on the health of organisms. Maritime Technol Res 6(2):265418. https://doi.org/10.33175/mtr.2024.265418

Bhardwaj LK, Rath P, Choudhury M (2023a) A comprehensive review on the classification, uses, sources of nanoparticles (NPs) and their toxicity on health. Aerosol Sci Eng 7(1):69–86

Bhardwaj LK, Rath P, Jain H, Purohit S, Yadav P, Singh V (2024a) The impact of climate-induced livelihood, health, and migration on women and girls: a review. Global insights on women empowerment and leadership, pp 100–118

Bhardwaj LK, Rath P, Tokas R (2024b) E-waste management in developing countries: current practices, challenges, disposal, and impact on human health and environment. Sustainable solutions for E-waste and development, pp 90–102

Bhardwaj LK, Rath P, Jain H, Choudhury M (2025a) Exploring the effects of e-waste on soil, water quality and human health. Discover Civil Eng 2(1):1–16

Bhardwaj LK, Rath P, Rehman F, Yadav P, Massahi T, Mittal R, Jain H (2025b) Status of micro/nano-plastic (MNP) pollution in the aquatic system. In: Societal and environmental ramifications of plastic pollution, pp 43–62. IGI Global

Bhardwaj LK, Rath P, Yadav P, Gupta U (2024c) Microplastic contamination, an emerging threat to the freshwater environment: a systematic review. Environ Syst Res 13(1):8

Bhardwaj LK, Sharma S, Jindal T (2023b) Estimation of physico-chemical and heavy metals in the lakes of Grovnes & Broknes Peninsula, Larsemann Hill. East Antarctica. Chem Africa 6(5):2677–2694

Bhardwaj LK, Singh VV, Dwivedi K, Rai S (2023) A comprehensive review on sources, types, impact and challenges of air pollution

Chen D, Cheng Y, Zhou N, Chen P, Wang Y, Li K, Huo S, Cheng P, Peng P, Zhang R, Wang L (2020) Photocatalytic degradation of organic pollutants using TiO_2-based photocatalysts: A review. J Clean Prod 268:121725

China SEPA (2022) Ministry of ecology and environment of the People's Republic of China. Soil environmental quality, risk control standard for soil contamination of agricultural land (GB15618-2018)

Clarke BO, Smith SR (2011) Review of 'emerging' organic contaminants in biosolids and assessment of international research priorities for the agricultural use of biosolids. Environ Int 37(1):226–247

Conder JM, Hoke RA, Wolf WD, Russell MH, Buck RC (2008) Are PFCAs bioaccumulative? A critical review and comparison with regulatory criteria and persistent lipophilic compounds. Environ Sci Technol 42(4):995–1003

Conrad CC, Hilchey KG (2011) A review of citizen science and community-based environmental monitoring: issues and opportunities. Environ Monit Assess 176:273–291

Costa LG (2006) Current issues in organophosphate toxicology. Clin Chim Acta 366(1–2):1–13

Cynthia J, Saroja MN, Sultana P, Senthil J (2019) IoT-based real time air pollution monitoring system. Int J Grid High Perform Comput (IJGHPC) 11(4):28–41

Das N, Chandran P (2011) Microbial degradation of petroleum hydrocarbon contaminants: an overview. Biotechnol Res Int 2011(1):941810

Das P, Arias EVA, Kambala V, Mallavarapu M, Naidu R (2013) Remediation of perfluorooctane sulfonate in contaminated soils by modified clay adsorbent—a risk-based approach. Water, Air, & Soil Pollution 224:1–14

Daughton CG (2016) Pharmaceuticals and the Environment (PiE): Evolution and impact of the published literature revealed by bibliometric analysis. Sci Total Environ 562:391–426

Decision EU (2015) 495/2015, Commission Implementing Decision (EU) 2015/495 of 20 March 2015 establishing a watch list of substances for Union-wide monitoring in the field of water policy pursuant to Directive 2008/105/EC of the European Parliament and of the Council. Off J Eur Union L 78:40–42

Dhara VR, Dhara R (2002) The Union Carbide disaster in Bhopal: a review of health effects. Arch Environ Health Int J 57(5):391–404

Dhimal M, Chirico F, Bista B, Sharma S, Chalise B, Dhimal ML, Ilesanmi OS, Trucillo P, Sofia D (2021) Impact of air pollution on global burden of disease in 2019. Processes 9(10):1719

Fuller R, Landrigan PJ, Balakrishnan K, Bathan G, Bose-O'Reilly S, Brauer M, Caravanos J, Chiles T, Cohen A, Corra L, Cropper M (2022) Pollution and health: a progress update. The Lancet Planetary Health 6(6):e535–e547

Funding R (2023) Drinking water state revolving fund

Galloway TS (2015) Micro-and nano-plastics and human health. Marine anthropogenic litter, pp 343–366

García-Lledó A, Rodríguez-Martín S, Tobías A, Alonso-Martín J, Ansede-Cascudo JC, de Abajo FJ (2020) Heat waves, ambient temperature, and risk of myocardial infarction: an ecological study in the Community of Madrid. Revista Española De Cardiología (English Edition) 73(4):300–306

Gavrilaș S, Ursachi CȘ, Perța-Crișan S, Munteanu FD (2022) Recent trends in biosensors for environmental quality monitoring. Sensors 22(4):1513

Gawlik-Kobylińska M, Gudzbeler G, Szklarski Ł, Kopp N, Koch-Eschweiler H, Urban M (2021) The EU-SENSE system for chemical hazards detection, identification, and monitoring. Appl Sci 11(21):10308

Gore AC, Chappell VA, Fenton SE, Flaws JA, Nadal A, Prins GS, Toppari J, Zoeller RT (2015) EDC-2: the Endocrine Society's second scientific statement on endocrine-disrupting chemicals. Endocr Rev 36(6):E1–E150

Harrad S, Drage DS, Sharkey M, Berresheim H (2019) Brominated flame retardants and perfluoroalkyl substances in landfill leachate from Ireland. Sci Total Environ 695:133810

Hayes EK, Gagnon GA (2024) From capture to detection: a critical review of passive sampling techniques for pathogen surveillance in water and wastewater. Water Res, 122024

Kanangire CK, Matano A, Dida GO, Anyona D (2016) A systematic review of effects of emerging pollutants on human health and livelihoods of population living in the Lake Victoria basin of Kenya. Kisumu, Kenya

Kumar A, Dhiman C, Kumar M, Kannappan N, Kumar D, Chourasia MK, Narayan KP (2025) Development of a quality by design based hybrid RP-HPLC method for Glimepiride: Bioanalytical and industrial applications. J Appl Pharmaceutical Sci

Kumar A, Rana R, Saklani R, Kumar M, Yadav PK, Tiwari A, Chourasia MK (2024) Technology transfer of a validated RP-HPLC method for the simultaneous estimation of andrographolide and paclitaxel in application to pharmaceutical nanoformulation. J Chromatogr Sci 62(4):356–363

Kumar A, Sakhare K, Bhattacharya D, Chattopadhyay R, Parikh P, Narayan KP, Mukherjee A (2022) Communication in non-communicable diseases (NCDs) and role of immunomodulatory nutraceuticals in their management. Front Nutr 9:966152

Kumar SS, Arivazhagan S, Rangarajan N (2013) Remote sensing and GIS applications in environmental sciences–a review. J Environ Nanotechnol 2(2):92–101

Kurowska P, Mlyczyńska E, Dawid M, Respekta N, Pich K, Serra L, Dupont J, Rak A (2022) Endocrine disruptor chemicals, adipokines and reproductive functions. Endocrine 78(2):205–218

Landrigan PJ, Fuller R, Acosta NJ, Adeyi O, Arnold R, Baldé AB, Bertollini R, Bose-O'Reilly S, Boufford JI, Breysse PN, Chiles T (2018) The Lancet Commission on pollution and health. The Lancet 391(10119):462–512

Lanphear BP, Hornung R, Khoury J, Yolton K, Baghurst P, Bellinger DC, Canfield RL, Dietrich KN, Bornschein R, Greene T, Rothenberg SJ (2019) Erratum: "Low-level environmental lead exposure and children's intellectual function: an international pooled analysis." Environ Health Perspect 127(9):099001

Lapworth DJ, Baran N, Stuart ME, Ward RS (2012) Emerging organic contaminants in groundwater: a review of sources, fate and occurrence. Environ Pollut 163:287–303

Lara-Martín PA, Li X, Bopp RF, Brownawell BJ (2010) Occurrence of alkyltrimethylammonium compounds in urban estuarine sediments: behentrimonium as a new emerging contaminant. Environ Sci Technol 44(19):7569–7575

Martine P, Rébé C (2019) Heat shock proteins and inflammasomes. Int J Mol Sci 20(18):4508

Medina E, Kim SH, Yun M, Choi WG (2021) Recapitulation of the function and role of ROS generated in response to heat stress in plants. Plants 10(2):371

Mishra A, Kumari M, Kumar R, Iqbal K, Thakur IS (2022) Persistent organic pollutants in the environment: Risk assessment, hazards, and mitigation strategies. Bioresource Technol Reports 19:101143

Naidu R, Espana VAA, Liu Y, Jit J (2016) Emerging contaminants in the environment: Risk-based analysis for better management. Chemosphere 154:350–357

Nel A, Xia T, Madler L, Li N (2006) Toxic potential of materials at the nanolevel. Science 311(5761):622–627

Oulton RL, Kohn T, Cwiertny DM (2010) Pharmaceuticals and personal care products in effluent matrices: a survey of transformation and removal during wastewater treatment and implications for wastewater management. J Environ Monit 12(11):1956–1978

Pamukcu S, Cheng L (eds) (2017) Underground sensing: Monitoring and hazard detection for environment and infrastructure. Academic Press

Poynton HC, Vulpe CD (2009) Ecotoxicogenomics: emerging technologies for emerging contaminants 1. JAWRA J American Water Resour Assoc 45(1):83–96

Puri M, Gandhi K, Kumar MS (2023) Emerging environmental contaminants: a global perspective on policies and regulations. J Environ Manage 332:117344
Rathi BS, Kumar PS, Show PL (2021) A review on effective removal of emerging contaminants from aquatic systems: current trends and scope for further research. J Hazard Mater 409:124413
Richardson SD (2012) Environmental mass spectrometry: emerging contaminants and current issues. Anal Chem 84(2):747–778
Scheurer M, Brauch HJ, Lange FT (2009) Analysis and occurrence of seven artificial sweeteners in German waste water and surface water and in soil aquifer treatment (SAT). Anal Bioanal Chem 394:1585–1594
Schieber M, Chandel NS (2014) ROS function in redox signaling and oxidative stress. Curr Biol 24(10):R453–R462
Serwecińska L (2020) Antimicrobials and antibiotic-resistant bacteria: a risk to the environment and to public health. Water 12(12):3313
Steinemann A (2004) Human exposure, health hazards, and environmental regulations. Environ Impact Assess Rev 24(7–8):695–710
Stuart M, Lapworth D, Crane E, Hart A (2012) Review of risk from potential emerging contaminants in UK groundwater. Sci Total Environ 416:1–21
Taylor F, Cartwright M, Ahmad R (2001) Analytical methods for environmental monitoring
Tollefsen KE, Nizzetto L, Huggett DB (2012) Presence, fate and effects of the intense sweetener sucralose in the aquatic environment. Sci Total Environ 438:510–516
Thompson RC, Olsen Y, Mitchell RP, Davis A, Rowland SJ, John AW, McGonigle D, Russell AE (2004) Lost at sea: where is all the plastic? Science 304(5672):838–838
Uğuz C, İşcan M, Togan İ (2009) Alkylphenols in the environment and their adverse effects on living organisms. Kocatepe Veterinary J 2(1):49–58
USEPA (2023) Clean water act overview. U.S. Environmental Protection Agency
van Eden W (2018) Immune tolerance therapies for autoimmune diseases based on heat shock protein T-cell epitopes. Philos Trans Royal Soc B: Biol Sci 373(1738):20160531
Vishwanathan SS, Hanaoka T, Garg A (2023) Impact of Glasgow climate pact and updated nationally determined contribution on mercury mitigation abiding by the Minamata convention in India. Environ Sci Technol 57(43):16265–16275
Vos T, Barber RM, Bell B, Bertozzi-Villa A, Biryukov S, Bolliger I, Charlson F, Davis A, Degenhardt L, Dicker D, Duan L (2015) Global, regional, and national incidence, prevalence, and years lived with disability for 301 acute and chronic diseases and injuries in 188 countries, 1990–2013: a systematic analysis for the Global Burden of Disease Study 2013. The Lancet 386(9995):743–800
Wang Z, Xie Z, Möller A, Mi W, Wolschke H, Ebinghaus R (2014) Atmospheric concentrations and gas/particle partitioning of neutral poly-and perfluoroalkyl substances in northern German coast. Atmos Environ 95:207–213
World Health Organization (WHO) (2020) Guidelines for personal protective equipment in pesticide application. World Health Organization
Wilson AD (2012) Review of electronic-nose technologies and algorithms to detect hazardous chemicals in the environment. Procedia Technol 1:453–463
World Health Organization (WHO) (2021) WHO global air quality guidelines: particulate matter (PM2. 5 and PM10), ozone, nitrogen dioxide, sulfur dioxide and carbon monoxide. World Health Organization
Ziani K, Ioniță-Mîndrican CB, Mititelu M, Neacşu SM, Negrei C, Moroşan E, Drăgănescu D, Preda OT (2023) Microplastics: a real global threat for environment and food safety: a state of the art review. Nutrients 15(3):617

Chapter 5
Emerging Contaminants from Agricultural Waste and Their Impact on Aquatic Ecosystems

Shreya Singh, Laxmi Kant Bhardwaj, Prangya Rath, and Shivendra Singh

Abstract Aquatic ecosystems are biological systems that have played an important role in managing the existence of the Earth. However, they are facing expanding defilement because of human actions like urbanization, industrialization, and agriculture. Water pollution, starting from the superfluous use of pesticides, excrements, untreated sewage, and other emerging contaminants (ECs) compromises the idea of maritime conditions and stances prosperity risks like the spread of powerful diseases and mortality. Water pollution can damage fish diversity, compromise immune structures, and lead to diseases like equilibrium rot and gill hurt. Agriculture is the base of human development which contributes to environmental degradation by agricultural waste production. Waste, arising from various farming practices, includes manure, crop residues, and chemical inputs like pesticides and fertilizers. Waste released in water bodies, like insecticide, pesticides, weighty metals, and modern waste that affect aquatic organisms, results in mortality and biological system disturbance. The toxic effect of using farming practices along with different human actions as viable and useless is known as squandering material. It contains a secure and liquid structure as ecological waste and causes damage to both humans and animals. Introducing a financial pathway leads to the direct degradation of contamination into useful substances. This investigation examined the influence of producing waste on seagoing circumstances, with an emphasis on believable organization and mechanical events.

S. Singh
Department of Agriculture, Ramlalit Singh Mahavidyalaya, Mirzapur, Uttar Pradesh, India

L. K. Bhardwaj
Amity Institute of Environmental Toxicology, Safety and Management, Amity University, Noida, Uttar Pradesh, India

P. Rath (✉)
Amity Institute of Environmental Science, Amity University, Noida, Uttar Pradesh, India
e-mail: prangyarath71@gmail.com

S. Singh
Department of Agriculture, Udai Pratap College, Bhojubir, Varanasi, Uttar Pradesh, India

H. Jain et al. (eds.), *Emerging Contaminants in the Aquatic Environment*, Advances in Wastewater Research, https://doi.org/10.1007/978-981-96-8548-6_5

Keywords Agricultural waste · Aquatic ecosystems · Emerging contaminants (ECs) · Pollution mitigation · Water pollution

Abbreviations

AgCs	Agrochemicals
AI	Artificial intelligence
AOPs	Advanced oxidation processes
ARGs	Antibiotic resistance genes
As	Arsenic
AWMS	Agricultural Waste Management System
BPA	Bisphenol A
Cd	Cadmium
CO_2	Carbon dioxide
Cr	Chromium
DDT	Dichlorodiphenyltrichloroethane
ECs	Emerging contaminants
EDCs	Endocrine-disrupting chemicals
ELISA	Enzyme-linked immunosorbent assay
FTIR	Fourier-transform infrared spectroscopy
GC	Gas chromatography
GC-ECD	Gas chromatography- electron capture detector
GC-MS	Gas chromatography-mass spectrometry
Hg	Mercury
HPLC	High-performance liquid chromatography
LC-MS	Liquid chromatography-mass spectrometry
LRRS	Low-resolution Raman spectroscopy
MF	Membrane filtration
MPs	Microplastics
NGOs	Non-governmental organizations
NPP	Net primary productivity
NPs	Nanoparticles
NPS	Non-point sources
O_2	Oxygen
PAHs	Polycyclic aromatic hydrocarbons
Pb	Lead
PBDEs	Polybrominated diphenyl ethers
PCBs	Polychlorinated biphenyls
PCPs	Personal care products
PFAS	Per- and poly-fluoroalkyl substances
POPs	Persistent organic pollutants
ppb	Parts per billion
PPCPs	Pharmaceuticals and personal care products

SDGs	Sustainable development goals
SWWTPs	Smart wastewater treatment plants
US	United States
UV	Ultraviolet
VOCs	Volatile organic compounds
WWTPs	Wastewater treatment plants

5.1 Introduction

Water environments covering more than two-thirds of the Earth's surface play major roles in moderating climate and offering inestimable valuable services to the ever-expanding global population. The marine and freshwater fellow is uniquely important for the sustenance of life on this planet as noted by Hain et al. (2014). Lakes, waterways, streams, wetlands, and other water bodies support the higher profile of global net primary productivity (NPP) and contribute more than 50% of global NPP. Seas and oceans are also capable of producing roughly half of the global biomass, sequestering about 50% of the carbon dioxide (CO_2) emissions due to human activities, and supporting the generation of about half of the atmospheric oxygen (O_2) from the other 50% (Bhat et al. 2017). As has been described above, these aquatic environments are extensive and crucially significant for the existence of the world population.

One of the special concerns is maintaining the decreased stability of these environments in order to ensure an environmentally friendly future for all living organisms on Earth. These different forms of ecosystems have been substantially degraded and polluted through the level of human interference. Pollution refers to activities that are carried out by human beings, or activities that come from human beings, which produce harm and, in this case, produce substances or energy within an environment. However, human beings further encroach on the expansion of the diverse activities in these ecosystems which are increasing regularly. Many human activities cause the accidental discharge of various ECs into water bodies, endangering living species and decreasing water quality, rendering them unsuitable to be utilized (Ekubo and Abowel 2011; Bhardwaj et al. 2019; Bhardwaj and Jindal 2020). ECs in agriculture are chemical compounds that are not monitored properly but have the potential to cause adverse ecological and human effects.

These chemical compounds are seen to occur both in terrestrial and aquatic environments and their intensity differs, and it is not limited to only organic matter but may also come from inorganic content (Mahesh et al. 2022). The organic contaminants include a variety of substances, such as agrochemicals (AgCs), endocrine-disrupting chemicals (EDCs), hormones, industrial chemicals, persistent organic pollutants (POPs), personal care products (PCPs), pesticides, pharmaceutical compounds, veterinary drugs (Bhardwaj et al. 2018; Impellitteri et al. 2023). Inorganic contaminants include microplastics (MPs) as well as heavy metals (Srikanth et al. 2019;

Alam et al. 2023; Bhardwaj et al. 2024a). These chemicals can be of synthetic origin or naturally occurring (Hossein 2019).

Toxic metals and nano-compounds are highly toxic to microorganisms (Bhardwaj et al. 2023a). Recent evaluations show that horticulture is rising as a significant supporter of worldwide water tainting. Cultivating tasks are liable for the arrival of a lot of ECs into water bodies, including AgCs, manures, natural matter, drug buildups, silt, and saline. This chapter focuses on the occurrence, sources, transport pathway, and detection methods of ECs in agricultural waste. This study also focuses on agricultural waste, its types, and the challenges and future direction in the detection of ECs.

5.2 Agriculture Waste, Types, and Their Sources

Agricultural waste is known as agricultural trash or agro waste. These wastes are residue substances used in the manufacturing and processing of agricultural products (Obi et al. 2016). These wastes include crops, livestock rearing, food, vegetables, and aquaculture. The total amount of waste from agriculture varies according to particulars. These wastes may be liquids, slurries, or solids and contain hazardous ECs that contaminate the environment (Karić et al. 2022). Different types of agricultural wastes are shown in Fig. 5.1. These wastes pose a toxicity risk to plants, animals, and humans through both direct and indirect routes. However, these substances may include elements that are helpful to people. The expenses involved with their collection, transportation, and processing are generally offset by the benefits of using agricultural waste. Table 5.1 highlights different agricultural wastes along with their environmental impacts as well as management practices used.

However exact numbers on rural waste creation are scant and assessments of agricultural waste generation are scarce, it is widely accepted that they contribute significantly to the overall waste volume in developed countries. Increased agricultural production has inevitably resulted in more livestock waste, crop waste, and agro-industrial byproducts. Moreover, in the event that emerging nations keep on growing their cultivating rehearses, crop buildup is expected to increase fundamentally on an overall scale. As indicated by existing assessments, about 998 million tons of farming waste is produced every year (Maji et al. 2020). Organic matter can account for up to 80% of total solid waste generated on a farm, including manure itself producing up to 5.27 kg per day per 1000 kg of live weight, based on wet weight (Abou Hussein and Sawan 2010). Currently, these wastes are treated utilizing the "3R" method, which emphasizes reduction, reuse, and recycling (Liu et al. 2017). The Agricultural Waste Management System (AWMS) was evaluated, with a focus on typical waste management solutions for chicken farms (Koul et al. 2022). Farms utilize six agricultural waste management functions. These procedures are concerned with the gathering, storage, management, processing, recycling, and disposal of agricultural waste (Abou Hussein and Sawan 2010).

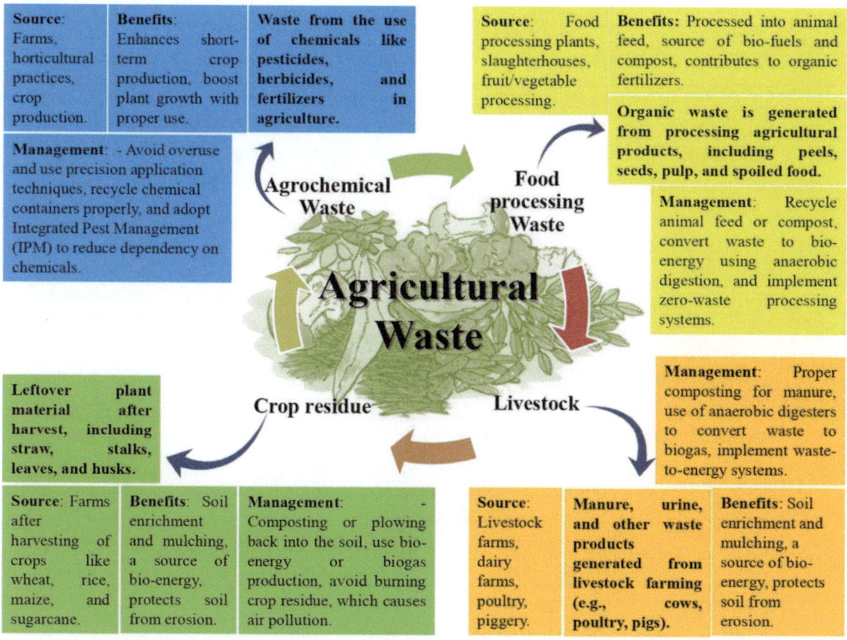

Fig. 5.1 Different types of agricultural waste

5.3 Sources of Emerging Contaminants (ECs) in Agriculture

The origins and routes of ECs within aquatic ecosystems exhibit considerable diversity and complexity (Zhang et al. 2024). Aquatic ecosystems have been reported to be repositories for pathogens. These pathogens enter aquatic ecosystems because of human activities, discharged industrial wastes, and chemical pollutants. Such pollutants contain detergents, heavy metals, microfibers, plastics, and different types of toxic materials (Hampel et al. 2015; Bhat et al. 2017). Per- and poly-fluoroalkyl substances (PFAS) are a class of synthetic chemicals that are also called forever chemicals (Dias et al. 2024).

PFAS are used for different types of applications, such as in firefighting foams, making non-stick cooking utensils, and preparing water-proof clothes. They are also used in a wide range of industrial processes. Manufacturing processes, chemical processing activities, and disposal of industrial wastes are the key sources of polybrominated diphenyl ethers (PBDEs), dioxins, industrial chemicals, and their byproducts. Nanoparticles (NPs) are used in the electronics industries, the medical field, and as components of other products (Bhardwaj et al. 2023b). There are three principal sources of ECs in agriculture: effluent water, livestock, manure, pesticides, agrochemicals, water irrigation, contaminated water bodies drainages, plastic waste, and fertilizers (Rathi et al. 2021).

Table 5.1 Agricultural waste with their environmental impact and management practices

S. no	Agricultural waste types	Description	Environmental impact	Management practices
1	Pesticide containers	Leftover pesticide holders are frequently disposed of into fields or lakes after use, adding to natural contamination	Foodborne diseases and lacking sanitation rehearse	Legitimate removal and reusing of pesticide compartments
2	Excessive fertilizer	Farmers frequently apply more compost than needed, prompting natural contamination through overflow and groundwater contamination	Contamination of surface water bodies	Carrying out preservation practices to lessen loss
3	Livestock waste	Includes strong waste like compost, wastewater by doing the washing, and air poisons like smelling salts and methane	Air pollution, soil fertility degradation, and water pollution from runoff and leaching	Appropriate waste drainage frameworks in animal offices
4	Aquaculture waste	Generate from aquaculture takes care of metabolic waste and waste from fish feces	Water contamination from disintegrated and suspended squander	Normal checking of water quality and waste levels
5	Slaughterhouse waste	Includes natural materials from slaughterhouse tasks, adding to ecological contamination when not appropriately managed	Soil, water, and air pollution from putrefaction processes	Consistency with administrative norms to squander the executives

5.3.1 Treated Wastewater and Effluents

In water-scar regions, the usage of treated wastewater is increasing for agricultural purposes. However, this water may still contain residues of ECs. ECs in treated wastewater and effluents originate from human activity and industrial processes. The mainstream pharmaceuticals and personal care products (PPCPs), MPs, and industrial chemicals involved are mostly found in wastewater effluents and these pollutants are not completely adsorbed in traditional wastewater treatment processes (Bhardwaj 2023).

5.3.2 Livestock and Manure

The veterinary uses of tetracycline and sulfonamides in the management of livestock have the ability to leach into the environment through the spreading of manure (Gaballah et al. 2021). Antibiotics can be used to encourage the transference of antibiotic resistance genes (ARGs) from a source to the environment, which is a threat to public health. Large and small animal veterinary facilities and animal feeding operations act as the main sources of ARGs (Hou et al. 2015). In developing countries, municipal solid waste is used as a compost that may have ECs. Humans are releasing most of the 443 antibiotic compounds, 99% antidepressants, and other pharmaceuticals into the environmental system (Bhardwaj et al. 2024b).

5.3.3 Pesticides, Agrochemicals (AgCs), and Fertilizers

POPs, for e.g., dichlorodiphenyltrichloroethane (DDT), and lindane are facing bans and restrictions (Bhardwaj and Jindal 2022). However, they may still be found as residual pollutants in the soil. Neonicotinoids and glyphosate-based herbicides exhibit environmental persistence and exert influence on soil microbiota (Kremer 2021). Biosolids have been found to comprise MPs, heavy metals, hormones, and POPs. By using biosolids as fertilizers, it becomes easier for such contaminants to be brought into the soil (Pozzebon and Seifert 2023). Considering the use of synthetic fertilizers, the following heavy metals can be brought into the soil: cadmium (Cd), arsenic (As), and chromium (Cr). These fertilizers are identified to contain these harmful substances, and their usage in large fields has been realized to result in the build-up of these substances (Atafar et al. 2010). The leading sources of these chemicals are the processes of farming and especially the discharge of the leftovers of these chemicals into water sources.

5.3.4 Irrigation and Runoff from Contaminated Water Bodies

Electrolytic conductivities in the water used for irrigation or runoff from water sources with water pollutants may stem from different sources. Rivers and lakes are some of the important water sources for irrigation uses. These systems may receive industrial wastewater or partially treated wastewater which contain ECs and hence pollute agricultural land. Bating crops by pesticide use in agricultural practices; could lead to contaminating groundwater for a long time (Srivastav 2020). Operations across agricultural industries let out environmental contaminants such as pesticides, fertilizers, and animal wastes. Municipal wastewater discharges, industrial wastewater, and urban runoff carry several ECs and can contaminate the aquatic ecosystem (Gupta et al. 2023). Artificial recharge may introduce ECs into groundwater (Renau-Pruñonosa et al. 2020).

5.3.5 Plastic Mulch and Agricultural Plastics

ECs in plastic mulch and agricultural plastics can be created from various traces. Plastic pipes, films, and other plastic products are used in agriculture and these products can break down into MPs (Gündoğdu et al. 2022). The size of these particles is <5 mm (Bhardwaj 2024). Phthalate and bisphenol A (BPA) are already a class of EDs that have the capability of entering water and soil through plastic products (Bhardwaj and Sharma 2021). Different alterations such as plasticizers, ultraviolet (UV) stabilizers, flame retardants, pigments, and dyes are used in the formulation process of plastic products. However, these additives are relatively dangerous materials and might leach into the environment (Bhardwaj 2022). Most of the plastics break down by the effects of UV and sunny weather, temperature fluctuations, and microbial action which produce MPs, nano-plastics, and other chemical compounds. Production of plastics involves its manufacture from monomers and other components such as oligomers and catalysts which on many occasions are left with the final products and may cause pollution.

Some ECs like heavy metals and POPs may be absorbed by the plastics when the plastics interact with the surrounding soil. Recycled agricultural plastics may introduce ECs into the environment. Some plastics are coated with hydrophobic coating agents to change their permeability rates, but these agents may cause pollution. Plastic storage and transport modes are vulnerable to exposure to these EDs to harm agricultural land through pollution.

5.3.6 Atmospheric Deposition

The following discussion explains that the daily emission of ECs into the atmosphere comes from both natural and human-initiated activities. The industrial emissions, agricultural application, transportation, waste digestion and disposal, and urban water overflow are part of the anthropogenic source while wildfires, volcanic activities, soil, and dust fallout are part of natural sources (Bhardwaj and Vikram 2023). Industrial pollutions include PFAS, volatile organic compounds (VOCs), mercury (Hg), lead (Pb), Cd, POPs, PPCPs, and NPs (Falandysz et al. 2024). Some of the chemical compositions used in agriculture include the use of organophosphates and pyrethroids released into the atmosphere (Ravula and Yenugu 2021). Exemplarily, ammonia emission, which occurs after fertilizer application, may cause eutrophication (Bhardwaj et al. 2023a). The emissions from the burning of fossil fuels such as car and plane engines and tire abrasion give out MPs, particulates, polycyclic aromatic hydrocarbons (PAHs), and black carbon which can spread through air transport systems (Bhardwaj et al. 2021). Waste incineration raises the possibility of air pollution in terms of the emission of dioxins, furans, and some heavy metals. The pollutants can be delivered to the agricultural fields through processes like direct fall, dust fall, and any atmospheric processes.

5.4 Transport Pathways of Emerging Contaminants (ECs)

ECs can move through surface water, groundwater, atmosphere, soil and sediment, bioaccumulation, and human pathways (Deere et al. 2021). Table 5.2 gives an overview of different types of ECs and their sources and pathways. These pollutants can affect water, soil, and biota. The behavior of ECs depends for the most part on their chemical properties, the environment, and additional impurities that might be present in the pathway. Facilities handling wastewater, agricultural drainage, and industries also directly introduce emerging pollutants to water in rivers, lakes, and oceans (Carere et al. 2021). Heavy rainfall can introduce ECs to surface water through untreated sewage and stormwater overflow. ECs can leach from contaminated soils, landfills, and agricultural fields into groundwater. Treated wastewater used for irrigation or groundwater recharge can introduce ECs. VOCs can be deposited onto water bodies or land after traveling long distances. NPs and MPs can become airborne and contribute to atmospheric contamination. Many ECs adsorb onto the soil particles and sediments. These contaminants may remain as such for a long time and may alter the lives of aquatic organisms. MPs and polychlorinated biphenyls (PCBs) are ingested by aquatic organisms and accumulate, also their concentration increases with their movement up the food chain. ECs admitted into plants by water and soils, and they are transported to all parts of the plants. ECs can enter the human body through contaminated drinking water, air, food, and PCPs. Skin disease and infection can be caused through exposure to ECs.

5.5 Impacts of Emerging Contaminants (ECs) on Aquatic Ecosystems

ECs contaminate water bodies and affect aquatic organisms. Personal hygiene products, MPs, NPs, and antibiotics used by humans can contaminate aquatic ecosystems (Hampel et al. 2015). Water pollution has major long-term economic ramifications for humans, leading to different health conditions such as dysentery, diarrhea, abortion, premature delivery, and stomach difficulties (Purnamitta 2012). The current state of contamination of water is exacerbated by the reduction in water flow in aquatic systems, which is mostly caused by crop irrigation requirements.

Table 5.2 Different types of emerging contaminants (ECs) with their sources and pathways

S. no	Emerging contaminants (ECs)	Sources	Pathways
1	Pharmaceuticals compounds	Industrial wastewater effluent, and human excretion	Surface water, and groundwater
2	Endocrine disrupter chemicals (EDCs)	Plastics, pesticides, as well as chemicals from industries	Surface water bodies, soil, as well as biota
3	Per- and poly-fluoroalkyl substances (PFAS)	Firefighting foams, non-sticky items, waterproof clothes, as well as processes of industries	Groundwater, as well as surface water bodies
4	Microplastics (MPs)	Plastic degradation, and textiles	Surface water, soil, and air
5	Pesticides	Agriculture, and urban landscaping	Agricultural runoff, and groundwater
6	Industrial chemicals	Manufacturing, and industrial discharge	Air, surface water, and soil
7	Antibiotic resistance genes (ARGs)	Medical and agricultural sectors	Wastewater, soil, and biota

Water contamination is a growing and significant problem in India (Sharma et al. 2020). Cultivation and horticultural practices overburden water accessibility and produce wastewater through the abuse of pesticides and composts (Ivanov et al. 2015). The residue of ECs remains as such for a long time and can constitute non-point sources (NPS) of pollution during storm and runoff events (Souza et al. 2020). Agriculture activity produces various compounds like aldrin, endrin, chlordane, As, and Cd which eventually end up in water bodies (Rahman and Debnath 2015). Khanna and Gupta (2018) studied the environmental effects of water pollution and its subsequent usage for irrigation. Nutrient runoff from agricultural and domestic resources transported by rivers to the sea promotes coastal eutrophication, resulting in algal blooms and hypoxia (Dungait et al. 2012). Water quality degradation presents a considerable number of issues for downstream water users, which are exacerbated by population increase, rising water demand, and agricultural and industrial activities.

Horticulture, in particular, contributes altogether to the debasement of water environments in numerous well-off countries. For example, the European Association reports that 38% of water habitats are subject to substantial agricultural stress. In a comparable manner, agricultural practices represent the primary source of pollution in the rivers and streams across the United States (US). The lake is positioned as the third largest in size and the second largest in terms of wetland area (Mateo-Sagasta et al. 2017).

Soil management for major agricultural production is one of the two primary causes of water pollution in China where nitrogen is found to have contaminated the groundwater (Mateo-Sagasta et al. 2017). In developing countries, the discharge of raw municipal and industrial effluents and the lack of proper management of it put forward certain dangers. These pollutants contribute to the degradation of aquatic environments, leading to harmful effects due to their toxicity (Hampel et al. 2015). Agrarian activities contribute essentially to water contamination by delivering compost, pesticides, and residue from fields into adjoining waterways. Adopting correct managerial practices is crucial for reducing agricultural chemical emissions and protecting water resources (Dalin et al. 2012). Preservation of the natural environment plays an important role in the provision of water that is fit for use within ecosystems and in economic practices such as domestic, agricultural, and industrial uses (Kroeze et al. 2012).

5.6 Detection Methods/Techniques for Emerging Contaminants (ECs)

There are several approaches to identifying and quantifying ECs in aquatic environments depending on the source, type, and concentration of the ecologies. Some of the following techniques are used: high-performance liquid chromatography (HPLC), gas chromatography (GC), liquid chromatography-mass spectrometry (LC-MS), gas chromatography-mass spectrometry (GC-MS), enzyme-linked immunosorbent assay (ELISA), Fourier-transform infrared spectroscopy (FTIR), and Raman spectroscopy, etc. The analytical techniques developed reveal high sensitivity, and this identifies the pollutants present in parts per billion (ppb) concentration (Stefanakis and Becker 2016). The use of chromatography enhances the assessment and determination of pollutants in water, soil, sediments, and biota samples. The first is the extraction of the pollutants from the samples, then passing the samples through chromatographic procedures.

ECs such as POPs, PFOS, and other volatile contaminants are analyzed through gas chromatography-electron capture detector/mass spectrometry (GC-ECD/GC-MS). Some POPs are analyzed by HPLC/LC-MS (Maurya et al. 2021). The structure and shape of MPs are analyzed by spectroscopy techniques. Raman spectroscopy and FTIR both are based on the analysis of the absorption and transmission spectra of compounds (Manivannan et al. 2022). Low-resolution Raman Spectroscopy (LRRS) is used for the detection of algae from aquatic environments (Adejimi et al. 2022). These techniques help in the estimation of the molecular structure of the chemical compound. ELISA is a biosensor-based technique used for the detection of specific EC such as antibiotics. These advanced techniques are very effective in the detection of ECs as compared to traditional techniques. MS is the best technique for the identification of unknown new pollutants based on their mass.

5.7 Mitigation and Management Strategies

Due to their persistent nature, ECs can disrupt the aquatic food chain (Clance et al. 2023). Effective mitigation and management strategies are essential to reduce or remove ECs from aquatic environments. Conventional treatments have been used for water treatment for a long time. These treatment processes contain primary, secondary, and tertiary steps. During primary treatment, large particles are removed physically through sedimentation. During secondary treatment, biological processes are used to degrade organic matter (Shah 2024a, b). However, studies have shown that the complete removal of ECs cannot be completely carried out by conventional methods. Studies have revealed that such contaminants persist even after traditional treatment processes and thus accumulate in aquatic ecosystems thereby posing significant risks to the health of humans as well as wildlife (Gogoi et al. 2018; Rasheed et al. 2019).

These inefficacies of conventional methodologies have necessitated the need for exploring more advanced and integrated technologies for addressing challenges that are generated due to ECs. Emerging technologies have been observed to provide promising solutions for effectively removing different ECs, especially from water resources (Varsha et al. 2022). Techniques such as advanced oxidation processes (AOPs), membrane filtration (MF), and adsorption methods possess significant potential. AOPs, like ozonation as well as photocatalysis, effectively decompose complex organic contaminants to lesser toxic constituents (Bilal et al. 2019). Technologies utilizing MF processes like nanofiltration as well as reverse osmosis have been studied to significantly remove different types of ECs. These are observed to be cost-prohibitive (Talib and Randhir 2016). Adsorption by using low-cost adsorbents, for e.g., biochar, and agricultural waste, has been studied to be a sustainable option removing ECs (Rathi and Kumar 2021; Varsha et al. 2022). Furthermore, methods using biological treatment including the usage of microalgae as well as enzyme-assisted biodegradation have been shown to be effective in degrading and removing ECs present in wastewater (Singh et al. 2021).

Such emerging technologies are currently at a naive stage and still require optimization. They are thus considered critical advancements in water treatment studies. ECs' effective management of aquatic ecosystems needs robust policies as well as regulatory frameworks. Existing regulations have been observed to be less effective in addressing different complexities that are associated with ECs. The primary reason for such shortage is due to a lack of comprehensive health standards and guidelines for addressing the issues related to these contaminants (Gogoi et al. 2018). Policymakers are required to develop and implement strict regulations which could encompass all stages of ECs, i.e., from their production and usage to their disposal and treatment.

This requires setting up permissible concentration limits for EC in different water resources. It also requires promoting the usage of sustainable management practices and would also encourage the adoption of other treatment technologies that are comparatively advanced. Public awareness as well as educating people through different campaigns are encouraged so as to reduce ECs release into the environment

(Talib and Randhir 2016; Deere et al. 2021). Collaborative interactions and efforts are required between different organizations, government agencies, industries, and the scientific community are important to establish policies and regulations that are effective and simultaneously safeguard the health of the environment as well as the public.

5.8 Challenges and Future Directions

Agricultural waste contains a mixture of ECs such as pesticides, heavy metals, POPs, and MPs that can have a hazardous impact on aquatic ecosystems. Due to the complex and diverse nature of ECs, the detection and quantification are complicated (Richardson and Kimura 2016). ECs resist degradation and can bioaccumulate in aquatic organisms then lead to toxicity in aquatic organisms (La Farre et al. 2008). The long-term impacts of ECs remain poorly understood at low concentrations and in combination with other stressors. Behavioral and reproductive changes are less studied in aquatic organisms but may have substantial implications for ecosystem health. Poor waste management practices in agriculture can contribute to the leaching of ECs into water bodies (Rashmi et al. 2020). Overdose of AgCs can increase the problem of contaminant contamination in regions with intensive farming practices. There are limited guidelines and regulations for many ECs because the data is limited on the occurrence of ECs in agricultural waste, fate, and their effects on aquatic organisms. Researchers focus on legacy pollutants in their studies as compared to newer contaminants due to the availability of data.

Advancing analytical techniques are required for the detection of ECs at trace levels. The employing of the predictive modeling instrument to assess the fate, transport, and ecological risks of ECs in aquatic ecosystems will be beneficial in future research. After incorporating the effects of climate change, they may alter contaminant behavior and affect aquatic ecosystem resilience. Sustainable agricultural practices such as the use of organic farming, and reduced pesticide application will be helpful to minimize the input of contaminants. Biopesticides and organic fertilizers may be the alternative to AgCs because they have less environmental impact. The Smart Wastewater Treatment Plants (SWWTPs) can remove ECs very efficiently. Phytoremediation and wetlands act as the two natural solutions to the removal of ECs prior to their escalation into the aquatic environment. Because the management of AgCs and pharmaceutical compounds pollutants may contribute to the pollution of aquatic water bodies in agriculture, guidelines on the safe use and disposal of these products should be provided. A comprehensive monitoring program together with ecotoxicological tests can be able to capture different kinds of water organisms and help to get a more extensive picture of the water pollution effects at the chronic and cumulative processes.

5.9 Recommendations

The environmental implications of toxic crop residue need to be evaluated, along with measures for their effective management. Integrating EC assessments in aquatic bodies into large-scale environmental evaluations and Sustainable Development Goals (SDGs) research is essential. The problem of ECs requires immediate action by means of stricter environmental rules and regulations, especially with reference to industrial-scale pollution. We recommend some suggestions to reduce or eliminate the pollution of ECs from the aquatic ecosystem.

- Artificial intelligence (AI) systems should be used with a waste management system for smart functioning.
- Regular monitoring of ECs should be done by the authority body to reduce the level of contamination.
- Scientist should focus on new emerging pollutants in their study.
- There is a need for people to use eco-friendly and rational AgCs and their substitutes.
- In addition, these authoritative/standard-setting organizations and non-governmental organizations (NGOs) should also carry out training for farmers in order to encourage them to adopt sustainable agricultural practices.
- The existing Wastewater Treatment Plants (WWTPs) should be changed to SWWTPs for the removal of ECs.
- Environmental policies and regulations related to the removal of ECs from aquatic environments should be updated.
- That is why informal and popular education of farmers, industries, and the common public regarding the hazards related to the ECs and the importance of practicing sustainable agriculture must be mandated.
- Research indicates that the removal of ECs requires synergy between researchers, policymakers, and agricultural practices.
- Collaboration among researchers, policymakers, and agricultural communities is essential for the development and implementation of effective strategies aimed at the removal of ECs.

5.10 Conclusions

Agricultural waste, therefore, means any waste product that is produced in the process of agriculture, and which is not utilized for human or animal feed. ECs are toxic to the environment and are found in agricultural wastes. The release of ECs from agricultural waste contaminates the aquatic ecosystem. The accumulation of ECs in aquatic ecosystems is a worldwide concern. The presence of ECs in effluents and in-land disposal has drawn the interest of policymakers and researchers owing to their probable hazard to human health and the ecosystem. The management and disposal of agricultural waste present an important challenge for the scientific community and authorities. Untreated or deficiently treated wastewater can cause serious water pollution that can be hazardous to aquatic organisms. Eventually, distinguishing sully processes including overflow, filtering, and direct release is basic for making powerful contamination control and counteraction methods to protect water assets and environments from perilous ECs. Endeavors to limit contamination utilizing better land the executives, wastewater treatment, and contamination control techniques are basic for safeguarding the biological system alongside general well-being. This study demonstrated that land-based human activities have far-reaching consequences for aquatic ecosystems, regardless of whether the pollution source is geographically remote within the impacted areas.

Acknowledgements The authors express their sincere gratitude to their respective institutes for their unwavering support, resources, and cooperation, which played a crucial role in the timely completion of this study. Their encouragement and constant assistance are highly appreciated.

References

Abou Hussein SD, Sawan OM (2010) The utilization of agricultural waste as one of the environmental issues in Egypt (a case study). J Appl Sci Res 6(8):1116–1124

Adejimi OE, Ignat T, Sadhasivam G, Zakin V, Schmilovitch ZE, Shapiro OH (2022) Low-Resolution Raman Spectroscopy for the detection of contaminant species in algal bioreactors. Sci Total Environ 809:151138

Alam S, Bhardwaj LK, Mallick R, Rai S (2023) Estimation of heavy metals and fluoride ion in vegetables grown nearby the stretch of River Yamuna, Delhi (NCR). India. Indian J Environ Protection 43(1):64–73

Atafar Z, Mesdaghinia A, Nouri J, Homaee M, Yunesian M, Ahmadimoghaddam M, Mahvi AH (2010) Effect of fertilizer application on soil heavy metal concentration. Environ Monit Assess 160:83–89

Bhardwaj LK, Jindal T (2022) Polar ecotoxicology: Sources and toxic effects of pollutants. New frontiers in environmental Toxicology, pp 9–14

Bhardwaj LK, Chauhan A, Ranjan A, Jindal T (2018) Persistent organic pollutants in biotic and abiotic components of Antarctic pristine environment. Earth Syst Environ 2:35–54

Bhardwaj LK, Sharma S, Ranjan A, Jindal T (2019) Persistent organic pollutants in lakes of Broknes peninsula at Larsemann Hills area, East Antarctica. Ecotoxicology 28:589–596

Bhardwaj LK, Jindal T (2020) Persistent organic pollutants in lakes of Grovnes Peninsula at Larsemann Hill area. East Antarctica. Earth Syst Environ 4(2):349–358

Bhardwaj LK, Sharma A (2021) Estimation of physico-chemical, trace metals, microbiological and phthalate in PET bottled water. Chemistry Africa 4(4):981–991

Bhardwaj LK, Vikram V (2023) Air pollution and its effect on human health. Futuristic trends in agriculture engineering & food sciences 3, 1. IIP Proceedings

Bhardwaj LK (2022) Evaluation of bis (2-ethylhexyl) phthalate (DEHP) in the PET bottled mineral water of different brands and impact of heat by GC–MS/MS. Chem Africa 5(4):929–942

Bhardwaj LK (2024) Occurrence of microplastics (MPs) in Antarctica and its impact on the health of organisms. Maritime Technol Res 6(2):265418

Bhardwaj LK, Kumar D, Kumar A (2023a) Phytoremediation potential of Ocimum sanctum: a sustainable approach for remediation of heavy metals. In: Phytoremediation potential of medicinal and aromatic plants, pp 46–57. CRC Press

Bhardwaj LK, Rath P, Choudhury M (2023b) A comprehensive review on the classification, uses, sources of nanoparticles (NPs) and their toxicity on health. Aerosol Sci Eng 7(1):69–86

Bhardwaj LK, Rath P, Bajpai S, Upadhyay D, Jain H, Kumar N, Sinha S (2024b) COVID-19 and the interplay with antibacterial drug resistance. In: Frontiers in combating antibacterial resistance: current perspectives and future horizons, pp 246–273. IGI Global

Bhardwaj LK, Rath P, Yadav P, Gupta U (2024b) Microplastic contamination, an emerging threat to the freshwater environment: a systematic review. Environ Syst Res 13(1):8

Bhardwaj LK, Sharma S, Jindal T (2021) Occurrence of polycyclic aromatic hydrocarbons (PAHs) in the lake water at Grovnes Peninsula Over East Antarctica. Chem Africa 4:965–980

Bhat SA, Meraj G, Yaseen S, Pandit AK (2017) Statistical assessment of water quality parameters for pollution source identification in Srinagar City, Kashmir: a case study. J Environ Prot 8(6):580–597

Bilal M, Adeel M, Rasheed T, Zhao Y, Iqbal HM (2019) Emerging contaminants of high concern and their enzyme-assisted biodegradation–a review. Environ Int 124:336–353

Carere M, Antoccia A, Buschini A, Frenzilli G, Marcon F, Andreoli C, Gorbi G, Suppa A, Montalbano S, Prota V, De Battistis F (2021) An integrated approach for chemical water quality assessment of an urban river stretch through Effect-Based Methods and emerging pollutants analysis with a focus on genotoxicity. J Environ Manage 300:113549

Clance LR, Ziegler SL, Fodrie FJ (2023) Contaminants disrupt aquatic food webs via decreased consumer efficiency. Sci Total Environ 859:160245

Dalin C, Konar M, Hanasaki N, Rinaldo A, Rodriguez-Iturbe I (2012) Evolution of the global virtual water trade network. Proc Natl Acad Sci USA 109:5989–5994

Deere JR, Streets S, Jankowski MD, Ferrey M, Chenaux-Ibrahim Y, Convertino M, Isaac EJ, Phelps NB, Primus A, Servadio JL, Singer RS (2021) A chemical prioritization process: applications to contaminants of emerging concern in freshwater ecosystems (Phase I). Sci Total Environ 772:146030

Dias D, Bons J, Kumar A, Kabir MH, Liang H (2024) Forever chemicals, per-and polyfluoroalkyl substances (PFAS). Lubrication. Lubricants 12(4):114

Dungait JA, Cardenas LM, Blackwell MS, Wu L, Withers PJ, Chadwick DR, Bol R, Murray PJ, Macdonald AJ, Whitmore AP, Goulding KW (2012) Advances in the understanding of nutrient dynamics and management in UK agriculture. Sci Total Environ 434:39–50

Ekubo AA, Abowel MN (2011) Impacts of crude oil pollution on aquatic organisms. Environ Pollut 12(2):189–205

Falandysz J, Liu G, Rutkowska M (2024) Analytical progress on emerging pollutants in the environment: An overview of the topics. TrAC trends in analytical chemistry, p 117719

Gaballah MS, Guo J, Sun H, Aboagye D, Sobhi M, Muhmood A, Dong R (2021) A review targeting veterinary antibiotics removal from livestock manure management systems and future outlook. Biores Technol 333:125069

Gogoi A, Mazumder P, Tyagi VK, Chaminda GT, An AK, Kumar M (2018) Occurrence and fate of emerging contaminants in water environment: a review. Groundw Sustain Dev 6:169–180

Gündoğdu R, Önder D, Gündoğdu S, Gwinnett C (2022) Plastics derived from disposable greenhouse plastic films and irrigation pipes in agricultural soils: a case study from Turkey. Environ Sci Pollut Res 29(58):87706–87716

Gupta P, Sharma A, Bhardwaj LK (2023) Solid waste management (SWM) and its effect on environment & human health. Futuristic trends in agriculture engineering & food sciences 3, 7. IIP Proceedings

Hain MP, Sigman DM, Haug GH (2014) Carbon dioxide effects of Antarctic stratification, North Atlantic Intermediate Water formation, and subantarctic nutrient drawdown during the last ice age: Diagnosis and synthesis in a geochemical box model. Global Biogeochem Cycles 28(8):757–773

Hampel M, Ellenbecker M, Rivera J, Brewer SK (2015) Aquatic pollution and its effects on living organisms. Environ Toxicol 3(2):45–67

Hossein M (2019) Toxicological aspects of emerging contaminants. Emerging and eco-friendly approaches for waste management, pp 33–58

Hou J, Wan W, Mao D, Wang C, Mu Q, Qin S, Luo Y (2015) Occurrence and distribution of sulfonamides, tetracyclines, quinolones, macrolides, and nitrofurans in livestock manure and amended soils of Northern China. Environ Sci Pollut Res 22:4545–4554

Impellitteri F, Multisanti CR, Rusanova P, Piccione G, Falco F, Faggio C (2023) Exploring the impact of contaminants of emerging concern on fish and invertebrates physiology in the Mediterranean Sea. Biology 12(6):767

Ivanov V, Stabnikov V, Ahmed Z, Dobrenko S, Saliuk A (2015) Production and applications of crude polyhydroxyalkanoate-containing bioplastic from the organic fraction of municipal solid waste. Int J Environ Sci Technol 12:725–738

Karić N, Maia AS, Teodorović A, Atanasova N, Langergraber G, Crini G, Ribeiro AR, Đolić M (2022) Bio-waste valorisation: Agricultural wastes as biosorbents for removal of (in) organic pollutants in wastewater treatment. Chem Eng J Adv 9:100239

Khanna R, Gupta S (2018) Agrochemicals as a potential cause of ground water pollution: a review. Int J Chem Stud 6(3):985–990

Koul B, Yakoob M, Shah MP (2022) Agricultural waste management strategies for environmental sustainability. Environ Res 206:112285

Kremer RJ (2021) Disruption of the soil microbiota by agricultural pesticides. In: Synthetic pesticide use in Africa, pp 147–164. CRC Press

Kroeze C, Bouwman L, Seitzinger S (2012) Modeling global nutrient export from watersheds. Curr Opin Environ Sustain 4(2):195–202

La Farre M, Pérez S, Kantiani L, Barceló D (2008) Fate and toxicity of emerging pollutants, their metabolites and transformation products in the aquatic environment. TrAC, Trends Anal Chem 27(11):991–1007

Liu L, Liang Y, Song Q, Li J (2017) A review of waste prevention through 3R under the concept of circular economy in China. J Mater Cycles Waste Manage 19:1314–1323

Mahesh N, Balakumar S, Danya U, Shyamalagowri S, Babu PS, Aravind J, Kamaraj M, Govarthanan M (2022) A review on mitigation of emerging contaminants in an aqueous environment using microbial bio-machines as sustainable tools: Progress and limitations. J Water Process Eng 47:102712

Maji S, Dwivedi DH, Singh N, Kishor S, Gond M (2020) Agricultural waste: Its impact on environment and management approaches. Emerging eco-friendly green technologies for wastewater treatment, pp 329–351

Manivannan B, Nallathambi G, Devasena T (2022) Alternative methods of monitoring emerging contaminants in water: a review. Environ Sci Process Impacts 24(11):2009–2031

Maurya P, Kumar A, Singh S, Nisha R, Pal RR, Chourasia MK, Saraf SA (2021) Bio-analytical method development for estimation of levofloxacin: application in estimation of drug in nano-formulations and pharmacokinetic studies. Indian J Pharmaceut Educ Res 55(3):S814–S824

Obi FO, Ugwuishiwu BO, Nwakaire JN (2016) Agricultural waste concept, generation, utilization and management. Niger J Technol 35(4):957–964

Pozzebon EA, Seifert L (2023) Emerging environmental health risks associated with the land application of biosolids: a scoping review. Environ Health 22(1):57

Purnamitta G (2012) Microbiological Aspects of Water: Key Criteria of Quality. Curr Res Biol Pharmaceut Sci 11(1):157–166

Rahman KA, Debnath SC (2015) Water logging and losses in ecosystem: a case analysis on DND embankment, Bangladesh. Int Res J Interdisciplinary Multidisciplinary Studies (IRJIMS) 1:27–33

Rasheed T, Bilal M, Nabeel F, Adeel M, Iqbal HM (2019) Environmentally-related contaminants of high concern: potential sources and analytical modalities for detection, quantification, and treatment. Environ Int 122:52–66

Rashmi I, Roy T, Kartika KS, Pal R, Coumar V, Kala S, Shinoji KC (2020) Organic and inorganic fertilizer contaminants in agriculture: Impact on soil and water resources. Contaminants in agriculture: sources, impacts and management, pp 3–41
Rathi BS, Kumar PS (2021) Application of adsorption process for effective removal of emerging contaminants from water and wastewater. Environ Pollut 280:116995
Rathi BS, Kumar PS, Show PL (2021) A review on effective removal of emerging contaminants from aquatic systems: Current trends and scope for further research. J Hazard Mater 409:124413
Ravula AR, Yenugu S (2021) Pyrethroid based pesticides–chemical and biological aspects. Crit Rev Toxicol 51(2):117–140
Renau-Pruñonosa A, García-Menéndez O, Ibáñez M, Vázquez-Suñé E, Boix C, Ballesteros BB, Hernández García M, Morell I, Hernández F (2020) Identification of aquifer recharge sources as the origin of emerging contaminants in intensive agricultural areas. La Plana de Castellón, Spain. Water 12(3):731
Richardson SD, Kimura SY (2016) Water analysis: emerging contaminants and current issues. Anal Chem 88(1):546–582
Shah MP (2024a) Trends in biological processes in industrial wastewater treatment. IOP Publishing
Shah MP (ed) (2024b) Emerging innovative trends in the application of biological processes for industrial wastewater treatment. Elsevier
Sharma R, Kumar R, Satapathy SC, Al-Ansari N, Singh KK, Mahapatra RP, Agarwal AK, Le HV, Pham BT (2020) Analysis of water pollution using different physicochemical parameters: a study of Yamuna River. Front Environ Sci 8:581591
Singh DV, Bhat RA, Upadhyay AK, Singh R, Singh DP (2021) Microalgae in aquatic environs: a sustainable approach for remediation of heavy metals and emerging contaminants. Environ Technol Innov 21:101340
Souza RM, Seibert D, Quesada HB, Bassetti FJ, Fagundes-Klen MR, Bergamasco R (2020) Occurrence, impacts and general aspects of pesticides in surface water: a review. Process Saf Environ Prot 135:22–37
Srikanth K, Sukesh K, Rao AR, Pavan G, Ravishankar GA (2019) Emerging contaminants effect on aquatic ecosystem: human health risks-a review. Agric Res Technol 19:556104
Srivastav AL (2020) Chemical fertilizers and pesticides: role in groundwater contamination. In: Agrochemicals detection, treatment and remediation, pp 143–159. Butterworth-Heinemann
Stefanakis AI, Becker JA (2016) A review of emerging contaminants in water: classification, sources, and potential risks. Impact of water pollution on human health and environmental sustainability, pp 55–80
Talib A, Randhir TO (2016) Managing emerging contaminants: status, impacts, and watershed-wide strategies. Exposure and Health 8:143–158
Varsha M, Kumar PS, Rathi BS (2022) A review on recent trends in the removal of emerging contaminants from aquatic environment using low-cost adsorbents. Chemosphere 287:132270
Zhang X, Liu X, Zhan C (2024) Emerging contaminants in aquatic ecosystems: sources, effects, and mitigation approaches. Int J Aquaculture, 14

Chapter 6
Microplastics Are Contaminants of Emerging Concern in Aquatic Environments: An Overview

Prangya Rath, Laxmi Kant Bhardwaj, Abhiram Kumar, and Kumar Pranav Narayan

Abstract Microplastics (MPs) have appeared as ubiquitous contaminants and pose a risk to aquatic environments as well as humans. They enter the environment from different sources, for e.g., discharge from industries, wastewater treatment plant (WWTP) emissions, breakdown of larger plastic debris, atmospheric deposition, and human activities. Because of the chemical composition of plastics, environments are exposed to both micro-sized as well as nano-sized particles, additives that get leached, as well as degradation byproducts, that living organisms can then take up. They are slowly degrading in soils as well as water which has generated concern among research communities as well as the public. They seriously threaten the aquatic ecosystems and cause physical and histo-physiological damage. Once they enter the environment, they accumulate and persist for a longer duration. They float on the water's surface. The transportation rate of these particles is higher in the water system. The detection of these particles depends on techniques of sampling, location, and analysis methods. This chapter explores the occurrence, sources, transport, health impact, identification, as well as treatment of MPs in aquatic environments.

Keywords Microplastics (MPs) · Emerging contaminants (ECs) · Human health · Aquatic environment

---

P. Rath (✉)
Amity Institute of Environmental Sciences, Amity University, Noida, Uttar Pradesh, India
e-mail: prangyarath71@gmail.com

L. K. Bhardwaj
Amity Institute of Environmental Toxicology, Safety and Management, Amity University, Noida, Uttar Pradesh, India

A. Kumar · K. P. Narayan
Birla Institute of Technology and Sciences Pilani, Hyderabad Campus, Telangana, India

H. Jain et al. (eds.), *Emerging Contaminants in the Aquatic Environment*, Advances in Wastewater Research, https://doi.org/10.1007/978-981-96-8548-6_6

Abbreviations

Al	Aluminum
BPA	Bisphenol A
DDT	Dichlorodiphenyltrichloroethane
ECs	Emerging contaminants
EDs	Endocrine disruptors
EDS	Energy dispersive X-ray spectroscopy
Fe	Iron
FPA	Focal plane array
FTIR	Fourier transform infrared spectroscopy
GC-MS	Gas chromatography-mass spectrometry
HCB	Hexachlorobenzene
HIS	Hyperspectral imaging
ICP-MS	Inductively coupled plasma mass spectrometry
MPs	Microplastics
PA	Polyamide
PAM	Polyacrylamide
PBDEs	Polybrominated diphenyl ether
PCPs	Personal care products
PE	Polyethylene
PET	Polyethylene terephthalate
PMPs	Primary microplastics
PP	Polypropylene
PS	Polystyrene
PVC	Polyvinyl chloride
SEM	Scanning electron microscopy
SMPs	Secondary microplastics
STPs	Sewage treatment plants
TBBPA	Tetrabromobisphenol A
TiO_2	Titanium dioxide
US	United States
WWTPs	Wastewater treatment plant

6.1 Introduction

Over the past 70 years, worldwide plastic production has substantially increased. It has been estimated to grow from approx. 1.5 million tons to around 359.0 million tons. Projections suggest that this figure could reach 500.0 million tons by 2025 (Lebreton and Andrady 2019; Hohn et al. 2020). In 2013, China's plastic production was ~63.0 million tons, representing a large portion of global production. When combined with the output from other Asian countries, the region's total reached about 114.0 million tons (Report of World Watch Institute 2015). Europe was the second-largest producer, with ~50.0 million tons, followed closely by North America, which produced 49.0 million tons (Real 2022). In comparison, Commonwealth countries, Latin America, Africa, and the Middle East together accounted for a smaller share, producing 37.0 million tons. India produces plastic at an average of approx. 5.6 million tons each year, making it one of the biggest consumers of plastics in the world (Laskar and Kumar 2019). Based on this pattern, scientists have predicted that there will be approx. 12,000 million metric tons of garbage plastic by 2050 that would be filling different landfills and severely affecting the environment.

The plastic usage has increased due to the increasing global population. Plastic is mainly utilized in electrical industries, packaging sectors, automotive industries, industries handling equipment of sports, building materials, food industries, cosmetic industry, medical industry, and construction industry (Bhardwaj and Sharma 2021). Some substances made of plastics contain pro-oxidants, which promote fragmentation. They possess the tendency to form MPs (Kershaw 2015). MPs are tiny pieces of plastic having a size smaller than 5 mm (Bhardwaj 2024; Bhardwaj et al. 2024). When its length is less than 1 μm, these are termed as nanoplastics. They are hydrophobic in nature and are emerging as a significant environmental pollutant globally. They vary in size, type, color, shape, as well as density, and such characteristics have been reported to be strongly associated with their fate, source, as well as toxicity (Eerkes-Medrano et al. 2015). They are considered as one of the sources causing air pollution in the form of dust as well as airborne fibrous particles.

6.1.1 Classification of Microplastics (MPs)

MPs have fallen into five groups: fragments, micro-pellets, films, fibers, as well as foams (Anderson et al. 2017). They are categorized as primary or secondary types based on their origin. Primary microplastics (PMPs) are intentionally manufactured at the microscale for use in products like cosmetics, detergents, insecticides, personal care items, pharmaceuticals, and synthetic textiles. Secondary microplastics (SMPs) result from the degradation of large-sized plastics, such as bags, bottles, clothes, fishing gear, tires, and food containers, through physical, chemical, or biological processes. They are also classifiable by their chemical composition, including materials like polyethylene (PE), polypropylene (PP), as well as polystyrene (PS). Figure 6.1 illustrates about classification of MPs.

Single-use plastics and microbeads have been banned by more than 60 nations (Giacovelli 2018). In the year 2015, the United States (US) enacted the Microbead-Free Waters Act. This act bans distribution as well as the production of different cosmetic products that have plastic microbeads (Kershaw 2015). Several researchers reported different types as well as MP concentrations in different aquatic bodies (Table 6.1). Koelmans et al. (2019) reported MPs' particles range from 0 to 103 particles/L in freshwater ecosystems. Jambeck et al. (2015) reported 4.8 to 12.7 million metric tons of waste of plastics from oceans and stated that the level of

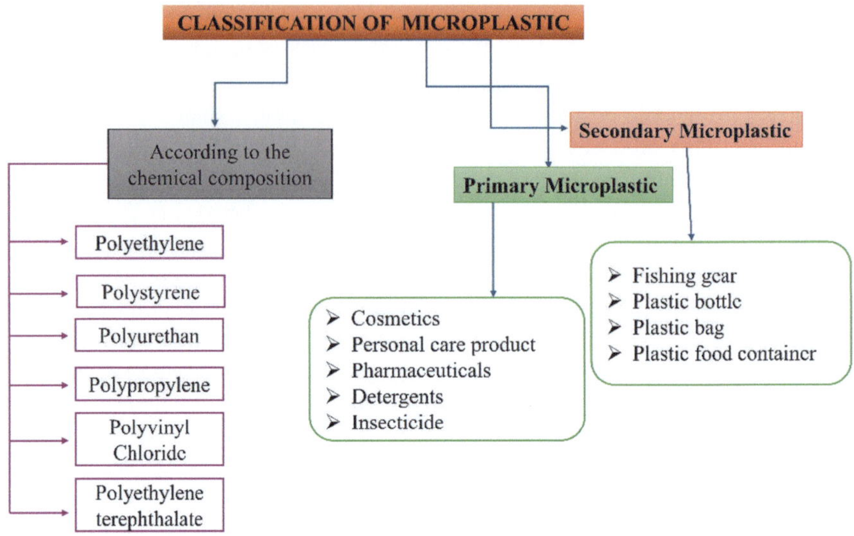

Fig. 6.1 Classification of microplastics (MPs) based on their origin

MP pollution has been observed to be higher in areas that are undeveloped because of a lack of sustainable and proper processes of waste management. Wang et al. (2017a, b) stated major types of MPs like polyethylene terephthalate (PET) and PP in the range between 1660.0 ± 639.1 and 8925 ± 1591 numbers/m^3 from the inland freshwaters of Wuhan, China. MPs were observed in the digestive systems of different invertebrate animals of the aquatic ecosystem. MPs having lower densities ($p < 1.0$ g/cm^3) float on the water surface and get ingested by mainly filter-feeding invertebrates (for e.g., *Daphnia magna*) as well as carnivorous fishes (for e.g., *Culter dabryi*, *Culter alburnus*) (Devriese et al. 2015; Schmid et al. 2018) while those having higher densities ($p > 1.0$ g/cm^3) were settled inside sediments as well as gets ingested by omnivorous fishes (e.g., *Sinibrama wui*) (Zhang et al. 2017). The occurrence, health impact, sources, transportation, identification, as well as treatment of MPs are described in detail in this chapter.

6.2 Sources and Transport of Microplastics (MPs)

MPs can enter aquatic environments from wastewater effluent, run-offs from sources that are land-based, overflows of combined sewers, deposition in the atmosphere, industrial effluents, production of drinking water, distributions, fragmentations as well as degradation of macro-plastics, human activities such as tourism, sewage treatment plants (STPs), and distribution (Dubaish and Liebezeit 2013; Cesa et al. 2017). They are found in various items, from cosmetics to synthetic clothes, and plastic bags. Based on stability, ease of discarding, and degradation resistance, MPs

Table 6.1 Occurrence of different types of microplastics (MPs) in aquatic environment

S. no	Sample type	Location	Detection method	Microplastics (MPs) conc	Type/color/size of microplastics (MPs)	References
1	Microbeads	Great Lakes	Scanning electron microscopy/ Energy dispersive X-ray spectroscopy (SEM/EDS)	0.043 particles/m^3	Blue, white, as well as gold; <1 × 10^{-3} m	Eriksen et al. (2013)
2	25% polyethylene (PE) and 19% polypropylene (PP)	Tamar Estuary, Southwest England	Sieving and Fourier transform infrared spectroscopy (FTIR)	0.028 particles/m^3	Black as well as yellow; 1–5 × 10^{-3} m polypropylene (PP) and <1 or 1–3 × 10^{-3} m nylon	Sadri and Thompson (2014)
3	Fragments and films	Lake Hovsgol, Mongolia	Sieving and light microscope	0.20 particles/m^3	Blue as well as white	Free et al. (2014)
4	Microfibers	Yangtze Estuary System, China	Stereomicroscope and floatation	4137.3 ± 2461.5 and 0.167 ± 0.138 numbers/m^3	Transparent, white, and black	Zhao et al. (2014)
5	Polystyrene (PS)	Pearl River Estuary, Hong Kong	Visual sorting and sieve	Highest 2098 ± 1705, Median 520 ± 688, lowest 94 ± 44 items/m^2		Fok and Cheung (2015)
6	Polypropylene (PP) and polyethylene (PE)	Urban estuaries, China	Agitation, filtrations, micro-Raman spectroscopy	Ranges from 10.6 -119.8%	Transparent, black, and white	Zhao et al. (2015)
7	Polystyrene (PS), polypropylene (PP), polyvinyl chloride (PVC), and polyethylene (PE)	Tibet Plateau Lake, China	Scanning electron microscopy (SEM) and Raman spectroscopy	8 ± 14 to 563 ± 1219 items/m^2	White, transparent, yellow, and blue	Zhang et al. (2016)

(continued)

Table 6.1 (continued)

S. no	Sample type	Location	Detection method	Microplastics (MPs) conc	Type/color/size of microplastics (MPs)	References
8	Cellophane, polyethylene (PE), polystyrene (PS), and polypropylene (PP)	Taihu Lake, China	Scanning electron microscopy/ energy dispersive X-ray spectroscopy (SEM/EDS), and Micro-Fourier transform infrared spectroscopy (μ-FTIR)	11.0–234.6 items/kg in sediment, and 3.4–25.8 $\times 10^3$ items/m^3 in water surface	Transparent, and white	Su et al. (2016)
9	Fibers and fragments	Lagoon-Channel of Bizerte (Northern Tunisia)	Stereomicroscopy	3000–18,000 items/kg, dry sediment	Black, blue, green, white, as well as red	Abidli et al. (2021)
10	43% microfibers, 22% polyethylene terephthalate (PET), 22% fluoro-polymer/teflon 7% polypropylene (PP), and 7% nitrocellulose	Hudson River, USA	Fourier transform infrared spectroscopy (FTIR)	0.625 to 2.45×10^3 fibers/m^3	Red, blue, transparent, and black	Miller et al. (2017)
11	91% fragments	River Thames, UK	Sieving, visual inspection, and Raman spectroscopy	$33.2 \pm 16.1 \times 10^3$ particles/100 kg sediment	Yellow, and red	Horton et al. (2017a)
12	Polyethylene (PE) and polypropylene (PP)	Urban surface waters, Wuhan, China	Stereoscopic microscope, Fourier transform infrared spectroscopy (FTIR), and scanning electron microscopy (SEM)	1660.0 ± 639.1 to 8925 ± 1591 numbers/m^3	50.4% to 86.9% transparent	Wang et al. (2017a, b)

(continued)

Table 6.1 (continued)

S. no	Sample type	Location	Detection method	Microplastics (MPs) conc	Type/color/size of microplastics (MPs)	References
13	Polypropylene (PP) and polyethylene (PE)	Beijiang River	Flotation, Fourier transform infrared spectroscopy (FTIR), and scanning electron microscopy (SEM)	178 ± 69 to 544 ± 107 items/kg sediment	Blue, and brown	Wang et al. (2017a, b)
14	Polyethylene (PE)	Vembanad Lake, Kerala, India	Sieving, wet peroxide oxidation, as well as Raman spectroscopy	252.80 ± 25.76 particles/m^2	Transparent as well as white	Sruthy and Ramasamy (2017)
15	Polyethylene (PE), polypropylene (PP), polystyrene (PS), polyethylene terephthalate (PET), polyvinyl chloride (PVC), acrylics, and polyamide (PA)	Japan, Thailand, Malaysia, as well as South Africa	Density separation and Fourier transform infrared spectroscopy (FTIR)	From 100 to 1900 pieces/kg of dry sediment	White, brown, as well as black	Matsuguma et al. (2017)
16	38.5% polystyrene (PS), 29.4% polypropylene (PP) and 21% polyethylene (PE)	Three Gorges Reservoir, China	Raman spectroscopy and Fourier transform infrared spectroscopy (FTIR)	In surface water (1597 to 12,611 numbers/m^3) and in sediments (25 to 300 numbers/kg wet weight)	Transparent	Di and Wang (2018)
17	76.2% fragments in Charleston Harbor and 77.5% fragments in Winyah Bay	South Carolina Estuaries	Sieving, hydrogen peroxide (H_2O_2) treatment, scanning electron microscopy (SEM), and Fourier transform infrared spectroscopy (FTIR)	In sediments of Charleston Harbor (413.8 ± 76.7) and in Winyah Bay, (221.0 ± 25.6) particles/m^2	Black, blue, colorless, gray, green, red, as well as white	Gray et al. (2018)

(continued)

Table 6.1 (continued)

S. no	Sample type	Location	Detection method	Microplastics (MPs) conc	Type/color/size of microplastics (MPs)	References
18	75.3% fibers in water and 68.7% fibers in sediment	Wind Farm, Yellow Sea, China	Sieving, density separation, and micro-Fourier transform infrared spectroscopy (μ-FTIR)	$2.58 \pm 1.14 \times 10^3$ items/kg (dry) in sediment, and 0.330 ± 0.278 items/m^3 in surface water	Colored, transparent, and black	Wang et al. (2018)
19	45% polyethylene (PE), 18% polystyrene (PS) and 15% polypropylene (PP)	Italian Subalpine Lakes	Fourier transform infrared spectroscopy (FTIR)	4000 to 57,000 particles/km^2		Sighicelli et al. (2018)
20	Polypropylene (PP)	Shanghai, China	Density separation, microscope, and Micro-Fourier transform infrared spectroscopy (μ-FTIR)	$80.2 \pm 59.4 \times 10^3$ items/100 kg dry weight	Red, blue, white, and transparent	Peng et al. (2018)
21	26.2% polyamide (PA) and 23.1% cellophane	Pearl River	Micro-Raman spectroscopy	19.86×10^3 MP/m^3 for urban and 8.90×10^3 MP/m^3 for the estuary	Film, fiber, and fragment	Yan et al. (2019)
22	37% polypropylene (PP) and 30% polyethylene (PE)	Poyang Lake	Micro-Raman spectroscopy	5 to 34×10^3 MP/m^3 for surface water while 54 to 506 MP/kg for sediments	Fiber	Yuan et al. (2019)
23	Polypropylene (PP) and polyethylene (PE)	Yongjiang River	Raman spectroscopy	0.5 to 7.7×10^3 MP/m^3 for surface water while 54 to 506 MP/kg for sediments	Fiber	Zhang et al. (2020)

(continued)

Table 6.1 (continued)

S. no	Sample type	Location	Detection method	Microplastics (MPs) conc	Type/color/size of microplastics (MPs)	References
24	Polypropylene (PP)	Danjiangkou Reservoir	Micro-Raman spectroscopy	0.47 to 15.02 × 10^3 MP/m^3 in surface water while 15 to 40 MP/kg in wastewater	Fiber	Di et al. (2019)
25	27.7% polystyrene (PS)	Suzhou River and Huangpu River	Micro-Fourier transform infrared spectroscopy (μ-FTIR)	0.08 to 7.4 × 10^3 MP/m^3	Fiber	Luo et al. (2019)
26	Polypropylene (PP)	Yangtze River	Raman spectroscopy	4.92 × 10^5 MP/km^2	Fragment	Xiong et al. (2019)
27	52.31% polypropylene (PP) and 27.39% polyethylene (PE)	Feilaixia Reservoir	Micro-Fourier transform infrared spectroscopy (μ-FTIR)	0.56 MP/m^3	Films	Tan et al. (2019)

sustain in the environment for a longer duration of time and get consumed by aquatic biota (Peiponen et al. 2019). They can originate through different sources in various forms like fiber, fragments, and foams.

Jiang (2018) reported that MPs primary sources include discharge from the house sewage, such as polymeric particles released from cleaning, cosmetic items, feedstocks utilized in manufacturing plastic items, as well as plastic powders and pellets that are utilized during air blasting. The progressive fragmentation of plastic items that are larger in size is a secondary source of MPs (Eriksen et al. 2014). Additionally, various types of elements come from different foundations. For instance, you have road superficial markings created from thermoplastic mixture paints, fibers originating because of artificial textiles, as well as bits of larger litter products like bottles made from plastics and packaging items (Horton et al. 2017a). The colors of MPs confirm its numerous sources (Yu et al. 2016). PMPs find their way into the environment through various channels. This can happen, for instance, when spills occur during manufacturing or transport, or when synthetic textiles are washed and release particles. Additionally, personal care products (PCPs) can end up in wastewater systems from our homes, contributing to the issue.

6.2.1 *Wastewater and Industrial Effluent*

It is a widely recognized contributor to the pollution of MPs in aquatic environments (Mintenig et al. 2017). It has been observed to be one of the important collection points of MPs that gets released in day-to-day life. Cosmetic microbeads, tiny bits that come from larger consumer products, along with synthetic fibers that wash off our clothes, are the main culprits entering our sewage systems. WWTP can serve as a route for these MPs to make their way into our water bodies. There are two ways this happens: directly and indirectly. In the direct route, MPs are released straight from the WWTP effluents, often in significant quantities. On the other hand, the indirect route involves MPs being released from the WWTP as they settle in the sludge (Gatidou et al. 2019). It is then utilized in green construction, as fertilizers, etc., in fields (Sun et al. 2019). 65 million particles of MPs were released daily in the WWTP effluents (Murphy et al. 2016). However, MPs released due to industrial activities have been observed in aquatic ecosystems.

6.2.2 *Run-Off from Land-Based and Combined Sewer Overflows*

MPs have been seen to originate from infrastructures, land-usage practices, runoffs from road surfaces, paints used for road markings, as well as debris from tire wear. Boucher and Friot (2017) have reported different inputs of MPs into aquatic ecosystems from sources that are land-based. MP fibers are considered as another main source that gets emitted from textiles because of processes of wear, tears, as well as washing (Henry et al. 2019). Dust generated during construction in the city is an example of land-based sources (Boucher and Friot 2017). Rainstorms, run-offs from agricultural sources, and activities of farming are reported as some of the potential causes of MPs in aquatic environments (Horton et al. 2017b). Heavy rainfall has been reported to be one of the main sources of MPs in aquatic ecosystems. This may be due to barriers getting temporarily bypassed during wastewater treatment.

6.2.3 Atmospheric Deposition

It has been reported that another potential contributor of MPs in aquatic ecosystems comes from dry as well as wet depositions, precipitations, and run-offs (Wright and Kelly 2017). MPs that come from urban as well as industrial dust can find their way into inland water bodies from the atmosphere (Abbasi et al. 2020). They are some of the indirect sources of MPs in aquatic environments. The airborne MPs come from various sources like waste incineration, buildings, industrial emissions, landfills, fertilizer use, and traffic. These particles have been studied to pose a direct threat to construction workers and young children, as they may inhale or ingest them on a regular basis (Dehghani et al. 2017). MPs in dust found on streets contain heavy metals and are toxic to aquatic ecosystems.

6.2.4 Drinking Water Production and Distribution

Treatment of drinking water gives a barrier to MPs. Certain parts of water treatment facilities contain plastic materials. As these plastic components break down, they create MPs that enter our drinking water (Mintenig et al. 2019). Bottle as well as its cap is another source of MPs in samples of drinking water (Schymanski et al. 2018).

6.2.5 Fragmentation and Degradation of Macroplastics

The debris of macroplastic is also represented as another important source of MP formation through fragmentations, and these MPs could enter aquatic environments. Data from research on macroplastic fragmentations as well as degradations in aquatic environments are limited. In UV light presence and higher temperatures, the debris of macroplastics fragmented into MPs in aquatic ecosystems (Andrady 2007).

The way MPs travel through the air is still unclear (Horton and Dixon 2018), and they are influenced by pollution from both land and water. For both animals and humans, the main way we come into contact with MPs is through our food (Wright

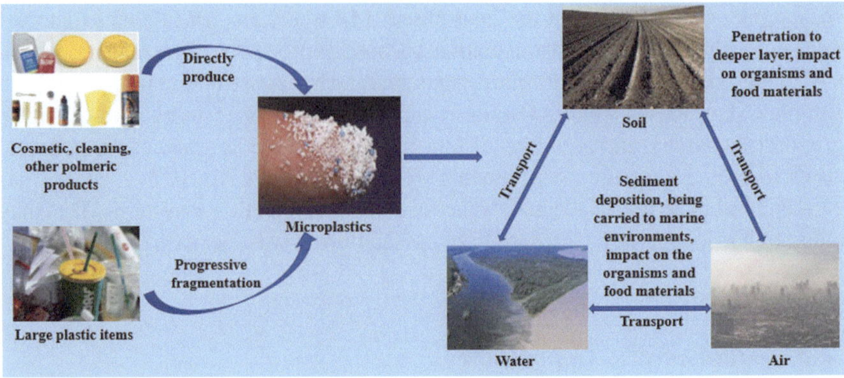

Fig. 6.2 Sources and transport pathways of microplastics (MPs)

and Kelly 2017). A study conducted by Lau and Wong (2000) found that PS and epoxy resins can be present in food items, which means they could potentially make their way into our bodies when we eat. The sources as well as transport pathways of MPs are well presented in Fig. 6.2.

6.3 Health Impact of Microplastics (MPs)

The pervasive presence of MPs in aquatic ecosystems raises significant ecological and health concerns. These hazards are present in three different forms, i.e., physical, chemical, as well as microbiological. Microbiological and chemical pollutants associated with MPs pose a low risk. After the interaction of MPs with various types of heavy metals as well as pollutants that are hydrophobic in nature, they can create a serious issue in aquatic ecosystems (Vedolin et al. 2018). MPs are transport agents for pollutants with toxic chemicals such as dichlorodiphenyltrichloroethane (DDT) as well as hexachlorobenzenes (HCB) (Laskar and Kumar 2019). The impacts on health, fate, as well as transportation of MPs, are still not well studied. Exposure to PVC or PE was able to damage the immune system of fish due to the oxidative stresses in fish leucocytes (Espinosa et al. 2018). Activities conducted by local humans are chiefly responsible for the accumulation of MPs in fish muscles (Akhbarizadeh et al. 2018).

Studies have pointed out that after ingesting MPs, fish experience fewer growth changes in terms of oxygen consumption, limited feeding capacity, shorter lifespans, and heightened activity of antioxidant-related enzymes (Windsor et al. 2019). With less food intake, fish produce less energy to support their life functions, which can lead to reproductive and neurological issues. Because of their tiny size, durability, and slow breakdown, MPs can impact aquatic organisms across multiple generations. These tiny particles are unintentionally consumed by aquatic organisms, such as plankton and fish marine mammals, potentially causing physical and chemical harm. Additionally, MPs can carry toxic pollutants, amplifying their environmental impact.

6.3.1 Physical Hazard

This hazard is found in tiny particles. MP toxicity depends on their physical traits like shapes, sizes, surface areas, as well as other surface features. MPs that are larger than 150 μm do not get absorbed in the human body, but smaller particles might be absorbed to a certain extent. In fact, the distributions as well as absorption of those really tiny particles could be even greater. Lu et al. (2016) reported about the toxicity of PS MPs on *Zebrafish.* PP fibers were observed to be more toxic as compared with PE spherical particles in *Hyalella Azteca* (Au et al. 2015).

6.3.2 Chemical Hazard

This hazard is present in the form of a polymer. Polybrominated diphenyl ether (PBDEs), phthalate esters, as well as tetrabromobisphenol A (TBBPA), are not covalently attached to polymers but are found in MPs. This hazard could easily migrate into the environment, and this migration depends on the molecular weights of compounds. Molecules having lower molecular weight migrate at a faster rate as compared to heavier molecules. Toxic chemicals like phthalates, bisphenol A (BPA), as well as PBDEs, are stuck on MPs and might induce toxic effects after ingestion by living organisms (Padervand et al. 2020). Such chemicals have been recognized as endocrine disruptors (EDs) and thus have been observed to exhibit toxicity upon their release. Volatile compounds found in plastics, like benzene, ethylbenzene, methylene chloride, styrene, as well as toluene, could lead to long-term health issues (Andrady 2017).

6.3.3 Microbiological Hazard

This hazard shows up as biofilm, which forms when microorganisms cling to the surfaces of MPs. In drinking water, it develops when these tiny organisms start to grow on pipes as well as other surfaces. Interestingly, these microorganisms have been observed to be non-pathogenic and tend to attach more quickly to surfaces that are hydrophobic and nonpolar as compared to hydrophilic ones.

6.4 Detection of Microplastics (MPs)

There are different methods for identification as well as characterization of MPs.

6.4.1 *Visual Identification*

By visually assessing the samples, we can effectively pinpoint which MPs come from field samples, and which were a result of contamination from the laboratory (Mathalon and Hill 2014). Doyle et al. (2011) stated that larger MPs could be detected by visual identification method. This method could not identify particles lesser than 1 mm (Lee et al. 2013).

6.4.2 *Density Separation*

The polymers are sorted based on their density differences. This technique is particularly effective for marine sediments since higher-density MPs have been observed to sink more readily as compared to their lighter plastic counterparts.

6.4.3 *Spectroscopic Techniques*

This technique is used for the identification of chemical structures of polymers by comparing its spectra of absorption or emission with that of reference spectra. This technique is well-established, faster, and reliable. Two techniques, Raman spectroscopy as well as FTIR, are considered as most popular for the identification as well as characterization of MPs. Raman spectroscopy is suitable for identifying particles of plastics which are common in marine environments (Lenz et al. 2015). Very small plastic particles <1 μm can be measured by Raman spectroscopy. Schymanski et al. (2018) have detected small particles (<20 μm) by utilizing high-resolution micro-Raman spectroscopy. FTIR can identify particles to about 10 to 20 μm when coupled with microscopy. It is a reliable technique for identifying different polymer types of MPs with lower cost as well as easy usage (Lusher et al. 2014). In FTIR, detection depends upon the stimulations of molecular vibrations caused by infrared radiations depending upon a substance's composition as well as molecular structure and wavelength.

6.4.4 Thermo-Analytical Technique

In this method, the sample undergoes pyrolysis in an inert environment prior to analysis. It is a useful technique for identifying different types of polymers, but it does need larger particle masses than the spectroscopic approach. Because of this, it's not the best choice for handling large quantities of samples. For identifying polymers and additives, gas chromatography-mass spectrometry (GC–MS) is employed, allowing for the analysis of one particle at a time in each run (Nuelle et al. 2014).

6.4.5 Chemical Method

This method is used to detect specific fragments of polymers or elements by the technique of inductively coupled plasma mass spectrometry (ICP-MS) (Braun et al. 2018).

6.4.6 Hyperspectral Imaging (HSI) Technique

Serranti et al. (2018) developed a hyperspectral imaging (HIS) technique to identify MPs easily. It could be utilized for the chemical characterization of MPs in marine environments, and it is a fast, reliable, as well as non-destructive technique. Sample preparation is not required in this method.

6.4.7 Combined Methods

In this method, both microscopic and spectroscopic methods are used to analyze many samples. The first spectroscopic method should be applied for MP identification in the samples, then a stereomicroscope could be utilized to count the MPs. A combination of methods could improve the identification of MPs in aquatic environments, while a single method makes the investigation of MPs difficult (Li et al. 2018a, b). Löder and Gerdts (2015) utilized the focal plane array (FPA) that depends upon μ-FTIR spectroscopy techniques for the identification of MPs, and this technique could detect plastic particles <500 μm.

6.5 Treatment of Microplastics (MPs)

Different techniques were used for the removal of MPs from aquatic ecosystems.

6.5.1 Sorption and Filtration Methods

6.5.1.1 By Adsorption on Green Algae

Sundbaek et al. (2018) studied how MP particles adhere to the surfaces of edible algae and seaweed. The key player in this sticking process is a gelatinous substance called alginate, which helps PS particles adhere to the seaweed (Martins et al. 2013). Alginate is released from seaweed cell walls in areas where it's been cut, and ~94.5% of the MPs tend to cling to these spots. The way MPs attach to the algae's surface is influenced by the charge on the surface of the particles. Interestingly, positively charged PS MPs are more readily absorbed by algae compared to their negatively charged counterparts (Nolte et al. 2017).

6.5.1.2 By Using Membrane Technology

MP removal efficiency is affected by membrane durability, concentration, as well as MP size. Li et al. (2018a, b) utilized membrane technology to remove MPs from water samples that are polluted. Horton and Dixon (2018) extracted filtrate of MPs in 20 minutes by reducing the turbidity of influent to the effluent. Padervand et al. (2020) designed a tool based on polymer coatings as an elongated mesh screen to remove MPs. Membrane bioreactors are the suitable technology for removing ~ 99.9 % of MP particles per liter (Lares et al. 2018).

6.5.2 Chemical Methods

Shirasaki et al. (2016) described coagulation as well as agglomeration processes for removing MPs from WWTPs. Ariza-Tarazona et al. (2019) utilized coagulants of iron (Fe) and aluminum (Al) salts and ultrafiltration techniques for PE MP removal. They used polyacrylamide (PAM) and $FeCl_3{\cdot}6H_2O$ as an enhancing coagulation agent. Akbal and Camcl (2011) utilized coagulants of metal hydroxide to remove MPs and stated that such coagulants were able to break up colloids and thereafter stabilize suspended MPs. After that, a sludge blanket was formed by the coagulants to trap suspended MPs from samples of wastewater. Li et al. (2010) reported that hydroxyl radicals promote the photochemical degradation of MPs in aqueous ecosystems. Tofa et al. (2019) studied photo-catalytic degradation of lower-density PE MPs in aquatic environments over rod-shaped ZnO NPs. Wang et al. (2019) reported that titanium dioxide (TiO_2)-based photocatalyst degradation of the PS MPs in solid and aqueous phases.

6.5.3 Biological Methods

By marine organisms, bacteria, and ingestion are popular biological methods for removing/degrading MPs from aquatic ecosystems.

6.5.3.1 By Marine Organisms

MPs are easily spread throughout the surface of oceans after traveling longer distances through wind due to their smaller size as well as negligible weight. They are considered as rich carbon sources for feeding marine organisms. Smaller-sized MPs get fragmented easily under different conditions of the environment. Dawson et al. (2018) studied fragmentations and ingestion of PE MPs through the Antarctic Krill (*Euphausiasuperba*) of Australia under the electron microscope. Cocca et al. (2020) used two native marine communities like *Agios* consortium as well as *Souda* consortium, for removing high-density PE MPs. Urbanek et al. (2018) studied about growth of microbial assemblages on MPs' surface, forming a new area that was prone to attachment of fungi and microalgae. After that, these microorganisms catalyzed MPs' metabolic reactions, thereby resulting in its biodegradation. Paço et al. (2017) stated that a native fungus called *Zalerion maritimum* utilized MPs as their source of nutrition, and they described the biodegradation capability of PE MPs based on variation of masses as well as sizes.

6.5.3.2 By Bacteria

Bacillus cereus and *Bacillus gottheilii* were utilized for the biodegradation of different MPs like PS, PP, PE, and PET (Auta et al. 2017). The biodegradation rate was assessed by measuring the weight loss of MPs. *B. gottheilii* reported a better potential MPs degrader.

6.5.3.3 By Ingestion

PP MPs were reported in the gut cavity of the *scleractinian* corals within 24 h after their ingestion (Hall et al. 2015). They stated that the capture and ingestion of MPs were in the corals with a rate of consumption of 50 μg plastics cm^{-2}/h. *Zooplankton* could ingest PS MPs and could keep MPs for up to 7 days after its entrance. Arossa et al. (2019) studied that the Red Sea giant clam can remove PE MPs from wastewater through sorption on the surface. The larger clams absorb higher concentrations of the MPs.

6.6 Recommendations

Effective plastic control measures should be ensured by water suppliers in cities, while processes of treatment of water need to be improved for removing MP particles. Although routine monitoring of MPs in water samples that are used for drinking purposes is not recommended because of less evidence indicating a human health concern, further studies should focus on MPs' prevention, awareness, and reduction methods. Understanding MP sources as well as pathways is crucial for preventing pollution, and research on river systems, watersheds, and MPs released by WWTPs can provide valuable insights for addressing aquatic environment contamination. Governments should implement policies for plastic pollution removal while considering the benefits of plastics before introducing new initiatives. Efforts should also be directed toward reducing plastic usage, banning some applications like microbeads found in cosmetics, and promoting biodegradable alternatives. Improving global management of plastic waste systems as well as encouraging the usage of bags that are biodegradable and materials that are made up of non-plastic substances through organizations that may be governmental or non-governmental can further mitigate the issue. Additionally, educational campaigns should raise awareness of plastic pollution and promote recycling and reuse. After removal from aquatic environments, the fate of MPs should also be monitored to understand their long-term impacts.

6.7 Conclusions

This chapter compiled detailed information about MP pollution in the aquatic environment which is considered a critical environmental issue. In regard to this, topics such as health impact, sources, transport, detection, and treatment for future works were discussed. The previous literature shows that various types of MPs get detected in different aquatic bodies. The MP's presence threatens aquatic ecosystems. The detection of different types of MPs was performed by using different techniques such as visual identification, SEM, FTIR, Raman Spectroscopy. In the context of color transparency, blue as well as white MPs were observed to be dominant; however, PE, PP, as well as fibers were the significant MPs. The levels of pollution due to MPs were high because of the activities of residents as well as industries located near the aquatic environment. Sorption and filtration, chemical, and biological methods were used for the treatment of MPs.

Acknowledgements The authors are thankful to Amity University Noida, and BITS Pilani Hyderabad for their continuous guidance and support for conducting this study.

References

Abbasi S, Keshavarzi B, Moore F, Hopke PK, Kelly FJ, Dominguez AO (2020) Elemental and magnetic analyses, source identification, and oxidative potential of airborne, passive, and street dust particles in Asaluyeh County. Iran. Sci Total Environ 707:136132

Abidli S, Akkari N, Lahbib Y, El Menif NT (2021) First evaluation of microplastics in two commercial fish species from the lagoons of Bizerte and Ghar El Melh (Northern Tunisia). Regional Studies Mar Sci 41:101581

Akbal F, Camcı S (2011) Copper, chromium and nickel removal from metal plating wastewater by electrocoagulation. Desalination 269(1–3):214–222

Akhbarizadeh R, Moore F, Keshavarzi B (2018) Investigating a probable relationship between microplastics and potentially toxic elements in fish muscles from northeast of Persian Gulf. Environ Pollut 232:154–163

Anderson PJ, Warrack S, Langen V, Challis JK, Hanson ML, Rennie MD (2017) Microplastic contamination in lake Winnipeg, Canada. Environ Pollut 225:223–231

Andrady AL (2007) Biodegradability of polymers. In: Physical properties of polymers handbook, pp 951–964. New York, NY: Springer New York

Andrady AL (2017) The plastic in microplastics: a review. Mar Pollut Bull 119(1):12–22

Ariza-Tarazona MC, Villarreal-Chiu JF, Barbieri V, Siligardi C, Cedillo-González EI (2019) New strategy for microplastic degradation: green photocatalysis using a protein-based porous N-TiO2 semiconductor. Ceram Int 45(7):9618–9624

Arossa S, Martin C, Rossbach S, Duarte CM (2019) Microplastic removal by Red Sea giant clam (Tridacna maxima). Environ Pollut 252:1257–1266

Au SY, Bruce TF, Bridges WC, Klaine SJ (2015) Responses of Hyalella azteca to acute and chronic microplastic exposures. Environ Toxicol Chem 34(11):2564–2572

Auta HS, Emenike CU, Fauziah SH (2017) Screening of Bacillus strains isolated from mangrove ecosystems in Peninsular Malaysia for microplastic degradation. Environ Pollut 231:1552–1559

Bhardwaj LK, Sharma A (2021) Estimation of physico-chemical, trace metals, microbiological and phthalate in PET bottled water. Chem Africa 4(4):981–991

Bhardwaj LK (2024) Occurrence of microplastics (MPs) in Antarctica and its impact on the health of organisms. Maritime Technol Res 6(2):265418–265418

Bhardwaj LK, Rath P, Yadav P, Gupta U (2024) Microplastic contamination, an emerging threat to the freshwater environment: a systematic review. Environ Syst Res 13(1):8

Boucher J, Friot D (2017) Primary microplastics in the oceans: a global evaluation of sources (Vol 10). Gland, Switzerland: Iucn

Braun U, Jekel M, Gerdts G, Ivleva N, Reiber J (2018) Microplastics analytics: sampling, preparation and detection methods. In: Discussion paper, BMBF research focus "Plastics in the Environment"

Cesa FS, Turra A, Baruque-Ramos J (2017) Synthetic fibers as microplastics in the marine environment: a review from textile perspective with a focus on domestic washings. Sci Total Environ 598:1116–1129

Cocca M, Di Pace E, Errico ME, Gentile G, Montarsolo A, Mossotti R, Avella M (eds) (2020) Proceedings of the 2nd international conference on microplastic pollution in the mediterranean sea. Springer Nature

Dawson AL, Kawaguchi S, King CK, Townsend KA, King R, Huston WM, Bengtson Nash SM (2018) Turning microplastics into nanoplastics through digestive fragmentation by Antarctic krill. Nat Commun 9(1):1001

Dehghani S, Moore F, Akhbarizadeh R (2017) Microplastic pollution in deposited urban dust, Tehran metropolis. Iran. Environ Sci Pollution Res 24:20360–20371

Devriese LI, Van der Meulen MD, Maes T, Bekaert K, Paul-Pont I, Frère L, Robbens J, Vethaak AD (2015) Microplastic contamination in brown shrimp (Crangon crangon, Linnaeus 1758) from coastal waters of the Southern North Sea and Channel area. Mar Pollut Bull 98(1–2):179–187

Di M, Wang J (2018) Microplastics in surface waters and sediments of the Three Gorges Reservoir, China. Sci Total Environ 616:1620–1627

Di M, Liu X, Wang W, Wang J (2019) Manuscript prepared for submission to environmental toxicology and pharmacology pollution in drinking water source areas: microplastics in the Danjiangkou Reservoir, China. Environ Toxicol Pharmacol 65:82–89

Doyle MJ, Watson W, Bowlin NM, Sheavly SB (2011) Plastic particles in coastal pelagic ecosystems of the Northeast Pacific ocean. Mar Environ Res 71(1):41–52

Dubaish F, Liebezeit G (2013) Suspended microplastics and black carbon particles in the Jade system, southern North Sea. Water Air Soil Pollut 224:1–8

Eerkes-Medrano D, Thompson RC, Aldridge DC (2015) Microplastics in freshwater systems: a review of the emerging threats, identification of knowledge gaps and prioritisation of research needs. Water Res 75:63–82

Eriksen M, Lebreton LC, Carson HS, Thiel M, Moore CJ, Borerro JC, Galgani F, Ryan PG, Reisser J (2014) Plastic pollution in the world's oceans: more than 5 trillion plastic pieces weighing over 250,000 tons afloat at sea. PLoS ONE 9(12):e111913

Eriksen M, Mason S, Wilson S, Box C, Zellers A, Edwards W, Farley H, Amato S (2013) Microplastic pollution in the surface waters of the Laurentian Great Lakes. Mar Pollut Bull 77(1–2):177–182

Espinosa C, Beltrán JMG, Esteban MA, Cuesta A (2018) In vitro effects of virgin microplastics on fish head-kidney leucocyte activities. Environ Pollut 235:30–38

Fok L, Cheung PK (2015) Hong Kong at the Pearl River Estuary: a hotspot of microplastic pollution. Mar Pollut Bull 99(1–2):112–118

Free CM, Jensen OP, Mason SA, Eriksen M, Williamson NJ, Boldgiv B (2014) High-levels of microplastic pollution in a large, remote, mountain lake. Mar Pollut Bull 85(1):156–163

Gatidou G, Arvaniti OS, Stasinakis AS (2019) Review on the occurrence and fate of microplastics in Sewage Treatment Plants. J Hazard Mater 367:504–512

Giacovelli C (2018) Single-use plastics: a roadmap for sustainability. International environmental technology Centre

Gray AD, Wertz H, Leads RR, Weinstein JE (2018) Microplastic in two South Carolina Estuaries: occurrence, distribution, and composition. Mar Pollut Bull 128:223–233

Hall NM, Berry KLE, Rintoul L, Hoogenboom MO (2015) Microplastic ingestion by scleractinian corals. Mar Biol 162(3):725–732

Henry B, Laitala K, Klepp IG (2019) Microfibres from apparel and home textiles: prospects for including microplastics in environmental sustainability assessment. Sci Total Environ 652:483–494

Hohn S, Acevedo-Trejos E, Abrams JF, De Moura JF, Spranz R, Merico A (2020) The long-term legacy of plastic mass production. Sci Total Environ 746:141115

Horton AA, Dixon SJ (2018) Microplastics: an introduction to environmental transport processes. Wiley Interdiscip Rev Water 5(2):e1268

Horton AA, Svendsen C, Williams RJ, Spurgeon DJ, Lahive E (2017a) Large microplastic particles in sediments of tributaries of the River Thames, UK–Abundance, sources and methods for effective quantification. Mar Pollut Bull 114(1):218–226

Horton AA, Walton A, Spurgeon DJ, Lahive E, Svendsen C (2017b) Microplastics in freshwater and terrestrial environments: Evaluating the current understanding to identify the knowledge gaps and future research priorities. Sci Total Environ 586:127–141

Jambeck JR, Geyer R, Wilcox C, Siegler TR, Perryman M, Andrady A, Narayan R, Law KL (2015) Plastic waste inputs from land into the ocean. Science 347(6223):768–771

Jiang JQ (2018) Occurrence of microplastics and its pollution in the environment: a review. Sustain Prod Consum 13:16–23

Kershaw PJ (2015) Biodegradable plastics & marine litter: misconceptions, concerns and impacts on marine environments. UNEP GPA

Koelmans AA, Nor NHM, Hermsen E, Kooi M, Mintenig SM, De France J (2019) Microplastics in freshwaters and drinking water: critical review and assessment of data quality. Water Res 155:410–422

Lares M, Ncibi MC, Sillanpää M, Sillanpää M (2018) Occurrence, identification and removal of microplastic particles and fibers in conventional activated sludge process and advanced MBR technology. Water Res 133:236–246

Laskar N, Kumar U (2019) Plastics and microplastics: a threat to environment. Environ Technol Innov 14:100352

Lau OW, Wong SK (2000) Contamination in food from packaging material. J Chromatogr A 882(1–2):255–270

Lebreton L, Andrady A (2019) Future scenarios of global plastic waste generation and disposal. Palgrave Commun 5(1):1–11

Lee J, Hong S, Song YK, Hong SH, Jang YC, Jang M, Heo NW, Han GM, Lee MJ, Kang D, Shim WJ (2013) Relationships among the abundances of plastic debris in different size classes on beaches in South Korea. Mar Pollut Bull 77(1–2):349–354

Lenz R, Enders K, Stedmon CA, Mackenzie DM, Nielsen TG (2015) A critical assessment of visual identification of marine microplastic using Raman spectroscopy for analysis improvement. Mar Pollut Bull 100(1):82–91

Li J, Liu H, Chen JP (2018a) Microplastics in freshwater systems: a review on occurrence, environmental effects, and methods for microplastics detection. Water Res 137:362–374

Li L, Xu G, Yu H, Xing J (2018b) Dynamic membrane for micro-particle removal in wastewater treatment: performance and influencing factors. Sci Total Environ 627:332–340

Li S, Xu S, He L, Xu F, Wang Y, Zhang L (2010) Photocatalytic degradation of polyethylene plastic with polypyrrole/TiO_2 nanocomposite as photocatalyst. Polym-Plast Technol Eng 49(4):400–406

Löder MG, Gerdts G (2015) Methodology used for the detection and identification of microplastics—a critical appraisal. Marine anthropogenic litter, pp 201–227

Lu Y, Zhang Y, Deng Y, Jiang W, Zhao Y, Geng J, Ding L, Ren H (2016) Uptake and accumulation of polystyrene microplastics in zebrafish (Danio rerio) and toxic effects in liver. Environ Sci Technol 50(7):4054–4060

Luo W, Su L, Craig NJ, Du F, Wu C, Shi H (2019) Comparison of microplastic pollution in different water bodies from urban creeks to coastal waters. Environ Pollut 246:174–182

Lusher AL, Burke A, O'Connor I, Officer R (2014) Microplastic pollution in the Northeast Atlantic Ocean: validated and opportunistic sampling. Mar Pollut Bull 88(1–2):325–333

Martins MJF, Mota CF, Pearson GA (2013) Sex-biased gene expression in the brown alga Fucus vesiculosus. BMC Genomics 14:1–14

Mathalon A, Hill P (2014) Microplastic fibers in the intertidal ecosystem surrounding Halifax Harbor. Nova Scotia. Marine Pollution Bulletin 81(1):69–79

Matsuguma Y, Takada H, Kumata H, Kanke H, Sakurai S, Suzuki T, Itoh M, Okazaki Y, Boonyatumanond R, Zakaria MP, Weerts S (2017) Microplastics in sediment cores from Asia and Africa as indicators of temporal trends in plastic pollution. Arch Environ Contam Toxicol 73:230–239

Miller RZ, Watts AJ, Winslow BO, Galloway TS, Barrows AP (2017) Mountains to the sea: river study of plastic and non-plastic microfiber pollution in the northeast USA. Mar Pollut Bull 124(1):245–251

Mintenig SM, Int-Veen I, Löder MG, Primpke S, Gerdts G (2017) Identification of microplastic in effluents of waste water treatment plants using focal plane array-based micro-Fourier-transform infrared imaging. Water Res 108:365–372

Mintenig SM, Löder MG, Primpke S, Gerdts G (2019) Low numbers of microplastics detected in drinking water from ground water sources. Sci Total Environ 648:631–635

Murphy F, Ewins C, Carbonnier F, Quinn B (2016) Wastewater treatment works (WwTW) as a source of microplastics in the aquatic environment. Environ Sci Technol 50(11):5800–5808

Nolte TM, Hartmann NB, Kleijn JM, Garnæs J, Van De Meent D, Hendriks AJ, Baun A (2017) The toxicity of plastic nanoparticles to green algae as influenced by surface modification, medium hardness and cellular adsorption. Aquat Toxicol 183:11–20
Nuelle MT, Dekiff JH, Remy D, Fries E (2014) A new analytical approach for monitoring microplastics in marine sediments. Environ Pollut 184:161–169
Paço A, Duarte K, da Costa JP, Santos PS, Pereira R, Pereira ME, Freitas AC, Duarte AC, Rocha-Santos TA (2017) Biodegradation of polyethylene microplastics by the marine fungus Zalerion maritimum. Sci Total Environ 586:10–15
Padervand M, Lichtfouse E, Robert D, Wang C (2020) Removal of microplastics from the environment. a review. Environ Chem Lett 18(3):807–828
Peiponen KE, Räty J, Ishaq U, Pélisset S, Ali R (2019) Outlook on optical identification of micro-and nanoplastics in aquatic environments. Chemosphere 214:424–429
Peng G, Xu P, Zhu B, Bai M, Li D (2018) Microplastics in freshwater river sediments in Shanghai, China: a case study of risk assessment in mega-cities. Environ Pollut 234:448–456
Real LEP (2022) Plastics statistics: production, recycling, and market data. Recycled materials for construction applications: plastic products and composites. Springer International Publishing, Cham, pp 103–113
Report of World Watch Institute (2015) Global plastic production rises, recycling lags. http://www.plastic-resource-center.com/wp-content/uploads/2018/11/Global-Plastic-Production-RisesRecycling-Lags.pdf
Sadri SS, Thompson RC (2014) On the quantity and composition of floating plastic debris entering and leaving the Tamar Estuary. Southwest England. Marine Pollution Bulletin 81(1):55–60
Schmid K, Winemiller KO, Chelazzi D, Cincinelli A, Dei L, Giarrizzo T (2018) First evidence of microplastic ingestion by fishes from the Amazon River estuary. Mar Pollut Bull 133:814–821
Schymanski D, Goldbeck C, Humpf HU, Fürst P (2018) Analysis of microplastics in water by micro-Raman spectroscopy: release of plastic particles from different packaging into mineral water. Water Res 129:154–162
Serranti S, Palmieri R, Bonifazi G, Cózar A (2018) Characterization of microplastic litter from oceans by an innovative approach based on hyperspectral imaging. Waste Manage 76:117–125
Shirasaki N, Matsushita T, Matsui Y, Marubayashi TJCEJ (2016) Effect of aluminum hydrolyte species on human enterovirus removal from water during the coagulation process. Chem Eng J 284:786–793
Sighicelli M, Pietrelli L, Lecce F, Iannilli V, Falconieri M, Coscia L, Di Vito S, Nuglio S, Zampetti G (2018) Microplastic pollution in the surface waters of Italian Subalpine Lakes. Environ Pollut 236:645–651
Sruthy S, Ramasamy EV (2017) Microplastic pollution in Vembanad Lake, Kerala, India: the first report of microplastics in lake and estuarine sediments in India. Environ Pollut 222:315–322
Su L, Xue Y, Li L, Yang D, Kolandhasamy P, Li D, Shi H (2016) Microplastics in taihu lake, China. Environ Pollut 216:711–719
Sun J, Dai X, Wang Q, Van Loosdrecht MC, Ni BJ (2019) Microplastics in wastewater treatment plants: detection, occurrence and removal. Water Res 152:21–37
Sundbæk KB, Koch IDW, Villaro CG, Rasmussen NS, Holdt SL, Hartmann NB (2018) Sorption of fluorescent polystyrene microplastic particles to edible seaweed Fucus vesiculosus. J Appl Phycol 30:2923–2927
Tan X, Yu X, Cai L, Wang J, Peng J (2019) Microplastics and associated PAHs in surface water from the Feilaixia Reservoir in the Beijiang River, China. Chemosphere 221:834–840
Tofa TS, Kunjali KL, Paul S, Dutta J (2019) Visible light photocatalytic degradation of microplastic residues with zinc oxide nanorods. Environ Chem Lett 17:1341–1346
Urbanek AK, Rymowicz W, Mirończuk AM (2018) Degradation of plastics and plastic-degrading bacteria in cold marine habitats. Appl Microbiol Biotechnol 102:7669–7678
Vedolin MC, Teophilo CYS, Turra A, Figueira RCL (2018) Spatial variability in the concentrations of metals in beached microplastics. Mar Pollut Bull 129(2):487–493

Wang J, Peng J, Tan Z, Gao Y, Zhan Z, Chen Q, Cai L (2017a) Microplastics in the surface sediments from the Beijiang River littoral zone: composition, abundance, surface textures and interaction with heavy metals. Chemosphere 171:248–258

Wang L, Kaeppler A, Fischer D, Simmchen J (2019) Photocatalytic TiO_2 micromotors for removal of microplastics and suspended matter. ACS Appl Mater Interfaces 11(36):32937–32944

Wang T, Zou X, Li B, Yao Y, Li J, Hui H, Yu W, Wang C (2018) Microplastics in a wind farm area: a case study at the Rudong Offshore Wind Farm, Yellow Sea, China. Mar Pollut Bull 128:466–474

Wang W, Ndungu AW, Li Z, Wang J (2017b) Microplastics pollution in inland freshwaters of China: a case study in urban surface waters of Wuhan, China. Sci Total Environ 575:1369–1374

Windsor FM, Tilley RM, Tyler CR, Ormerod SJ (2019) Microplastic ingestion by riverine macroinvertebrates. Sci Total Environ 646:68–74

Wright SL, Kelly FJ (2017) Plastic and human health: a micro issue? Environ Sci Technol 51(12):6634–6647

Xiong X, Wu C, Elser JJ, Mei Z, Hao Y (2019) Occurrence and fate of microplastic debris in middle and lower reaches of the Yangtze River–from inland to the sea. Sci Total Environ 659:66–73

Yan M, Nie H, Xu K, He Y, Hu Y, Huang Y, Wang J (2019) Microplastic abundance, distribution and composition in the Pearl River along Guangzhou city and Pearl River estuary, China. Chemosphere 217:879–886

Yu X, Peng J, Wang J, Wang K, Bao S (2016) Occurrence of microplastics in the beach sand of the Chinese inner sea: the Bohai Sea. Environ Pollut 214:722–730

Yuan W, Liu X, Wang W, Di M, Wang J (2019) Microplastic abundance, distribution and composition in water, sediments, and wild fish from Poyang Lake, China. Ecotoxicol Environ Saf 170:180–187

Zhang K, Su J, Xiong X, Wu X, Wu C, Liu J (2016) Microplastic pollution of lakeshore sediments from remote lakes in Tibet plateau, China. Environ Pollut 219:450–455

Zhang K, Xiong X, Hu H, Wu C, Bi Y, Wu Y, Zhou B, Lam PK, Liu J (2017) Occurrence and characteristics of microplastic pollution in Xiangxi Bay of Three Gorges Reservoir. China. Environ Sci Technol 51(7):3794–3801

Zhang X, Leng Y, Liu X, Huang K, Wang J (2020) Microplastics' pollution and risk assessment in an urban river: a case study in the Yongjiang River, Nanning City. South China. Exposure and Health 12(2):141–151

Zhao S, Zhu L, Li D (2015) Microplastic in three urban estuaries, China. Environ Pollut 206:597–604

Zhao S, Zhu L, Wang T, Li D (2014) Suspended microplastics in the surface water of the Yangtze Estuary System, China: first observations on occurrence, distribution. Mar Pollut Bull 86(1–2):562–568

Chapter 7
Advanced Monitoring and Detection Techniques for Emerging Contaminants in Aquatic Environments

Arittrabha Sengupta, Jeniva Jasmine, Bhavya Chaudhary, Kritika Singh, Sahil Sharma, Harsh Sable, and Rohan Samir Kumar Sachan

Abstract This chapter investigates the recent developments in analyzing and detecting emerging contaminants (ECs) in aquatic habitats. For the detection and evaluation of contaminants, this chapter focuses on the incorporation of evolved systemic techniques or situation-based measures to detect pollutants in aquatic environments. For the assessment of contaminants in aquatic environments, there is the use of techniques such as bioassays and biomarkers which provide useful information about the hazardous effects and possible risks caused by ECs. Simultaneously, high-resolution mass spectrometry (HRMS) techniques, inclusive of liquid chromatography integrated with HRMS (LC-HRMS), this technique delivers remarkable sensitivity and specificity for the assessment of contaminants which allows the detection of a wide range of micropollutants in aquatic environments and on the other hand it also detects derived products present in minute quantity. These advanced technologies markedly enhance the precision of EC evaluation, providing extensive information about their occurrence, dispersal, and environmental effects. The study emphasizes the crucial role of these advanced techniques in strengthening management programs while taking the view of risk evaluation, which also helps safeguard the aquatic environment and human health by forming effective management strategies. By the incorporation of evolved analytical measures with biological evaluative

A. Sengupta · J. Jasmine · K. Singh · R. S. K. Sachan
School of Bioengineering and Biosciences, Lovely Professional University, Phagwara, Punjab, India

B. Chaudhary
Department of Genomic Sciences, Central Institute of Kerala, Periye, Kerala, India

S. Sharma
School of Allied and HealthCare Sciences, GNA University, Phagwara, Punjab, India

H. Sable
School of Allied Health Science, Sharda University, Greater Noida, Uttar Pradesh, India

R. S. K. Sachan (✉)
Department of Medical Laboratory Sciences, Chandigarh School of Business, CGC, Jhanjeri, Mohali 140307, Punjab, India
e-mail: rohan.sachan009@gmail.com

H. Jain et al. (eds.), *Emerging Contaminants in the Aquatic Environment*, Advances in Wastewater Research, https://doi.org/10.1007/978-981-96-8548-6_7

techniques, researchers can easily identify and reduce the risks of ECs while ensuring biosafety in reaction to these prevalent environmental threats.

Keywords Emerging contaminants (ECs) · Aquatic environment · Human health · Advanced analytical techniques · High-resolution mass spectrometry

7.1 Introduction

The varied byproducts from pharmaceutical industries, cosmetics, agricultural discharge, and various pollutants from industries can affect the equilibrium of the aquatic environment (Rathi et al. 2021; Chakraborty et al. 2023). These substances are usually present in minute quantities and have significantly impacted both aquatic life and human health (Karbalaei et al. 2018; Issac and Kandasubramanian 2021). Conventional analytical methods which are essential for the detection of some specific contaminants often fail to meet the requirements of detection and quantification of novel contaminants which are present in low amounts but impart significant long-term effect that usually affects the aquatic environment for acceptable survival (Zielinski et al. 2009).

The varying chemical nature of contaminants and usually lower concentrations in aquatic environments have appeared to be one of the major problems in monitoring ECs. Conventional methods, for instance, wet chemistry and bioassays, can be used but they take a longer duration, require great physical efforts, and detection of various substances that are present in the trace is not achievable due to lack of sensitivity in techniques. Furthermore, the effective monitoring options are not to the point as the emergence of new contaminants at unprecedented speed makes it difficult to keep pace with regulatory requirements.

To combat these challenges, there is the development of advanced analytical and management or regulatory techniques by scientists and engineers which meets the essential requirements that are not met by the conventional techniques. Various analytical techniques, including liquid chromatography-mass spectrometry (LC–MS), and gas chromatography-mass spectrometry (GC–MS) have been effective in the detection of a broader range of pollutants, these mentioned techniques offer higher specificity and accuracy (Luo et al. 2014). Bioassays have improved their effectiveness and sensitivity because earlier it was very time-consuming and have been upgraded by high-performance screening, as well as genetically modified organisms offer the detection of various biological processes with high speed and sensitivity (Wu et al. 2019). There is the use of passive samplers, which can be employed by placing for a broadened period of time which is essential to provide a more effective measure of exposure to contaminants over the desired period of time which permits the researcher to measure contaminants in trace amounts (Richardson and Kimura 2016). Also, to improve the awareness of the concern about the depleting water quality there are citizen science programs that are helpful in expanding the efforts to collect data (Brack et al. 2015).

By the integration of these advanced analytical techniques with the available conventional methods and by utilizing the data investigation methods, it is practicable to form comprehensive monitoring programs that have the potential to find as well as quantify ECs in aquatic systems. This detailed knowledge is crucial for the assessment of any hazard to human health and the environment, adherence to regulatory decisions, and also the formation of successful plans for the mitigation of various risks. The current development in systematic techniques for the analysis and the growing accessibility of efficient resources of computing have been effective in further enhancement of our management and monitoring capability for ECs in aquatic ecosystems.

7.1.1 Environmental Occurrence and Distribution of Emerging Contaminants (ECs)

ECs consist of a broader range of compounds that vary from synthetic to naturally forming compounds, including self-care products, pharmaceutical compounds, hormones that have the capability to affect public health, chemicals that prevent burning or slow the spread of fire, etc. These are naturally present in very low levels within the environment but have severe consequences for humans and biota. The consumption or utilization of these substances is on the rise—making their presence in all environmental matrices critical. They are a class of chemicals ubiquitous in the environment primarily due to common human practices like domestic, agricultural, and other industrial operations.

Though it is entirely not new to the environment, advancements in analytical techniques have helped in disclosing EC's presence in levels as low as parts per trillion (ppt) in air, soil, and water (Tang et al. 2019). They include more than 3,000 kinds of chemicals and their products or derivatives of the various groups of substances, including pesticides, fertilizers, toxic metals, microplastics (MPs), active pharmaceutical ingredients (APIs), and also the natural/synthetic hormonal compounds among other chemicals (Mishra et al. 2023).

Therefore, ECs are pseudo-persistent compounds due to their low polarity and half-life in the vicinity of hours plus concentrations often happening at ng/L levels or below that. Most of the ECs present are resistant to conventional treatments occurring in urban wastewater treatment plants (WWTPs), and thus they can be a source of introduction into the environment (Zhang et al. 2023).

7.1.2 Sources of Emerging Contaminants (ECs)

Sources of ECs are many and varied, hence the reason they are always ubiquitously present in ecosystems (Naidu et al. 2016) (Fig. 7.1). The sources include point sources including WWTPs and industrial discharges as well as hospitals and other healthcare facilities (Khan et al. 2022). WWTPs that treat sewages from various sources like households, industries, and hospitals do not effectively eliminate these chemical substances from water resulting in their discharge to the water bodies. Some of these are APIs, consumer products (CP), biological persistent chemicals (BPCs), and endocrine-disrupting chemicals (EDCs). Pharmaceuticals and chemical industries emit ECs including solvents, plasticizers, and per- and poly-fluoroalkyl substances (PFASs) into the water bodies of the industrial facilities (Wilkinson et al. 2016). Hospitals are also capable of discharging pharmaceutical compounds and diagnostic chemicals including those that go through various treatment stages. There are nonpoint sources that are more general and can be difficult to tackle such as those from agriculture and urban water bodies. During the application of pesticides, herbicides, and veterinary drugs, ECs are used as ingredients. During raining phase or irrigation procedures, ECs find their way into water bodies (Bhardwaj and Jindal 2019, 2020; Bhardwaj et al. 2019). These are washed off their surfaces into the stormwater drains, and hence the bodies of water such as rivers and lakes, especially in cases of storms (Valdez-Carrillo et al. 2020). The depositing of ECs continues in the atmosphere; volatile organic compounds (VOCs) and particulate matter (PM), emitted by industries and vehicles, may easily travel several kilometers before they are deposited on the surface of ground or water (Kumar et al. 2022; Bhardwaj and Vikram 2024; Bhardwaj et al. 2024a).

Landfills and waste disposal sites complement the problem through leachate which is normally a liquid that complicates groundwater or surface water systems. Furthermore, marine debris especially plastics acts as a source and vector of ECs and transports them to environments all over the globe (Feng et al. 2023). These complications are made worse by the effects of climate change; the occurrence of heavy rain that increases the runoff or high temperatures that may alter the degradation rates of the ECs. Knowledge about the above sources is vital in crafting measures of supervision, suppression, and proper management of the risks posed by ECs as a way of protecting the environment and human health.

7.1.3 Pathways of Emerging Contaminants (ECs) into the Environment

ECs are released into the environment through several paths or avenues, and each contributes to their dispersion and possible impact on the environment and the people around them (Rasheed et al. 2019). The first major route is the transport of ECs through the surface water, which carries these substances from different locations to rivers, lakes, or streams (Dzara et al. 2021). This is sometimes done through the effluent release from WWTPs that fails to remove all pollutant compounds hence polluting the water surface. Also, these water bodies are threatened by direct input of ECs and other pollutants from agricultural and urban fields (Bo et al. 2015). For

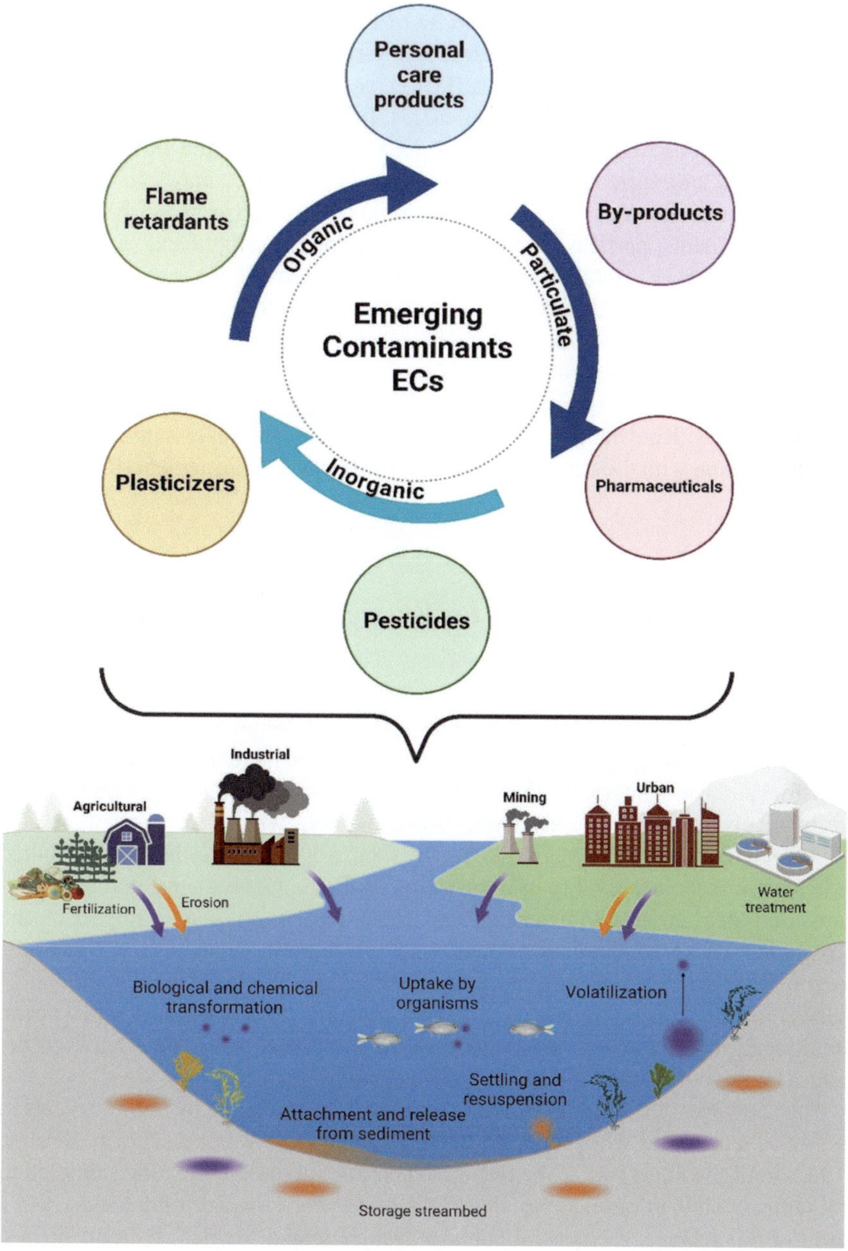

Fig. 7.1 Source of emerging contaminants (ECs)

instance, agricultural sources may dispose of ECs into water sources through rain or irrigation water, while urban sources discharge chemicals used in households, products from personal care products (PCPs), and MPs from roads and pavements into the stormwater systems (Lei et al. 2023). Intervention occurs when hazardous substances such as ECs pollute groundwater through percolation from the surface which may include agricultural lands or even dump sites which may lead to pollution of drinking water. Some of the ECs are capable of being adsorbed or precipitating onto the sediments, particularly those that are sparingly soluble and as such, form bed deposits in river and lake beds and which can re-suspend the contaminants once again in the water when stirred (Schnoebelen et al. 2006). Another deposition method is also present in that VOCs and PM can be transported through the air and deposited on land or water surfaces. Furthermore, through the carrying of pollutants and debris including MPs, ECs influence marine ecosystems through the transportation of pollutants and debris. Knowledge of these pathways is critical in assessing the scope of EC's interaction with the environment and in formulating management strategies for the impact.

7.2 Effect-Based Measures for Emerging Contaminant (EC) Detection

ECs are chemicals found in the environment that lack regulation or monitoring, on account of their being considered novels. This category includes everything from industrial and nanomaterials (NMs) to PPCPs (Richardson 2009). Since their effects are relatively new, they are still undergoing research, which is why they are also considered emerging. The main reason why regulatory measures have been limited to this point is the availability of information. Still, the fact remains, that the pollutants are potent enough to enter the food chain and leave a lasting impact on the ecosystem. These contaminants, along with agricultural and industrial discharge, stem from pharmaceutical wastes (Khan et al. 2022). However, the recognition of these compounds as possible environmental concerns dates back only a few years. Traditional pollutants were the focus for a large portion of the time between the Industrial Revolution and the late twentieth century.

This, however, has resulted in delays as it was made possible by improvements in analytical techniques which delayed the development of awareness and knowledge of ECs until the latter part of the twentieth century. Increased scientific study led to a significant shift in public opinion of ECs by the early twenty-first century. As a result of this change, the general public's awareness and worries about medicines, EDCs, and NMs have grown (Metcalfe 2013). Concerns about the presence and complex mixture combinations of these contaminants, their hazardous nature, and the challenges in assessing and managing the risks they present are increasing at a faster rate.

7.2.1 Detection and Quantification of Emerging Contaminants (ECs)

Detection and quantification of ECs, often at trace levels, would require advanced procedures specifically designed for this purpose, which can detect and analyze such contaminants in environmental samples. It is only by achieving excellent sensitivity and selectivity that one can identify and quantify the ECs against a very broad range of other chemicals. The process begins with careful sample preparation, such as the use of microextraction by packed sorbent (MEPS) or solid-phase microextraction (SPME) to concentrate ECs while reducing interference from the sample matrix, as well as solid-phase extraction (SPE), liquid–liquid extraction (LLE), ultrasound-assisted extraction (UAE), and so on (Martín-Pozo et al. 2019). The main separation techniques and accurate quantification of substances in this process are by using gas chromatography (GC) and liquid chromatography (LC) along with various mass spectrometry (MS) such as tandem MS, high-resolution MS, and several others. It is an essential analytical technique. Nuclear magnetic resonance spectroscopy (NMR) and capillary electrophoresis are even more sophisticated in improving EC analysis and meet certain other purposes, namely, identification of structure and study of ionic species. To ensure that it is reliable and in conformity with legal obligations, this blanket approach also entails screening of nontarget and suspect samples for unknown or potentially hazardous chemicals. It is supported by strict quality control and technique validation. Finally, sophisticated data analysis techniques are applied to process and decode the complex datasets generated, thereby ensuring that the conclusions drawn are valid and meaningful. Besides complying strictly with the legislative conditions, this holistic approach increases the understanding of ECs in the environment (Shyamalagowri et al. 2023).

7.2.2 Technological Advances in Detection

Advanced techniques are emerging and discussed in this chapter to overcome the challenges of the detection and quantification of ECs. More innovative approaches have been developed to detect pollutants with higher sensitivity and at lower concentrations than traditional methods, as shown in Table 7.1. The unknown pollutants could now be detected with molecular methods and MS for multi-residue analysis. These sensing technologies are continuously being developed to enable quick on-site screening. The culmination of all these new analytical techniques that are in practice has enhanced our ability to detect pollutants of growing concern and obtain critical information regarding their prevalence in the environment.

Table 7.1 Different analytical techniques for particular emerging contaminants (ECs)

Methods	Emerging contaminants (ECs)	References
Gas chromatography-mass spectrometry (GC–MS)	Naproxen, diclofenac, fenofibrate, nonylphenol, bisphenol A (BPA), triclosan, methyl triclosan, tri-n-butyl phosphate, tris (2-chloroethyl) phosphate, tris (2-chloroisopropyl) phosphate, tris (1,3-dichloro-2-propyl) phosphate, geosmin, 2-methylisoborneol, glycol ethers, benzene, toluene, ethylbenzene, xylene, butylated hydroxytoluene, butylated hydroxyanisole, Estrone, 17β-estradiol, estriol, 17a-ethinylestradiol, clofibric acid, ibuprofen	Cincotta et al. 2018; Bristow et al. 2023; Suvar et al. 2024
Enzyme-linked immunosorbent assay (ELISA)	Cetirizine, caffeine, pesticides, herbicides, heavy metals, hormone disruptors, sulfamethazine, BPA, sulfadiazine, sulfamethoxazole, ractopamine, pentachlorophenol, *Salmonella*, *Staphylococcus*, 17α-ethinylestradiol, 17β-estradiol, carbamazepine	Silva et al. 2020a, b; Aslam et al. 2023; Wang et al. 2023
Liquid chromatography-mass spectrometry (LC–MS)	Paracetamol, ibuprofen, olanzapine, simvastatin, albendazole, hydrolyzed dyes, trichothecene mycotoxins, imidacloprid (neonicotinoids), estriol, codeine, oxazepam, 2,4-dinitrotoluene, 1,3,5-trinitroperhydro-1,3,5-triazine, BPA, triclosan, caffeine, carbamazepine, lincomycin, perfluorooctanoic acid, perfluorononanoic acid, perfluorohexane sulfonic acid, perfluorooctane sulfonic acid	He et al. 2019; Safwat et al. 2020; Encarnação et al. 2021; Aslam et al. 2023; Nobile et al. 2024; Soliz et al. 2024; Tang et al. 2024
Nuclear magnetic resonance spectroscopy (NMR)	Polyfluoroalkyl substances, olmesartan acid, heparin	Camdzic et al. 2021; Colombo et al. 2022
Fourier transform infrared spectroscopy (FTIR)	Tetracycline, polycyclic aromatic hydrocarbons (PAHs), microplastics (MPs), BPA	Khoubi et al. 2024; Xie et al. 2024; Bhardwaj 2024; Bhardwaj et al. 2024b

7.2.2.1 Gas Chromatography-Mass Spectrometry (GC–MS)

Suvar et al. (2024) showed the potential of GC–MS in quantifying volatile and semi-volatile molecules during the study on the application of the technique in fire investigation for the detection of hazardous organic substances. For example, Bristow et al. (2023) developed a micro-SPE technique combined with GC–MS for the determination of 2-methylisoborneol and geosmin in drinking water. It achieved a limit of detection (LOD) values of 2.0 ng/g and 4.3 ng/g, respectively. Though GC–MS is very sensitive and specific for the determination and quantification of ECs, it has a few drawbacks in this aspect. One of the prime limitations includes the detection of the recently discovered contaminants, which have the properties of high molecular weight and polarity, and this may require derivatization of chemicals in the analytical techniques including GC–MS, further making sample preparation complex and also increasing the chances of error. Still, if the identification and quantification of ECs are the objective, GC–MS cannot be applied freely, although it has excellent sensitivity and selectivity.

7.2.2.2 Liquid Chromatography-Mass Spectrometry (LC–MS)

Among various detection methodologies, the techniques of LC, LC–MS, and LC–MS/MS are employed for the detection of pollutants in trace amounts. The present developments have highlighted their functionality in the analysis of residues in marine water which has strong potential to analyze and identify varied ECs including various contaminants like phthalates, BPA, and pharmaceutical remnants. The reverse-phase high-performance liquid chromatography (HPLC) technique, as described by Encarnação et al. (2021) is critical for the analysis with enhanced precision and permits the detection of varied contaminants formed during the decontamination of microalgae. There is the introduction of the novel method by He et al. (2019) which is known as the fast freeze-centrifugation extraction measure for the quantification of 11 ECs in three crops includes—corn, wheat, and tomato which is further analyzed by LC–MS, it shows the LOD within the range of 0.01 to 2.0 ng/g.

Maggee et al. (2023) assert that the sensitivity with which LC–MS can identify analytes in intricate matrices has made it an essential method in analytical chemistry and environmental monitoring. LC–MS has some limitations even with its remarkable EC detection sensitivity and selectivity. One of the major disadvantages of LC–MS is its susceptibility to matrix effects, which may adversely affect the accuracy and reproducibility of results. These effects occur when other compounds in the sample interfere with the ionization efficiency of the analytes, leading to enhancement or suppression that might affect quantitation. Besides, LC–MS has the property of analysis of different compounds but may not work effectively with some of the substances like ECs; this is because of their chemical composition and properties. For example, if the compounds have a larger size or highly polarized molecules, they require more care or derivatization which makes the analysis complicated.

7.2.2.3 Enzyme-Linked Immunosorbent Assay (ELISA)

ELISA is an extensively established analytical method that has been used to identify some of the pollutants present in different types of samples. The technology relies on the principle of antibody-antigen interactions, and this allows for the accurate identification and quantification of target analytes like specific pesticides, hormones, and heavy metals. Recent applications of ELISA included water quality analysis in order to detect the possible dangerous materials, such as herbicides, insecticides, and compounds disturbing the endocrine system. For the detection or quantification of different pharmaceutical compounds such as caffeine, carbamazepine, cetirizine,

17β-estradiol, and 17α-ethinylestradiol in samples of water, Silva et al. (2020a, b) utilized ELISA which has already proven its efficiency in the analysis of the mentioned compounds above. Commercially available kits of ELISA have turned out to be a better analytical technique for getting a speedy and accurate assessment of compounds, which are critical as they have resulted in threats to the environment and have given rise to issues concerning safety. ELISA has proved to be crucial for identifying pesticide residues and hormone disruptors because it gives greater sensitivity, specificity, accuracy, and precision. ELISA has been used mainly for diverse environmental contaminants for the control of pollutants and is very much required in the application field in food safety sectors which involves rigorous observation of the removal of pollutants. Furthermore, ELISA is less sensitive and specific than the above-mentioned techniques which have the potential to detect pollutants at much lesser concentrations. Such techniques include advanced analytical techniques like LC–MS or GC–MS (Gomes et al. 2020; Rout et al. 2021; Tong et al. 2022; Samal et al. 2022; Miller et al. 2023; Puri et al. 2023; Espíndola et al. 2024). In addition to that, the unknown or unclassified endocrine disrupters (EDs) are not detected since each target contaminant requires antibodies. Additionally, sample preparation techniques and matrix effects may interfere with the quantitative outcome of ELISA, thus causing differences in data accuracy and reproducibility (Aslam et al. 2023).

7.2.2.4 Sophisticated Spectroscopic Techniques

For the quantification and analysis of organic contaminants, there are two essential methods: FTIR and NMR. NMR spectroscopy has been found to be quite promising for the analysis of trace contaminants in different types of samples. NMR appears to be useful for the analysis and differentiation of structurally similar pollutants. According to Camdzic et al. (2021) and Colombo et al. (2022), heparin, poly-fluoroalkyl compounds, as well as olmesartan acid can be detected by NMR. FTIR spectroscopy is very useful for the presence as well as nano plastics (NPs) count (Bhardwaj et al. 2025). According to Chen et al. (2020), NMR and FTIR are fundamental techniques for characterizing complex organic pollutants in diverse materials.

Although NMR yields very good information about the molecular structure of compounds trace pollutants in environmental samples can often be very challenging to analyze because it demands very large sample sizes and very high concentrations of target compounds for achieving sufficient signal intensity. Routine application of NMR to environmental monitoring is also limited due to its often-lower sensitivity compared with MS methods and higher expense and time demands. Although FTIR has several advantages, including rapid analysis and non-destructive research, it is mainly used for identifying the functional groups in molecules rather than detailed structural information. The main problem with the FTIR technique is that it can be significantly affected by sample complexity and interference due to ubiquitous components like water in environmental samples, which may mask the spectra of the pollutant. Different microscopy techniques have been applied in the identification of MPs and NPs, such as dissection, polarized, fluorescence, scanning electron, and atomic force microscopy. However, these methods have often proven to be insufficient, mainly due to the minuteness of the particles; this problem has been especially recently overcome by combining chemical analysis methods, including spectroscopy, with microscopy techniques. The development of analytical techniques also led to the development of new methods that were not earlier possible. Direct visualization is possible by applying holographic imaging in a microscope setup of the MPs in unfiltered water, together with separation from diatoms, and further classification according to the kind, size, and shape of different kinds of plastics. New analytical tools are going to help solve the present problems that come with the identification of MPs and NPs, whether they are incorporated into or used in combination with conventional and innovative microscopy techniques. These innovations can enhance accuracy and detail in the detection and characterization of MPs and NPs (Cai et al. 2021; Mariano et al. 2021).

7.3 High-Resolution Mass Spectrometry (HRMS) Techniques

HRMS has revolutionized the field of environmental analysis by introducing unprecedented sensitivity, specificity, and structural resolution of detecting and monitoring a variety of contaminants like EDCs, pharmaceutical waste products, cosmetics, and pesticides applied for farming purposes, which might be harmful to the environment (Richardson and Kimura 2016). The capability of the HRMS technique would be the exact mass measurement and structural elucidation required in the detection and monitoring of these unknown or uncharacterized entities that seem useful to biosafety (Jorge et al. 2016).

7.3.1 Types of High-Resolution Mass Spectrometry (HRMS)

7.3.1.1 Liquid Chromatography-High-Resolution Mass Spectrometry (LC-HRMS)

LC-HRMS is one of the most sensitive detection techniques that are used to quantify and estimate ECs in aquatic ecosystems. This blend of methodologies provides a synergistic effect which often comprises enhanced sensitivity, specificity, and structural elucidation, and also the selective detection of contaminants which provides the capability for monitoring and detection of diverse varieties of contaminants which is essential for the management of aquatic systems. LC-HRMS helps tap into the individual analytical characteristics of each technique that work toward greater precision in measuring masses. This technique allows the analysis of intricate mixtures of compounds, which lends precision in detection and quantification, irrespective of the fact that there may be agents that can interfere with a mixture of compounds. This technique selectively analyzes the compounds to measure them appropriately.

7.3.1.2 Gas Chromatography-High-Resolution Mass Spectrometry (GC-HRMS)

GC-HRMS has become a very efficient detector for the purpose of EC analysis, known to exist within small quantities in any aquatic ecosystem. This combination of techniques allows exploitation of that combined effect for the analysis which has required sensitivity, specificity, and structural elucidation which revolutionize the analytical technique and make effective the identification and surveillance of a broad variety of volatile and semi-volatile contaminants. Coupling the individual strengths of GC with high mass accuracy and HRMS enables synergistic effects. This technique allows researchers to find complex combinations of volatile and semi-volatile compounds with a very high degree of accuracy for identification and quantification and even if some interfering agents could be there, this method selectively analyzes the compound (Wang et al. 2007; Gómez et al. 2009; Polyakova et al. 2013).

7.3.1.3 Direct Infusion-High-Resolution Mass Spectrometry (DI-HRMS)

DI-HRMS has been identified as one of the key analytical techniques that can be used to analyze and evaluate ECs in aquatic systems. The technique provides critical information on the composition and evaluation of molecular pollutants that have the potential to harm the aquatic environment, the technique provides a fast and straightforward approach to the evaluation of complex mixtures in the water (Pezzatti et al. 2020).

7.4 Integrating Advanced Analytical Methods

This radically changes our ability to detect, quantify, and identify the presence of pollutants in water bodies due to advanced analytical techniques that include MS, spectroscopy, electrochemical techniques, and biosensors. Such advanced sensitivity, specificity, and efficiency cannot be offered by other conventional analytical methods (Grieshaber et al. 2008). These technologies now make possible the detection and measurement of trace levels of contaminants in even the most complex matrices. Such enhanced analytical ability has indeed been quite crucial in addressing some of the latest environmental issues that have arisen lately, including identifying ECs, determining long-term trends in pollution, and mitigation strategies. Advanced techniques have further helped in the development of effective water quality monitoring networks, preventing harmful substances to human health and aquatic ecosystems. The latter part of the discussion will now cite to present the basic advanced analytical methods being utilized in the monitoring of aquatic contaminants.

7.4.1 Advanced Analytical Methods

7.4.1.1 Mass Spectrometry (MS)

MS is a very powerful analytical technique that has changed the surface of chemical analysis. It is a universal device for the detection and measurement of a wide range of molecules from simple molecular species to the most complex macromolecules. The principle of MS essentially comprises the process of ionizing molecules and then separating the formed ions according to the mass-to-charge (M/Z) ratio. The separation of ions can be performed by means of the forces of an electric and magnetic field in a mass spectrometer. The concentration of the ions is measured, and the presence is detected since the ions are released when separation takes place. Such measurement may reflect the molecular weight, configuration, and composition of the analyte. For wider application in different scientific fields including chemistry, biochemistry, ecology, as well as forensic sciences, MS usage has been observed to be sensitive, selective, and resolving analytical technique (Faull et al. 2008).

- **High-Performance Liquid Chromatography-Mass Spectrometry (HPLC–MS)**

HPLC–MS application has become essential for monitoring aquatic pollutants. It offers preciseness, sensitivity, specificity, and efficiency in pollutant monitoring. It is one powerful analytical technique that combines the capabilities of HPLC in separation and MS in the detection and quantification powers, thus offering a wide analysis range of various contaminants in complicated environmental matrices. HPLC–MS is particularly suitable for the determination and quantification of non-volatile and thermally labile compounds such as pharmaceuticals, pesticides, EDCs, and other ECs. MS is highly sensitive. Interfering substances still allow detection at trace levels. In addition, MS offers specificity in the identification and quantification of the individual compounds in complex mixtures. HPLC–MS has played a very important role in enhancing our understanding of aquatic pollution and assisted in designing useful monitoring programs. HPLC–MS gives reliable and precise data on EC concentrations. Thus, it plays a major role in protecting human health. It also protects aquatic ecosystems from harmful impacts that are caused due to pollution. (Avar et al. 2016).

- **Gas Chromatography-Mass Spectrometry (GC–MS)**

GC–MS is effective toward different compounds including VOCs, PAHs, pesticides, as well as organic compounds with volatile and semi-volatile activity. The analytical apparatus separates the analytes based on their boiling points, which then get detected through accurate quantitation measures by the MS part of the equipment.

Some advantages can be quoted with respect to the GC–MS technique for aquatic contaminant monitoring. These include (1) enhanced sensitivity toward trace levels of contaminants, (2) better specificity toward the identification and quantification of individual compounds in complex mixtures, (3) ability to analyze a wide range of volatile and semi-volatile organic compounds, and (4) possibility for automated analysis that may permit high-throughput sample processing. GC–MS has enriched our knowledge in connection with water pollution. It helps in decision-making at regulatory levels. It gives scope for effectively executing remedial measures (Mondal et al. 2018; Pozzer et al. 2022).

- **Inductively Coupled Plasma-Mass Spectrometry (ICP-MS)**

This technique is highly adaptable for determining inorganic pollutants, heavy metals, metalloids, as well as rare earth elements. The ICP part of the system generates a high-temperature plasma that atomizes and ionizes the material, whereas the MS part is used for accurate identification and quantitation of the elements. Some of the diverse advantages that emerge from the use of ICP-MS for the analysis of aquatic pollutants include: (1) ultra-low detection limits for a wide range of elements; (2) high precision and accuracy in the quantitation process; (3) ability to analyze complex matrices such as seawater and wastewater; and (4) ability to carry out multi-element determination which is useful in the identification of diverse species from the same sample simultaneously. ICP-MS has highly contributed to the understanding of aquatic pollution, which helps in the formation of potent remedial techniques as well as contributing to regulatory choices (Tegladza et al. 2020; Barbieri et al. 2023).

7.4.1.2 Spectroscopy

Spectroscopy is one of the developed analytical techniques that deals with the study of the interactivity of various compounds with radiation of any pattern of the electromagnetic spectrum. Spectroscopy is applied to monitor aquatic pollutants. It identifies the nature as well as the amount of pollutants that are found in aquatic ecosystems. It works on the principle of absorption, emission, or scattering of light by different types of compounds. It includes other types of techniques, including ultraviolet–visible spectroscopy (UV–Vis), infrared spectroscopy (IR), and Raman spectroscopy. Their applications have been able to identify and quantify a wide range of contaminants from heavy metals and organic compounds and from other ECs. UV–Vis spectroscopy is rather sensitive in the determination of colored species but IR spectroscopy is actually used in identifying functional groups in organic compounds. Raman spectroscopy is another kind of analysis with non-invasive as well as rapid examination in both organic as well as inorganic species of pollutants. Spectroscopic techniques are largely used in amalgamation with any other analytical methods like chromatography for greater sensitivities and specificity in detection. With this application of spectroscopy in the monitoring of aquatic contaminants, our knowledge improved of problems associated with water quality; and helped us design appropriate measures of pollution management (Zulkifli et al. 2018; Li et al. 2020; Ong et al. 2020).

7.4.1.3 Electrochemical Techniques

Electrochemical technologies are now becoming practical tools for monitoring aquatic contaminants as they provide multiple techniques for pollutant detection and quantification in water bodies. Such methods exploit the electrical nature of pollutants and their interaction with electrodes to yield excellent information regarding the presence and concentrations of these impurities. Ion-selective electrodes (ISEs) and potentiometric sensors are most commonly used for detecting specific ions such as heavy metals, nutrients, and anions. These sensors have been desirable for use over other sensors because they are easily carried and transportable and also, they are economical as well as work in real time (Gupta et al. 2011). In addition to those voltametric methods of estimation, such as cyclic voltammetry and differential pulse voltammetry can be employed to measure the redox reaction of pollutants and their quantity. Electrochemical methods are applicable for the constant determination of the quality of water since the above measurements can be carried out very rapidly in situ. Using the sensitivity, selectivity, and mobility of electrochemical techniques, researchers and environmental managers can sensitively assess levels of pollutants in aquatic environments and implement proper remediation measures (Hernandez-Vargas et al. 2018).

7.4.1.4 Biosensors

Biosensors have emerged as useful tools for monitoring water pollutants, integrating biological with physical transducer elements. These technologies allow for the unique detection and determination of contaminants within aquatic systems due to their specificity and sensitivity for biological identification elements. Biological elements that are employed in biosensors include enzymes, antibodies, and nucleic acids, which provide for selective detection of contaminants, such as heavy metals, organic pollutants, and other ECs. Biosensors can be classified into two categories, which are direct and indirect detection systems. This depends on the nature of the biological component applied and the transducer. Direct detection takes place where the pollutant forms direct contact with the biological component, while indirect detection relies on some form of secondary response or signal amplification mechanism. Biosensors are advantageous in determining aquatic contaminants since they can possess increased sensitivity, specificity, and flexibility in analyzing in real time. Biosensors can be miniaturized and integrated into portable devices for on-site monitoring and early warning systems for water quality issues (Ejeian et al. 2018).

7.4.2 Benefits of Advanced Analytical Techniques in Aquatic Contaminant Monitoring

Advanced analytical techniques would be best described as the modern improvement that has transformed field monitoring of aquatic contamination, providing incredible sensitivity, specificity, and efficiency in determining and quantifying various pollutants. Recent advanced techniques such as MS, spectroscopy, electrochemical approaches, and biosensors have enabled researchers to detect traces of pollutants in complex environmental matrices at trace levels. These new technologies have become vital tools for protecting human health and ecosystems against the harmful effects of pollution by delivering quick, precise, and reproducible measurements. The ability to detect and quantify contaminants at increasingly lower levels has enabled lead detection and avoidance early on, thus enabling prompt intervention to mitigate potential hazards. The higher specificity of these technologies ensures accurate identification of specific pollutants, even within complex mixtures, and is crucial in establishing the origin, transport pathways, and potential impacts of pollution. In addition, advanced analytical methods often increase efficiency through automation and the ability to analyze several pollutants at once, thereby reducing analysis time and costs. The integration of these advantages has transformed the monitoring of aquatic pollutants, enabling scientists to obtain extensive and reliable data that informs decision-making and supports effective pollution control measures. Some of their main advantages are as follows:

- **Higher Sensitivity**

The superior sensitivity of cutting-edge analytical technologies is one of the significant gains in the monitoring of aquatic pollutants since it allows monitoring at very low concentrations (Rodriguez-Mozaz et al. 2007). In this way, it facilitates new emerging pollutants being discovered and pollution trends over very long periods being characterized, even as contamination levels reach below conventional LOD (Khan et al. 2022). It is through such technologies that trace-level detection of pollutants becomes possible. This once again gives the overall ability to have an early warning system that allows for timely intervention before further pollution and probable health risks are set in (Zielinski et al. 2009). The ability to trace contaminants at low levels is necessary for understanding environmental mobility and the transport of contaminants in aquatic environments, and in assessing the success of controlling approaches to pollution (Rathi et al. 2021). High sensitivity has also led to more stringent standards and policies of water quality to ensure public health and the environment.

- **Resolution**

Advanced analytical tools are able to increase the specificity of the monitoring of aquatic contamination, and thus, it is possible to identify and measure individual pollutants even in complex mixtures (Wille et al. 2012). This specificity would help in identifying the sources, transport mechanisms, and potential impacts of various contaminants. These techniques can differentiate closely related molecules that may provide valuable information on individual contaminants in water bodies, therefore enabling targeted monitoring and remediation efforts (Altenburger et al. 2015). The specificity of new analytical procedures reduces false positives and negatives and therefore ensures the reliability of the data (Leeflang et al. 2008). Since precise determination of concentrations is required, this will imply full regulatory compliance along with sound decision-making that arises from this fundamental aspect because it determines risk assessment or provides the right measure for ameliorative action to be taken (Linkov et al. 2006). Advanced analytical methods used for monitoring contaminants in water show advantages in quicker and more efficient analyses of several samples at the same time (Rodriguez-Mozaz et al. 2007).

This efficiency is realized through automation, downsizing, and the capability to monitor simultaneously several pollutants (Silva Burato et al. 2020). These technologies allow for regular monitoring of water quality through reduced analysis time and costs, therefore allowing fast pollution incident identification and determination of measures used in pollution control effectiveness (Glasgow et al. 2004). The real-time monitoring systems may be upgraded using modern analytical approaches to improve the efficacy of providing critical information for decision-making and emergency response (Thibaud et al. 2018). The technologies have better efficacy, hence enhancing the aquatic pollutant monitoring programs that are efficient and cost-effective, therefore protecting public health and the environment (Gray and Shimshack 2011). Automation is an important advantage of advanced analytical techniques in monitoring aquatic contamination because it enhances efficiency and reduces human error, thus facilitating high-throughput analysis (Rodriguez-Mozaz et al. 2007). Automated systems can handle the same sample preparation, analysis, and data processing so laboratory people would have the time to focus on other necessary tasks (Lippi et al. 2011). This reduces the time needed for analysis while ensuring uniformity and repeatability of results (Zwanenburg 2019). In addition, automation reduces human error, which is very significant when dealing with sensitive and complex studies (Dekker 2017). The automation of routine processes can enable laboratories to increase their capacity to process larger volumes of samples, thus increasing the frequency of water quality monitoring and speeding up the data analysis. Generally, automation enhances the efficiency and effectiveness of aquatic pollution monitoring systems, which saves public health and the environment (Glasgow et al. 2004). In a nutshell, state-of-the-art analytical methods in aquatic contaminant monitoring have fitted researchers with all the tools to be able to recognize, quantify, and understand different polluters. What is also important as regards their sensitivity and specificity as well as efficiency in data quality is critical for keeping both public health and aquatic resources sustainable.

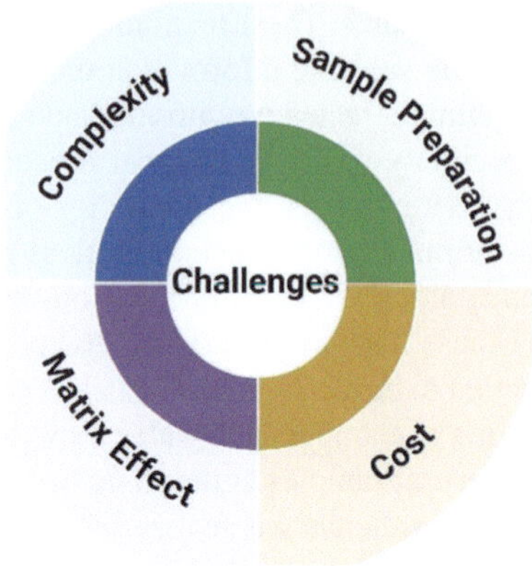

Fig. 7.2 Challenges to the advanced analytical techniques utilized in pollutant monitoring

7.4.3 Challenges and Considerations

The characteristics of aquatic habitats are complex features that pose many challenges to the advanced analytical techniques utilized in pollutant monitoring (Zulkifli et al. 2018) (Fig. 7.2). Such variability in contaminants, the complex matrix they exist in, and their very low concentrations often require sophisticated methodologies and careful analysis (Geissen et al. 2015). Matrix effects from dissolved organic matter, suspended particles, and salts can also interfere with the analysis of contaminants, resulting in false results and a loss of sensitivity (Richardson and Ternes 2018).

The vast spectrum of contaminants such as heavy metals, organic pollutants, medicines, and other ECs calls for specific analytical techniques and intricate sample preparation (Neale et al. 2021). Many of the pollutants are trace levels, and hence, identification and quantification are a tough task that demands sophisticated methodologies with high sensitivity and LOD (Zulkifli et al. 2018). The continued emergence of new pollutants, such as MPs and NMs, adds to the complexity of the issue, requiring novel analytical techniques and the development of new standards. However, spatial and temporal heterogeneity in the concentration of pollutants in the aquatic environment would lead to sampling problems and result in misinterpretation of data (Richardson 2009). Powerful yet flexible analytical methodologies that can withstand the complexity of the matrix, trace levels of pollutant detection, and changing landscape in the aquatic pollution arena need to be developed (Thakur and Kumar 2022). For instance, highly innovative analytical methods can greatly limit the observance of aquatic pollutant monitoring programs since their costs are high (Richardson 2009). Obtaining and maintaining apparatus specialists in MS, spectrophotometers, and electrochemical analyzers could be expensive since they require significant capital inputs (Crocombe 2018). In addition, the use of most of this apparatus often requires the consumption of expensive reagents, standards, and consumables. Furthermore, the training and expertise needed to operate and analyze data produced by these advanced methods can be costly, especially for smaller laboratories or organizations that have limited resources. Advanced analytical technologies are expensive, which severely limits their availability and accessibility to developing countries or regions with severely restricted budgets (Cameron et al. 2009). Such decisions regarding analytical techniques should hence be considered with precautions, measuring the demand for accurate and reliable data against the constraints that accompany available resources and budget stipulations (Drummond et al. 2008). Sample preparation involves critical steps but presents a great challenge with advanced techniques used in determining the aquatic pollutants.

The complexity of matrices in water ecosystems, often incorporating various dissolved chemical species, particulate matter, and biological compounds, requires sample preparation that may be elaborate, to ensure a high degree of accuracy and validity in the ensuing analysis (Mitra 2004). Time-consuming, labor-intensive, and prone to contamination sampling preparation methods sometimes compromise data quality (Nováková and Vl ková 2009). Furthermore, some analytical techniques require specific sample preparation techniques, and certain methods necessitate protocols, such as filtration, extraction, or concentration, and therefore make the process more complex. The selection of an appropriate sample preparation technique is essential through proper assessment to ensure that the targeted contaminants are effectively extracted and concentrated with minimal interference resulting from the introduction of extraneous contaminants or matrix effects (Trufelli et al. 2011). Suboptimal sample preparation can lead to wrong results and decrease the sensitivity of analytical techniques, which can pose a threat to the integrity of aquatic pollutant monitoring programs (Pérez-Fernández et al. 2017).

- **Matrix Effects**

While strong interference from additional compounds present in water samples makes contaminant detection challenging for advanced analytical techniques, it decreases their selectivity. Many factors, such as dissolved organic matter, suspended particles, salts, and other pollutants, may cause these effects (Richardson 2009). In MS, it occurs through ionization; in spectroscopy, through light absorption; and in sensor electrochemical responses, giving false positives and lower sensitivity values (Niu et al. 2018). Aquatic matrices are complex and often a mixture of diverse chemicals. This makes it difficult to predict and control the matrix effects (Niessen et al. 2006). Therefore, appropriate analytical techniques must be used, and careful sample preparation is essential to control the matrix effects so that accurate and reliable identification of contaminants is made (Hajšlová and Zrostlikova 2003). One of the big challenges in the monitoring of aquatic pollutants is the understanding and mitigation of matrix effects, which always require continuous research and the development of new analytical techniques (Richardson 2009).

7.5 Recommendations

Alternative detection and monitoring methods should use state-of-the-art technology and multidisciplinary approaches to tackle the problem of new pollutants in aquatic ecosystems. The implementation of biosensors, which use living organisms to identify pollutants with pinpoint accuracy is recommended. Furthermore, cutting-edge mass spectrometry methods may provide unmatched precision when it comes to detecting minute pollutants. These methods include high-resolution and tandem mass spectrometry, (HRMS and MS/MS), respectively. Improved predictive analytics and interpretation of real-time data may be achieved by the integration of environmental monitoring systems with machine learning algorithms. Local stakeholders may be engaged in sample collection and data exchange through community-based monitoring initiatives, which can supplement these technologies. There must be investment in scalable, low-cost technologies that are suited for developing locations and the promotion of uniform protocols for assessing new contaminants. With global cooperation and these solutions, detection skills may be greatly enhanced, and sustainable water management practices can be better informed.

7.6 Conclusion and Future Directions

Advanced monitoring and detection techniques are required for the mitigation of levels of ECs in aquatic habitats since these have reached considerable contamination levels threatening human health and ecosystems. The study discusses the need for effect-based metrics in providing early warnings for potential biological consequences and the efficiency of HRMS techniques that offer unmatched sensitivity and selectivity in the identification and quantification of ECs. Combining these advanced techniques with conventional analytical methods will be necessary for improving our capability to identify and assess the hazards associated with ECs. Understanding the environmental availability and distribution of ECs forms the basis for a proper monitoring framework. This information could be obtained using targeted sampling, as well as through the use of passive samplers that provide information regarding long-term exposures to pollutants.

Besides, the ecological and public health effects of ECs have to be examined with care so that the development of risk evaluation and management strategies can be formed.

Investment is required in novel techniques for the research and development (R&D) of new techniques. We need to increase data sharing and collaboration across researchers and regulatory bodies. And finally, public awareness of the importance of maintaining good quality water needs to be increased. Implementing all these measures ensures that our monitoring programs are better equipped to respond to the challenges an ever-changing aquatic environment is bound to bring along and, thereby, protect both the ecosystems and the human population. Several critical areas require further research and development to further the field of EC monitoring:

- **Miniaturization and Automation**: It is necessary to develop small, low-cost, automated analytical systems to enable more and more persistent sampling, particularly for remote or inaccessible areas.
- **Real-Time Monitoring**: Real-time monitoring techniques, inclusive of continuous flow analysis and online sensors, may be helpful in providing immediate alarms on pollutant spills or sudden increases in the amount of pollutants.
- **Integration of Sources**: Data integration from multiple sources, like traditional monitoring, citizen science, and remote sensing, could provide complete information on distribution and trends in ECs.
- **Advanced Data Analysis**: Machine learning (ML) and artificial intelligence (AI) methods are useful for identifying different patterns, trends prediction, and ranking the monitoring activities.
- **Development of New Analytical Methods**: There is a requirement for developing novel analytical techniques that possess the measuring capabilities and simultaneously quantify emerging contaminants. These methods are highly sensitive, selective, and efficient.
- **Assessment of ECs**: Since novel pollutants are regularly emerging, it is required to have rapid as well as efficient techniques for identifying and assessing hazardous ECs.
- **Public Outreach and Education**: Increased general public awareness and efforts toward ecological issues would lead to a better-informed and more engaged public. It is by pursuing these future efforts that we are in a better position to monitor and regulate the continuously growing problems in ecosystems, thus protecting human health and environmental integrity for generations.

Acknowledgements The authors hope to express sincere gratitude to their respective institutes for their assistance and resources, which were very vital in the prompt writing of this chapter.

Abbreviations

AI	Artificial intelligence
APIs	Active pharmaceutical ingredients
BPA	Bisphenol A
BPCs	Biological persistent chemicals
CP	Consumer products
DI-HRMS	Direct infusion-high resolution-mass spectrometry
ECs	Emerging contaminants
EDs	Endocrine disrupters
EDCs	Endocrine-disrupting chemicals
ELISA	Enzyme-linked immunosorbent assay

FTIR	Fourier transform infrared spectroscopy
GC	Gas chromatography
GC-HRMS	Gas chromatography-high resolution-mass spectrometry
GC–MS	Gas chromatography-mass spectrometry
HPLC	High-performance liquid chromatography
HPLC-MS	High-performance liquid chromatography-mass spectrometry
HRMS	High-resolution mass spectrometry
ICP-MS	Inductively coupled plasma-mass spectrometry
ISEs	Ion-selective electrodes
IR	Infrared spectroscopy
LC	Liquid chromatography
LC-HRMS	Liquid chromatography-high resolution-mass spectrometry
LC-MS	Liquid chromatography-mass spectrometry
LLE	Liquid-liquid extraction
LOD	Limit of detection
MEPS	Microextraction by packed sorbent
ML	Machine learning
MPs	Microplastics
MS	Mass spectrometry
NMR	Nuclear magnetic resonance spectroscopy
NMs	Nanomaterials
NPs	Nano plastics
PAHs	Polycyclic aromatic hydrocarbons
PCPs	Personal care products
PFASs	Per- and poly-fluoroalkyl substances
PM	Particulate matter
PPCPs	Pharmaceuticals and personal care products
ppt	Parts per trillion
R&D	Research and development
SPE	Solid-phase extraction
SPME	Solid-phase microextraction
UAE	Ultrasound-assisted extraction
UV-Vis	Visible spectroscopy
VOCs	Volatile organic compounds
WWTPs	Wastewater treatment plants

References

Altenburger R, Ait-Aissa S, Antczak P, Backhaus T, Barceló D, Seiler TB, Brion F, Busch W, Chipman K, de Alda ML, de Aragão Umbuzeiro G (2015) Future water quality monitoring—adapting tools to deal with mixtures of pollutants in water resource management. Sci Total Environ 512:540–551

Aslam S, Zhang Z, Muhammad A (2023) A review on validation of enzyme linked immunosorbent assay (Elisa) Techniques for detection and quantification of different contaminant in aquatic environment. J Bioresour Manag 10(3):6

Avar P, Maász G, Takács P, Lovas S, Zrínyi Z, Svigruha R, Takátsy A, Tóth LG, Pirger Z (2016) HPLC-MS/MS analysis of steroid hormones in environmental water samples. Drug Test Anal 8(1):123–127

Barbieri F, Zavattini B, Marchetto D, de Ferri L, Pojana G (2023) Surface sea water monitoring of micro-and mesoplastics in a Northern adriatic coastal area. In: 18 th international conference on chemistry and the environment-book of abstracts. European Chemical Society, pp 506–506

Bhardwaj LK, Jindal T (2019) Contamination of lakes in Broknes peninsula, East Antarctica through the pesticides and PAHs. v 31(7):1574–1580

Bhardwaj LK, Jindal T (2020) Persistent organic pollutants in lakes of Grovnes Peninsula at Larsemann Hill area, East Antarctica. Earth Syst Environ 4(2):349–358

Bhardwaj L, Sharma S, Ranjan A, Jindal T (2019) Persistent organic pollutants in lakes of Broknes peninsula at Larsemann Hills area, East Antarctica. Ecotoxicol 28:589–596

Bhardwaj LK, Vikram V (2024) Air pollution and its effect on human health. Futur Trends Agric Eng & Food Sci 3. IIP Proceedings. e-ISBN: 978–93–5747–344–6. https://doi.org/10.58532/V3BCAG4P1CH1

Bhardwaj LK, Rath P, Yadav P, Gupta U (2024b) Microplastic contamination, an emerging threat to the freshwater environment: a systematic review. Environ Syst Res 13(1):8

Bhardwaj LK, Singh VV, Dwivedi K, Rai S (2024a) A comprehensive review on sources, types, impact and challenges of air pollution. Int J Sustain Energy Environ 1:46–50. https://www.ijsee.net/show-165-2117-1.html

Bhardwaj LK, Rath P, Rehman F, Yadav P, Massahi T, Mittal R, Jain H (2025) Status of micro/nanoplastic (MNP) pollution in the aquatic system. In: Societal and environmental ramifications of plastic pollution. IGI Global, pp 43–62

Bhardwaj LK (2024) Occurrence of microplastics (MPs) in Antarctica and its impact on the health of organisms. Maritime Technol Res 6(2):265418 [online]. https://doi.org/10.33175/mtr.2024.265418

Bo Z, Mao S, Han ZJ, Cen K, Chen J, Ostrikov KK (2015) Emerging energy and environmental applications of vertically-oriented graphenes. Chem Soc Rev 44(8):2108–2121

Brack W, Altenburger R, Schüürmann G, Krauss M, Herráez DL, van Gils J, Slobodnik J, Munthe J, Gawlik BM, van Wezel A, Schriks M (2015) The SOLUTIONS project: challenges and responses for present and future emerging pollutants in land and water resources management. Sci Total Environ 503:22–31

Bristow RL, Haworth-Duff A, Young IS, Myers P, Hampson MR, Williams J, Maher S (2023) An automated micro solid phase extraction gas chromatography–mass spectrometry (μSPE-GC–MS) detection method for geosmin and 2-methylisoborneol in drinking water. Sci Rep 13(1):1768

Cai H, Xu EG, Du F, Li R, Liu J, Shi H (2021) Analysis of environmental nanoplastics: progress and challenges. Chem Eng J 410:128208

Camdzic D, Dickman RA, Aga DS (2021) Total and class-specific analysis of per-and polyfluoroalkyl substances in environmental samples using nuclear magnetic resonance spectroscopy. J Hazard Mater Lett 2:100023

Cameron A, Ewen M, Ross-Degnan D, Ball D, Laing R (2009) Medicine prices, availability, and affordability in 36 developing and middle-income countries: a secondary analysis. Lancet 373(9659):240–249

Chakraborty A, Adhikary S, Bhattacharya S, Dutta S, Chatterjee S, Banerjee D, Ganguly A, Rajak P (2023) Pharmaceuticals and personal care products as emerging environmental contaminants: prevalence, toxicity, and remedial approaches. ACS Chem Health & Saf 30(6):362–388

Chen Y, Wen D, Pei J, Fei Y, Ouyang D, Zhang H, Luo Y (2020) Identification and quantification of microplastics using Fourier-transform infrared spectroscopy: current status and future prospects. Curr Opin Environ Sci & Health 18:14–19

Cincotta F, Verzera A, Tripodi G, Condurso C (2018) Volatile emerging contaminants in melon fruits, analysed by HS-SPME-GC-MS. Food Addit & Contam: Part A 35(3):512–518

Colombo E, Mauri L, Marinozzi M, Rudd TR, Yates EA, Ballabio D, Guerrini M (2022) NMR spectroscopy and chemometric models to detect a specific non-porcine ruminant contaminant in pharmaceutical heparin. J Pharm Biomed Anal 214:114724

Crocombe RA (2018) Portable spectroscopy. Appl Spectrosc 72(12):1701–1751

Dekker S (2017) The field guide to understanding 'human error'. CRC Press

Drummond MF, Schwartz JS, Jönsson B, Luce BR, Neumann PJ, Siebert U, Sullivan SD (2008) Key principles for the improved conduct of health technology assessments for resource allocation decisions. Int J Technol Assess Health Care 24(3):244–258

Dzara MJ, Godoy AO, Odgaard M, Zulevi B, Serov A, Jankovic J, Pylypenko S (2021) Physicochemical properties of ECS supports and Pt/ECS catalysts. ACS Appl Energy Mater 4(9):9111–9123

Ejeian F, Etedali P, Mansouri-Tehrani HA, Soozanipour A, Low ZX, Asadnia M, Taheri-Kafrani A, Razmjou A (2018) Biosensors for wastewater monitoring: a review. Biosens Bioelectron 118:66–79

Encarnação T, Santos D, Ferreira S, Valente AJ, Pereira JC, Campos MG, Burrows HD, Pais AA (2021) Removal of imidacloprid from water by microalgae Nannochloropsis sp. and its determination by a validated RP-HPLC method. Bull Environ Contam Toxicol 107:131–139

Espíndola JC, Scaccia N, Barbosa Segundo I, Diniz DDS, Diniz JU, Mierzwa JC (2024) Evaluation of the pathway of contaminants in the environment: a case study of different aquatic environmental compartments. Sustain 16(10):3927

Faull KF, Dooley AN, Halgand F, Shoemaker LD, Norris AJ, Ryan CM, Laganowsky A, Johnson JV, Katz JE (2008) An introduction to the basic principles and concepts of mass spectrometry. Compr Anal Chem 52:1–46

Feng W, Deng Y, Yang F, Miao Q, Ngien SK (2023) Systematic review of contaminants of emerging concern (CECs): distribution, risks, and implications for water quality and health. Water 15(22):3922

Geissen V, Mol H, Klumpp E, Umlauf G, Nadal M, van der Ploeg M, van de Zee SE, Ritsema CJ (2015) Emerging pollutants in the environment: a challenge for water resource management. Int Soil Water Conserv Res 3(1):57–65

Glasgow RE, Marcus AC, Bull SS, Wilson KM (2004) Disseminating effective cancer screening interventions. Cancer: Interdiscip Int J Am Cancer Soc 101(S5):1239–1250

Gomes IB, Maillard JY, Simões LC, Simões M (2020) Emerging contaminants affect the microbiome of water systems—strategies for their mitigation. NPJ Clean Water 3(1):39

Gómez MJ, Gómez-Ramos MM, Agüera A, Mezcua M, Herrera S, Fernández-Alba AR (2009) A new gas chromatography/mass spectrometry method for the simultaneous analysis of target and non-target organic contaminants in waters. J Chromatogr A 1216(18):4071–4082

Gray WB, Shimshack JP (2011) The effectiveness of environmental monitoring and enforcement: a review of the empirical evidence. Rev Environ Econ Policy

Grieshaber D, MacKenzie R, Vörös J, Reimhult E (2008) Electrochemical biosensors-sensor principles and architectures. Sens 8(3):1400–1458

Gupta VK, Ganjali MR, Norouzi P, Khani H, Nayak A, Agarwal S (2011) Electrochemical analysis of some toxic metals by ion–selective electrodes. Crit Rev Anal Chem 41(4):282–313

Hajšlová J, Zrostlıkova J (2003) Matrix effects in (ultra) trace analysis of pesticide residues in food and biotic matrices. J Chromatogr A 1000(1–2):181–197

He X, Zhang H, Xue R, Liu W, Bagheri M, Limmer MA, Burken JG, Shi H (2019) Green analysis: high throughput analysis of emerging pollutants in plant sap by freeze–thaw–centrifugal membrane filtration sample preparation—HPLC-MS/MS analysis. J Agric Food Chem 67(46):12927–12935

Hernandez-Vargas G, Sosa-Hernández JE, Saldarriaga-Hernandez S, Villalba-Rodríguez AM, Parra-Saldivar R, Iqbal HM (2018) Electrochemical biosensors: a solution to pollution detection with reference to environmental contaminants. Biosens 8(2):29

Issac MN, Kandasubramanian B (2021) Effect of microplastics in water and aquatic systems. Environ Sci Pollut Res 28:19544–19562
Jorge TF, Mata AT, António C (2016) Mass spectrometry as a quantitative tool in plant metabolomics. Philos Trans R Soc a: Math, Phys Eng Sci 374(2079):20150370
Karbalaei S, Hanachi P, Walker TR, Cole M (2018) Occurrence, sources, human health impacts and mitigation of microplastic pollution. Environ Sci Pollut Res 25:36046–36063
Khan S, Naushad M, Govarthanan M, Iqbal J, Alfadul SM (2022) Emerging contaminants of high concern for the environment: current trends and future research. Environ Res 207:112609
Khoubi J, Ghiasvand A, Bahrami A, Shahna FG, Farhadian M (2024) Needle trap devices packed with an imine-based 2D COF: an innovative tool for the sampling of polycyclic aromatic hydrocarbons in air. Microchem J 199:110018
Kumar R, Vuppaladadiyam AK, Antunes E, Whelan A, Fearon R, Sheehan M, Reeves L (2022) Emerging contaminants in biosolids: presence, fate and analytical techniques. Emerg Contam 8:162–194
Leeflang MM, Deeks JJ, Gatsonis C, Bossuyt PM (2008) Systematic reviews of diagnostic test accuracy. Ann Intern Med 149(12):889–897
Lei L, Pang R, Han Z, Wu D, Xie B, Su Y (2023) Current applications and future impact of machine learning in emerging contaminants: a review. Crit Rev Environ Sci Technol 53(20):1817–1835
Li L, Wang Y, Zhang W, Yu S, Wang X, Gao N (2020) New advances in fluorescence excitation-emission matrix spectroscopy for the characterization of dissolved organic matter in drinking water treatment: a review. Chem Eng J 381:122676
Linkov I, Satterstrom FK, Kiker G, Batchelor C, Bridges T, Ferguson E (2006) From comparative risk assessment to multi-criteria decision analysis and adaptive management: recent developments and applications. Environ Int 32(8):1072–1093
Lippi G, Chance JJ, Church S, Dazzi P, Fontana R, Giavarina D, Grankvist K, Huisman W, Kouri T, Palicka V, Plebani M (2011) Preanalytical quality improvement: from dream to reality. Clin Chem Lab Med 49(7):1113–1126
Luo Y, Guo W, Ngo HH, Nghiem LD, Hai FI, Zhang J, Liang S, Wang XC (2014) A review on the occurrence of micropollutants in the aquatic environment and their fate and removal during wastewater treatment. Sci Total Environ 473:619–641
Magee HY, Maurer MM, Cobos A, Adhikari S, Halden RU (2023) Impact of LC-MS/MS quantification methods on analytical results by example of nationwide US environmental samples. Sci Total Environ 892:164425
Mariano S, Tacconi S, Fidaleo M, Rossi M, Dini L (2021) Micro and nanoplastics identification: classic methods and innovative detection techniques. Front Toxicol 3:636640
Martín-Pozo L, de Alarcón-Gómez B, Rodríguez-Gómez R, García-Córcoles MT, Çipa M, Zafra-Gómez A (2019) Analytical methods for the determination of emerging contaminants in sewage sludge samples. A review. Talanta 192:508–533
Metcalfe CD (2013) Pharmaceutical contaminants of emerging concern in the environment Pharmaceutical contaminants of emerging concern in the environment. Environ Toxicol & Chem 32(8)
Miller DM, Abels K, Guo J, Williams KS, Liu MJ, Tarpeh WA (2023) Electrochemical wastewater refining: a vision for circular chemical manufacturing. J Am Chem Soc 145(36):19422–19439
Mishra RK, Mentha SS, Misra Y, Dwivedi N (2023) Emerging pollutants of severe environmental concern in water and wastewater: a comprehensive review on current developments and future research. Water-Energy Nexus
Mitra S (ed) (2004) Sample preparation techniques in analytical chemistry. John Wiley & Sons
Mondal R, Mukherjee A, Biswas S, Kole RK (2018) GC-MS/MS determination and ecological risk assessment of pesticides in aquatic system: a case study in Hooghly River basin in West Bengal, India. Chemosphere 206:217–230
Naidu R, Espana VAA, Liu Y, Jit J (2016) Emerging contaminants in the environment: risk-based analysis for better management. Chemosphere 154:350–357
Neale P, Leusch F, Escher B (2021) Bioanalytical tools in water quality assessment. IWA publishing

Niessen WMA, Manini P, Andreoli R (2006) Matrix effects in quantitative pesticide analysis using liquid chromatography–mass spectrometry. Mass Spectrom Rev 25(6):881–899
Niu Z, Zhang W, Yu C, Zhang J, Wen Y (2018) Recent advances in biological sample preparation methods coupled with chromatography, spectrometry and electrochemistry analysis techniques. TrAC, Trends Anal Chem 102:123–146
Nobile M, Chiesa LM, Villa RE, Danesi L, Arioli F, Panseri S (2024) Occurrence of perfluoroalkyl substances in canned tuna and their impact on food safety. Food Control 159:110301
Nováková L, Vlčková H (2009) A review of current trends and advances in modern bio-analytical methods: chromatography and sample preparation. Anal Chim Acta 656(1–2):8–35
Ong TT, Blanch EW, Jones OA (2020) Surface enhanced Raman spectroscopy in environmental analysis, monitoring and assessment. Sci Total Environ 720:137601
Pérez-Fernández V, Rocca LM, Tomai P, Fanali S, Gentili A (2017) Recent advancements and future trends in environmental analysis: sample preparation, liquid chromatography and mass spectrometry. Anal Chim Acta 983:9–41
Pezzatti J, Boccard J, Codesido S, Gagnebin Y, Joshi A, Picard D, González-Ruiz V, Rudaz S (2020) Implementation of liquid chromatography–high resolution mass spectrometry methods for untargeted metabolomic analyses of biological samples: a tutorial. Anal Chim Acta 1105:28–44
Polyakova OV, Mazur DM, Artaev VB, Lebedev AT (2013) Determination of polycyclic aromatic hydrocarbons in water by gas chromatography/mass spectrometry with accelerated sample preparation. J Anal Chem 68:1099–1103
Pozzer AC, Gómez PA, Weiss J (2022) Volatile organic compounds in aquatic ecosystems–detection, origin, significance and applications. Sci Total Environ 838:156155
Puri M, Gandhi K, Kumar MS (2023) Emerging environmental contaminants: a global perspective on policies and regulations. J Environ Manage 332:117344
Rasheed T, Bilal M, Nabeel F, Adeel M, Iqbal HM (2019) Environmentally-related contaminants of high concern: potential sources and analytical modalities for detection, quantification, and treatment. Environ Int 122:52–66
Rathi BS, Kumar PS, Vo DVN (2021) Critical review on hazardous pollutants in water environment: occurrence, monitoring, fate, removal technologies and risk assessment. Sci Total Environ 797:149134
Richardson SD (2009) Water analysis: emerging contaminants and current issues. Anal Chem 81(12):4645–4677
Richardson SD, Kimura SY (2016) Water analysis: emerging contaminants and current issues. Anal Chem 88(1):546–582
Richardson SD, Ternes TA (2018) Water analysis: emerging contaminants and current issues. Anal Chem 90(1):398–428
Rodriguez-Mozaz S, de Alda MJL, Barceló D (2007) Advantages and limitations of on-line solid phase extraction coupled to liquid chromatography–mass spectrometry technologies versus biosensors for monitoring of emerging contaminants in water. J Chromatogr A 1152(1–2):97–115
Rout PR, Zhang TC, Bhunia P, Surampalli RY (2021) Treatment technologies for emerging contaminants in wastewater treatment plants: a review. Sci Total Environ 753:141990
Safwat N, Ayad M, Farouk M (2020) Development of an eco-friendly reversed-phase HPLC method for the simultaneous determination of three emerging contaminants in water samples. Arch Pharm Sci Ain Shams Univ 4(1):79–91
Samal K, Mahapatra S, Ali MH (2022) Pharmaceutical wastewater as emerging contaminants (EC): treatment technologies, impact on environment and human health. Energy Nexus 6:100076
Schnoebelen D, Kolpin D, Barber L, Furlong E, Meyer M, Skopec M (2006) Chemicals of concern. Iowa's first field research area for emerging contaminants, the Iowa policy project
Scott RPW (2012) Liquid chromatography. Reese-Scott Partnership

Shyamalagowri S, Shanthi N, Manjunathan J, Kamaraj M, Manikandan A, Aravind J (2023) Techniques for the detection and quantification of emerging contaminants. Phys Sci Rev 8(9):2191–2218

Silva CP, Carvalho T, Schneider RJ, Esteves VI, Lima DL (2020a) ELISA as an effective tool to determine spatial and seasonal occurrence of emerging contaminants in the aquatic environment. Anal Methods 12(19):2517–2526

da Silva S, Burato J, Vargas Medina DA, de Toffoli AL, Vasconcelos Soares Maciel E, Mauro Lanças F (2020b) Recent advances and trends in miniaturized sample preparation techniques. J Sep Sci 43(1):202–225

Soliz DL, Garcinuño RM, Paniagua González G, Bravo JC, Fernández Hernando P (2024) Assessing matrix solid-phase dispersion extraction strategies for determining bisphenols and phthalates in gilthead sea bream samples. Foods 13(3):413

Suvar NS, Prodan M, Ghicioi E, Toplician A (2024) Identification of hazardous organic substances for fire investigation with portable GC-MS. In: MATEC web of conferences, vol 389. EDP Sciences, p 00020

Tang Y, Yin M, Yang W, Li H, Zhong Y, Mo L, Liang Y, Ma X, Sun X (2019) Emerging pollutants in water environment: occurrence, monitoring, fate, and risk assessment. Water Environ Res 91(10):984–991

Tang G, Chen Y, Lin S, Li X (2024) The photo-and microbial degradation kinetics and pathways of sulfadoxine in seawater elucidated by liquid chromatography coupled with time-of-flight mass spectrometry. Chemosphere 351:141225

Tegladza ID, Qi T, Chen T, Alorku K, Tang S, Shen W, Kong D, Yuan A, Liu J, Lee HK (2020) Direct immersion single-drop microextraction of semi-volatile organic compounds in environmental samples: a review. J Hazard Mater 393:122403

Thakur A, Kumar A (2022) Recent advances on rapid detection and remediation of environmental pollutants utilizing nanomaterials-based (bio) sensors. Sci Total Environ 834:155219

Thibaud M, Chi H, Zhou W, Piramuthu S (2018) Internet of Things (IoT) in high-risk environment, health and safety (EHS) industries: a comprehensive review. Decis Support Syst 108:79–95

Tong X, Mohapatra S, Zhang J, Tran NH, You L, He Y, Gin KYH (2022) Source, fate, transport and modelling of selected emerging contaminants in the aquatic environment: current status and future perspectives. Water Res 217:118418

Trufelli H, Palma P, Famiglini G, Cappiello A (2011) An overview of matrix effects in liquid chromatography–mass spectrometry. Mass Spectrom Rev 30(3):491–509

Valdez-Carrillo M, Abrell L, Ramírez-Hernández J, Reyes-López JA, Carreón-Diazconti C (2020) Pharmaceuticals as emerging contaminants in the aquatic environment of Latin America: a review. Environ Sci Pollut Res 27:44863–44891

Wang S, Zhao P, Min G, Fang G (2007) Multi-residue determination of pesticides in water using multi-walled carbon nanotubes solid-phase extraction and gas chromatography–mass spectrometry. J Chromatogr A 1165(1–2):166–171

Wang Y, Wang M, Zhou L, Geng X, Xu Z, Zhang H (2023) Development of a competitive ELISA based on estrogen receptor and weak competitive molecule for the screening of potential estrogens in foods. Food Chem 401:134084

Wilkinson JL, Hooda PS, Barker J, Barton S, Swinden J (2016) Ecotoxic pharmaceuticals, personal care products, and other emerging contaminants: a review of environmental, receptor-mediated, developmental, and epigenetic toxicity with discussion of proposed toxicity to humans. Crit Rev Environ Sci Technol 46(4):336–381

Wille K, De Brabander HF, Vanhaecke L, De Wulf E, Van Caeter P, Janssen CR (2012) Coupled chromatographic and mass-spectrometric techniques for the analysis of emerging pollutants in the aquatic environment. TrAC, Trends Anal Chem 35:87–108

Wu D, Zhou Y, Lu G, Hu K, Yao J, Shen X, Wei L (2019) The occurrence and risks of selected emerging pollutants in drinking water source areas in Henan, China. Int J Environ Res Public Health 16(21):4109

Xie M, Liang M, Liu C, Xu Z, Yu Y, Xu J, You S, Wang D, Rad S (2024) Peroxymonosulfate activation by CuMn-LDH for the degradation of bisphenol a: effect, mechanism, and pathway. Ecotoxicol Environ Saf 270:115929

Zhang H, Shen N, Li Y, Hu C, Yuan P (2023) Source, transport, and toxicity of emerging contaminants in aquatic environments: a review on recent studies. Environ Sci Pollut Res 30(58):121420–121437

Zielinski O, Busch JA, Cembella AD, Daly KL, Engelbrektsson J, Hannides AK, Schmidt H (2009) Detecting marine hazardous substances and organisms: sensors for pollutants, toxins, and pathogens. Ocean Sci 5(3):329–349

Zulkifli SN, Rahim HA, Lau WJ (2018) Detection of contaminants in water supply: a review on state-of-the-art monitoring technologies and their applications. Sens Actuators, B Chem 255:2657–2689

Zwanenburg A (2019) Radiomics in nuclear medicine: robustness, reproducibility, standardization, and how to avoid data analysis traps and replication crisis. Eur J Nucl Med Mol Imaging 46(13):2638–2655

Chapter 8
Applications of Graphene Nanocomposites in Advanced Detection and Removal of Emerging Water Contaminants

Souravi Bardhan and Dipak Kr Chanda

Abstract Emerging pollutants cover a wide range of substances, including drugs, personal care products (PCPs), softeners, hormones, microplastics (MPs)/nanoplastics (NPs), pesticides, perfluorinated organic compounds, endocrine disruptors (EDs), etc., which cause heavy damage to the health of humans, animals, as well as the environment. Because they are often only found in small amounts in water with a complicated composition, their exact detection, and removal pose a big challenge. Graphene, a material that is primarily a two-dimensional carbon atom sheet, and its nanocomposites are considered to be an up-and-coming solution for the efficient and cost-effective detection and removal of pesticides and other similar pollutants. The unique lattice structure of graphene and its tailored surface chemistry, the easy surface functionalization and pore formations, and the high mobility of electrons are the main points to consider for the purification of these new contaminants of the water bodies by the technology. This chapter deals with the main idea that depends on graphene and its various derivatives, such as graphene quantum dots, graphene oxide (GO), and reduced graphene oxide (rGO), and their functionalization with a variety of chemical moieties like metals, aptamers, molecularly imprinted polymers, etc., for improving the selectivity toward target pollutants. Additionally, this chapter also examines the potential of chip technology, which will be critical in the near future, as well as the new methods and materials recently innovated in the present technological field of environmental concerns.

Keywords Graphene · Graphene nanocomposites · Emerging pollutants · Electrochemical sensing · Quenching · Adsorption · Photocatalysis

S. Bardhan (✉)
Department of Physics, Jadavpur University, Kolkata, West Bengal, India
e-mail: bardhan666srv@gmail.com

Department of Environmental Science, Netaji Nagar College for Women, University of Calcutta, Kolkata, West Bengal, India

D. K. Chanda
Matrexia Advanced Technologies Pvt. Ltd., Hoogly, West Bengal, India

H. Jain et al. (eds.), *Emerging Contaminants in the Aquatic Environment*, Advances in Wastewater Research, https://doi.org/10.1007/978-981-96-8548-6_8

List of Abbreviations

AChE	Acetylcholinesterase
Ag	Silver
AOPs	Advanced oxidation processes
As	Arsenic
Au	Gold
BBP	Butyl benzyl phthalate
BHA	Butylated hydroxyanisole
BHT	Butylated hydroxytoluene
BPA	Bisphenol A
Cd	Cadmium
CNT	Carbon nanotubes
CPE	Carbon paste electrode
Cr	Chromium
CVD	Chemical vapor deposition
DBP	Di-n-butyl phthalate
DEHP	Di(2-ethylhexyl) phthalate
DEP	Diethyl phthalate
DNA	Deoxyribonucleic acid
ECs	Emerging contaminants
EDCs	Endocrine-disrupting compounds
EDs	Endocrine disrupters
EP	Ethylparaben
FTIR	Fourier Transform Infrared
GC	Gas chromatography
GCE	Glassy carbon electrode
GO	Graphene oxide
GQDS	Graphene quantum dots
Hg	Mercury
HPLC	High-performance liquid chromatography
IBP	Isobutyl paraben
IoT	Internet of Things
LC	Liquid chromatography
LC-ESI-HRMS	Liquid chromatography-electrospray ionization high-resolution mass spectrometry
LOD	Limit of detection
MF	Membrane filtration
MIPs	Molecularly imprinted polymers
MPs	Microplastics
MS	Mass spectroscopy
MWW	Municipal wastewater
nM	NanoMolar
NMs	Nanomaterials

NPs	Nanoplastics
NSAIDs	Non-steroidal anti-inflammatory drugs
NTS	Non-targeted screening
PANI	Polyaniline
Pb	Lead
PCPs	Personal care products
PE	Polyethylene
PET	Polyethylene terephthalate
PFOA	Prfluorooctanoic acid
PFOCs	Perfluorinated organic compounds
PFOS	Perfluorooctane sulfonate
PLA	Polylactic Acid
PMRs	Photocatalytic membrane reactors
PMMA	Poly (methyl methacrylate)
POPs	Persistent organic pollutants
PP	Polypropylene
PP	Propylparaben
PS	Polystyrene
PVA	Polyvinyl alcohol
rGO	Reduced graphene oxide
ROS	Reactive oxidative species
RO	Reverse osmosis
SPR	Surface plasmon resonance
TiO_2	Titanium oxide
UV	Ultraviolet
WWTPs	Wastewater treatment plants
ZnO	Zinc oxide

8.1 Introduction

Water is one of the essential elements necessary for life to thrive on the earth (Alam et al. 2023; Bhardwaj et al. 2025a). However, it's being polluted in perpetuity by an increased level of organic and inorganic pollutants emanating from various industrial and commercial sectors and sources of houses and agricultural land. This is rendering the water unfit for consumption along with posing a threat to the biota. The problem of water contamination with emerging contaminants (ECs) is also recent, and it causes a lot of health and environmental risks even at low levels. ECs include different kinds of PCPs and pharmaceutical compounds, endocrine-disrupting compounds (EDCs), food additives, flame retardants, wood preservatives, disinfectants, and pesticides as well as tiny MPs/NPs (Gogoi et al. 2018; Kumar et al. 2022). They are transported to the aquatic system by industry, commerce, agriculture, and the private sector and are the sources that are shown in Fig. 8.1.

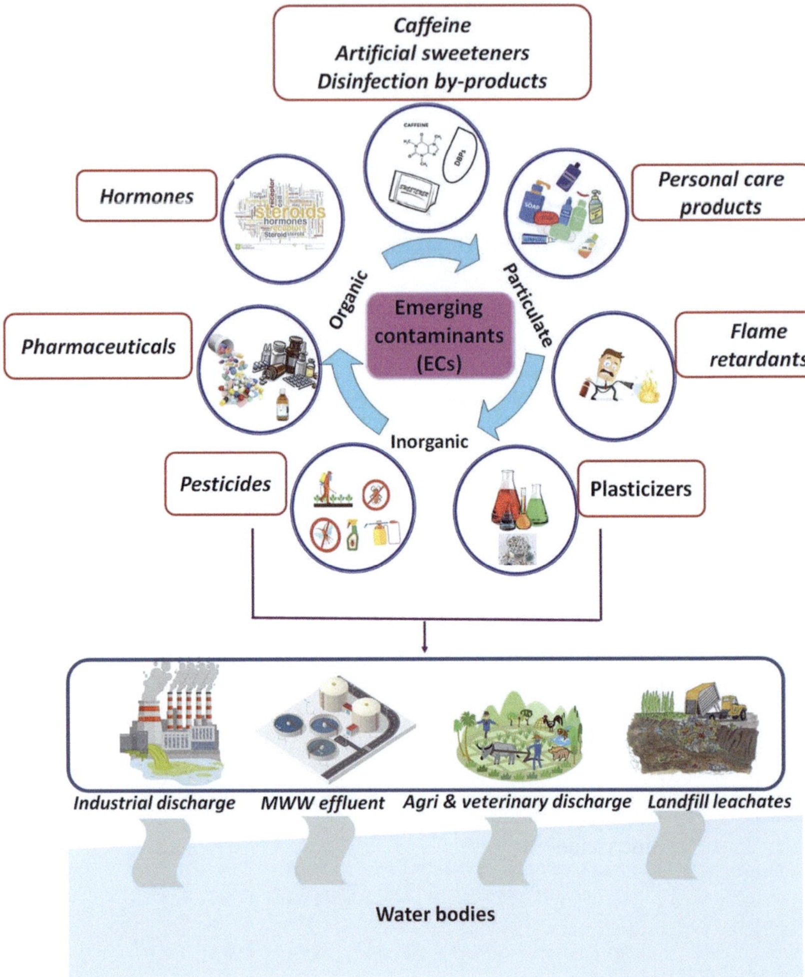

Fig. 8.1 Pictorial representation depicting various categories of emerging contaminants (ECs) and the pathways through which they enter aquatic systems

They not only cause physical damage but also bring poisonous substances into the food chain of the highest living organisms such as human beings (Gomes et al. 2020). Advancements in pharmaceuticals introduce new chemicals to nature, most of which often resist degradation and can bioaccumulate in aquatic organisms. For example, chemicals such as caffeine, butylated hydroxytoluene (BHT), diclofenac, chloroform, ibuprofen, naproxen, methyl dihydro jasmonate, and butylated hydroxyanisole (BHA) which were detected in irrigation water were found in fishes and crops (Calderón-Preciado et al. 2011). Moreover, bacteria become associated with resistance to several antibiotics, thereby giving rise to antibiotic-resistant bacteria, which may pose a huge risk to health globally (Urban-Chmiel et al. 2022). Unbeknown to us, hormones found in the water and bisphenol A (BPA) also contribute to hormonal disruptions in fish, as well as other wildlife species (Jasrotia et al. 2021).

Additionally, the immune system, hormonal system, and reproductive function of any fish exposed to BPA will suffer, resulting in a decline in their population. The consequences of exposure to ECs include interruptions of hormone balance, cancer, and neurological disorders. Therefore, urgent action for their lecturing is requested (Pan et al. 2024). Conventional water treatment methods cannot address them properly since their chemical properties are complex, and they exist in trace concentrations. As global demand for water increases, good management of ECs is central to protecting human and environmental health.

Here, MWW means municipal wastewater and agri means agricultural discharge (Manivannan et al. 2022). Reprinted with permission from The Royal Society of Chemistry 2022.

To combat hazards from ECs in water bodies, proper detection, primarily at sources, can alleviate the issues. Due to their diverse chemical structures and very low concentrations, effective analytical techniques are required in the detection of ECs in water. In the majority of cases noted, water pollution is being monitored through the use of electrodes, electrochemical sensors, fluorescent optical sensors, and colorimetric sensors, as well as through gas chromatography (GC) and liquid chromatography (LC) methods (Pena-Pereira et al. 2021). With the combination of mass spectrometry (MS) with the said techniques not only the degree of concentration is improved, but also the sensitivity of the systems methods, allowing pollutants to be measured even at low concentrations and in fairly complex water matrices (Wille et al. 2012). Spectroscopy-including Fourier Transform Infrared Spectroscopy (FTIR) and various other methods have been applied in the detection of MPs (Hanvey et al. 2017; Bhardwaj et al. 2025b) as well as the composition of other pollutants. The development of biosensors represents an even faster and more inexpensive technology for the effective detection of specific contaminants like antibiotics and hormones. In this method, biological recognition elements are utilized in the identification of the target molecules (Majdinasab et al. 2020). Moreover, to further enhance the ability to detect unknown contaminants in complex aquatic environments, newer technologies, such as non-targeted screening (NTS) with reversed-phase liquid chromatography-electrospray ionization high-resolution mass spectrometry (LC-ESI-HRMS) are adopted (Malm et al. 2024).

Other than the detection of ECs, their removal from water is also becoming challenging. Among various methods of pollution removal, adsorption, that is, the deposition of a substance onto the surface of another substance, is the most employed. It has gained wide acceptance due to its ease, cost-effectiveness, and efficiency (Rathi and Kumar 2021). Other common methods include the application of membrane filtration (MF), nanofiltration, and reverse osmosis (RO) (Anis et al. 2019). Although this technique is effective in pollutant removal, it has drawbacks such as high cost, energy consumption, and operational costs. Advanced oxidation processes (AOPs), such as ozonation and ultraviolet (UV) irradiation, involve the chemical degradation of ECs to less harmful substances through reactive radicals. Various biological treatments and microbial degradation are also performed to break down biodegradable pollutants into simpler forms (Ahmed et al. 2017), yet they require threshold conditions for optimum performance. Thus, each of these methods has its limitations, and hybrid systems that combine multiple techniques are often necessary to achieve comprehensive removal. For the last few decades, to overcome the drawbacks and to improve performance and efficiency much progress has been made in developing various materials along with advanced techniques to combat pollution from ECs (Shahid et al. 2021).

Among the various classes of materials, graphene, and its derivatives have gained immense popularity worldwide as a potential candidate in the water treatment sector. The unique properties of graphene and its excellent conductivity and sensitivity enable it to be used as a platform for sophisticated sensors that can detect even trace levels of contaminants. Graphene-based biosensors, for example, can identify specific ECs like hormones and antibiotics with high precision (Li et al. 2024). These are miniaturized, fast, and low-cost sensors that are perfectly used for real-time monitoring. In addition, graphene is compatible with other nanomaterials (NMs), allowing for the fabrication of hybrid systems with dual functions of detection and treatment, presenting integrated solutions for the current water crisis. This chapter explores some state-of-the-art solutions that include novel separation or detection methods and removal methods using graphene-based NMs.

8.2 Graphene Nanocomposites as a Promising Candidate in Water Treatment

Since 2004, there has been an exponential growth of various applications of graphene nanotechnology platforms (Geim 2009; Bhardwaj et al. 2023), especially emerging as innovative solutions for detecting and removing pollutants. As a two-dimensional carbon allotrope with sp^2-hybridized atoms in a honeycomb lattice (Fig. 8.2a), graphene offers a high specific surface area, superior electrical conductivity, impressive mechanical strength, and excellent chemical stability. Moreover, the honeycomb lattices comprising the hexagonal arrangement of carbon atoms in graphene provide a stable yet lightweight framework and exceptional surface-area-to-volume ratio making it one of the thinnest yet strongest materials which is beneficial for adsorption and catalytic applications.

These attributes make graphene an ideal candidate for designing nanocomposites with enhanced water treatment functionality (Marques et al. 2011). Graphene's high surface area and π-electron-richness offer a large number of sites to bind pollutants through π-π stacking, hydrogen bonding, van der Waals, and ionic bonds. It is electrically conductive which broadens the view of the usage in electrochemical sensing and catalytic degradation applications (Perreault et al. 2015). Each of the derivatives presents distinct features enabling a wide range of approaches for water treatment. Besides pristine graphene and its derivatives, further enhancement includes the use

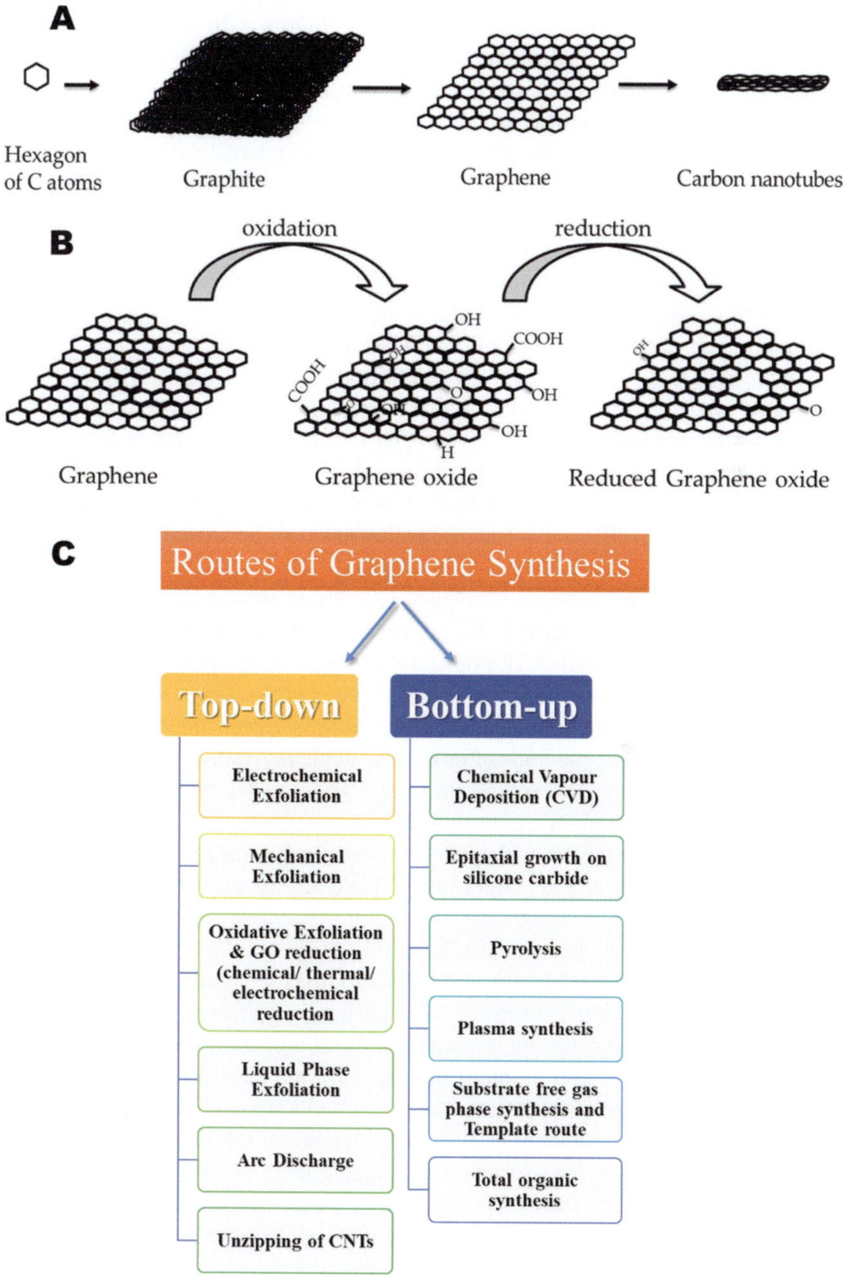

Fig. 8.2 **a** Various forms of carbon materials; **b** various derivatives of graphene obtained by conversion of graphene-to-graphene oxide and to reduced graphene oxide (GO). Reproduced from Memisoglu et al. (2023). Open access, **c** various synthesis routes of development of graphene

of graphene combined with composite materials. Figure 8.2b illustrates graphene derivatives, such as GO, rGO, and graphene quantum dots (GQDs), which have the appropriate properties required for detecting and removing target pollutants effectively. GO is the most water-dispersible derivative of graphene containing oxygen-functional groups to improve the adsorption of polar pollutants. rGO is demonstrated to be better than most other graphene derivatives in terms of electrical conductivity which is related to the fact that it emerges well for an electrochemical sensor application. GO and rGO are highly associated with the aggregation of pollutants through π-π stacking and hydrogen bonding and were thus speculated to be beneficial in the treatment of complex organic pollutants (Dong et al. 2017).

Contaminant detection using GQDs is possible due to their small size and fluorescent properties (Durodola et al. 2022). Additionally, they can be surface-modified to produce graphene-based nanocomposites. Graphene-containing nanocomposites are composite materials containing graphene and other active materials, such as metals, metal oxides, and polymers, designed to generate hybrid systems with improved water treatment performance (Ramírez et al. 2021). Such composites exhibit combinatorial effects that enhance pollutant adsorption, catalytic degradation, and sensitivity in detection.

Various graphene synthesis techniques, using top-down and bottom-up approaches, are illustrated in Fig. 8.2c. These techniques are useful for research and nanocomposite fabrication. Bottom-up approaches, Chemical Vapor Deposition (CVD), or epitaxial growth require molecular precursors using a nickel catalyst to ensure an orderly preparation of quality graphene in control of structural ordering (Devi et al. 2024). Those methods offer ideal scalable layers for higher-added value in electronics and advanced composite fields. Top-down processes will be relatively economical with minimal complexity, limited scaling, and inferior controls due to limited scalability with inadequate precursory accuracy as well due to process controls. Future improvements target higher yield and quality and lower costs for industrial use.

Moreover, to improve the reactivity, selectivity, dispersibility, and sensitivity, and to achieve better results in the removal of pollutants, the functionalization of graphene nanocomposites involving modifying their surface is performed (Chang and Wu 2013). According to the strong electron-accepting nature, large specific surface area, and extremely high electron mobility of graphene, the adsorption mechanism for several pollutants is likely to take place on the doped graphene surface by changing the reactant's electronic structures; in order to facilitate the reactivity, selectivity, dispersibility, and sensitivity and get the excellent effect in pollutant removal, functionalization of graphene nanocomposites, which means surface-modification of the nanoparticle, is carried out to improve the catalytic performance (Pooresmaeil and Namazi 2024).

Furthermore, emerging pollutants are often difficult to detect in water due to their modified structures or trace quantities, hence advancements in nanocomposite synthesis are necessary accordingly. Due to its high surface area, tunable chemistry, and excellent electrical properties, functionalized graphene is a platform for pollutant sensing. Covalent approaches involve directly getting more functional groups or molecules to covalently bind to the surface or edges of graphene to tune its electronic properties (Mao et al. 2013). Methods of carboxylation and hydroxylation in which –COOH and –OH groups are incorporated could help toward a better interaction with metal ions like mercury ions (Hg^{2+}) thus enhancing water dispersibility; amination technology to functionalize the sensor with amine ($-NH_2$) groups for detecting ammonia or other similar polluting agents, while polymerization or combing graphene with polymers, e.g., polyaniline (PANI) would aid in its enhanced sensitivity for heavy metals and organic contaminants (Peng and Zhang 2021).

Non-covalent functionalization methods retain the sp^2 hybridization and electrical conductivity of graphene and rely on π-π stacking, van der Waals forces, or hydrogen bonding interactions. It includes methods, such as biomolecule attachment of binding enzymes or aptamers through π-π interactions (Skaltsas et al. 2015) to improve sensitivity and selectivity toward various pollutants like hormones, EDCs, antibiotics, and heavy metals. Another method is the surfactant coating technique which uses surfactants to stabilize graphene and improve dispersibility in aqueous media for pollutant sensing (Qamar et al. 2024). Due to its outstanding properties, functionalized graphene composites like GO-metal hybrids or rGO-polymer systems are highly sensitive to antibiotics, pesticides, and perfluorinated compound detections.

8.2.1 Graphene Oxide (GO)

The graphene oxide or GO derivative of graphene is known for its oxygen-rich functional groups like epoxy, hydroxyl, and carboxyl moieties which also impart a hydrophilic nature. The functional groups enhance dispersibility in water and thus result in strong interactions with polar and ionic contaminants. For instance, GO is effective in the adsorption of heavy metals, like lead (Pb^{2+}), Hg^{2+}, and cadmium (Cd^{2+}), by the functional group-rich structure of GO. Investigations show that GO can be capable of taking up as much as 256.6 mg/g Hg when optimal conditions have been attained (Dreyer et al. 2010). GO is mainly excellent for adsorbing polar pollutants, such as EDCs and pharmaceuticals, in that their respective functional groups promote hydrogen bonding and electrostatic attraction (Thakur and Kandasubramanian 2019). In the meantime, its amphiphilic property also facilitates GO in absorbing organic pollution agents, which includes some dyes and pharmaceutical agents through π-π type interaction and hydrogen bonding.

GO is also outstanding in its catalytic and photocatalytic applications with metal oxides like titanium dioxide (TiO_2), or zinc oxide (ZnO). These composites utilize the charge-separation ability of GO, besides its ability to enhance light absorption, which leads to the efficient degradation of organic pollutants into less harmful byproducts (Zubair et al. 2024). Such composites are particularly effective against dyes, pharmaceuticals, and EDCs making them versatile in water treatment systems.

8.2.2 Reduced Graphene Oxide (rGO)

The rGO is formed from the reduction of GO by the removal of oxygen-containing groups with the retention of graphene's hydrophobicity and excellent electrical conductivity. These characteristics make rGO a great material for electrochemical detection as well as the removal of nonpolar contaminants. Due to high conductivity, rGO-based electrodes have shown great sensitivity in the detection of trace heavy metals as well as organic pollutants (Chamoli et al. 2020). Its higher conductivity also allows its incorporation into electrochemical sensors for trace-level detection of pharmaceuticals and pesticides (Rowley-Neale et al. 2018).

For the adsorption-based water treatment method, reduced GO presents itself as the most effective material in removing aromatic hydrocarbons, pesticides, and other persistent organic pollutants (POPs). The negative point of the immediate application of rGO is improved $-NH_2$ or thiol group bonding which enables specific contaminants such as Hg^{2+} to be bound through coordination bonding. These modified rGO composites have demonstrated adsorption capacities that are higher than those of conventional adsorbents (Monjezi et al. 2024). In addition, rGO-based nanocomposites which have been doped with silver (Ag) or iron oxide (Fe_3O_4) nanoparticles exhibit superior catalytic activity.

8.2.3 Graphene Quantum Dots (GQDs)

GQDs, which are mainly nanoscale fragments of graphene, exhibit excellent photoluminescent properties useful in optical sensing applications. This is because of the phenomenon of quantum confinement which causes size-dependent fluorescence. They have tunable fluorescence emissions, so they can accurately detect water contaminants, which include heavy metals such as Hg and Pb, as well as organic pollutants such as pesticides and pharmaceuticals (Durodola et al. 2022). Such properties make it possible for GQDs to use fluorescence quenching or enhancement mechanisms to detect food additives or disinfectants. Their nanoscale dimensions and adjustable optical properties further enhance their importance in the detection of MPs/NPs. Molecular recognition elements exist to further the functionalization of GQDs and make the detection of specific contaminants in complex aqueous samples possible. GQDs also have degradation functions of the pollutant. When combined with metal oxides, the GQD photocatalytic degradation can improve its efficiency by effectively absorbing more light and the charge separation. This made them successful in breaking the organic pollutants under visible light, making GQDs position as multifunctional material in integrated water treatment systems (Pan et al. 2015).

8.2.4 Graphene-Based Membranes

Graphene is a most promising material for the enrichment of polymer nanocomposites and improved material performance attributed to its outstanding mechanical, thermal, and electrical characteristics. Even at a small percentage, its addition considerably boosts the strength, thermal stability, and conductivity of host polymers. Graphene has been integrated into diverse polymers such as polyvinyl alcohol (PVA), poly (methyl methacrylate) (PMMA), and polylactic acid (PLA) through solution casting, melt blending, and in-situ polymerization method (Ganguly 2022). Further, functionalized graphene-like GO or rGO may provide improved compatibility with the hydrophilic or hydrophobic polymers individually.

Graphene and its derivatives also have been integrated into membranes to provide advanced water filtration systems. GO-based membranes can selectively separate the pollutants, depending on the size, charge, and hydrophobicity. The obtained membranes are known to exhibit high water flux with an ability to prevent fouling and remove salts, heavy metals, and organic contamination (Perreault et al. 2015). rGO membranes feature such performance for the removal of oil and grease in the wastewaters which are suitable for industries' effluents treatment. Graphene membranes can integrate adsorption, filtration, and catalytic functions, hence, paving the way for hybrid systems that treat many contaminants simultaneously. For example, graphene-TiO_2 membranes operate with photocatalytic degradation while physically filtering out contaminants, achieving complete removal of dyestuffs and pathogens from the dirty water (Fu and Gray 2024).

8.3 Applications of Graphene Nanocomposites in the Detection of Emerging Contaminants (ECs)

Accurate and reliable detection of emerging pollutants is a prerequisite for effective water treatment. Graphene-based nanocomposites have emerged with wonderful promise in the development of advanced sensors to monitor contaminant levels at trace and ultra-trace levels in water. These possess higher electrical conductivity, high surface area, and surface chemical tunability to facilitate building highly sensitive and selective detection systems. Graphene nanocomposites have contributed immensely to electrochemical detection systems. Its high conductivity is capable of allowing the electron to be transferred fast (Lawal 2015). In such a manner, electrochemical sensors make perfect materials with graphene as a medium because they function on principles whereby changes in electrical signals emanate from the interactions of functionalized graphene surfaces with contaminants.

For instance, GO doped with metal nanoparticles or enzymes was proven to be very sensitive toward the detection of heavy metals such as Pb^{2+}, Cd^{2+}, and Hg^{2+} (Fig. 8.3). In one study, reduced GO was used as a conductive platform in an electrochemical sensor to detect Hg^{2+} at nanomolar concentrations (Ma et al. 2021). The oxygen-functional groups of GO act as binding sites to the metal ions, while its high conductivity amplifies signals for improving sensitivity and accuracy. Functionalized graphene material, including chelation agents or molecular recognition agents, is capable of performing electrochemical detection of organophosphate pesticides that could result in fast detection and accurate results (Mahajan and Patil 2022). The use of graphene in electrochemical sensors improves performance metrics but also reduces manufacturing costs, making them ideal for large-scale monitoring applications in the environment.

Other than electrochemical sensing, graphene nanocomposites have been used in optical detection techniques particularly fluorescence-based and surface plasmon resonance (SPR) systems. GQDs have been widely used for the detection of organic pollutants and heavy metals because of their tunable fluorescence properties. Their nanoscale dimensions and quantum confinement effects enable fluorescence quenching or enhancement upon interaction with specific contaminants. For example, it has been reported that GQD-based sensors can detect trace levels of Hg and arsenic (As) in water through fluorescence quenching (Anh et al. 2017). Such sensors are highly sensitive and selective even in a complex matrix of water with competing ions. Graphene-metal nanocomposites, including graphene-Ag and graphene-gold (Au) hybrids, were also applied as detection platforms in SPR-based detection systems (Nurrohman and Chiu 2021). This allows the sensor to monitor real-time changes in the refractive index caused by pollutant binding. The synergistic effect between graphene and metal nanoparticles has improved the stability and sensitivity to make these sensors applicable to pharmaceuticals, pesticides, as well as other POPs.

Another material developed from graphene, biosensing technologies have gained significant limelight. Biosensors are prepared by incorporating biological recognition elements like enzymes, antibodies, or aptamers with graphene and detecting target contaminants (Venkatesan et al. 2024). It offers optimal immobilization conditions for biomolecules without losing their activity, enabling electron transfer; hence, the surface area coupled with conductivity enhances the value of the matrix. This approach encompasses the application of functionalized GO doped with enzyme acetylcholinesterase (AChE) for use in a detection platform on the basis of organophosphate pesticides, which act through inhibitory effects toward AChE's enzymatic activity in ways causing measurable electrochemical changes (Pratap Singh Raman et al. 2024). GQDs, functionalized using deoxyribonucleic acid (DNA) aptamers have been employed in antibiotic residue detection along with other pharmaceutical residues with the principal basis being through selective aptamer binding onto target molecules. The use of graphene in biosensors increases detection sensitivity but also permits miniaturization, portability, and real-time monitoring. Thus, the use of these devices is more practical for field applications.

Emerging innovations in graphene-based sensing technologies deal with the integration of this material into multifunctional and hybrid systems. Thus, for example, using microfluidic platforms combining graphene sensors allows real-time on-site monitoring of contamination in water. Such techniques are based on very few volumes

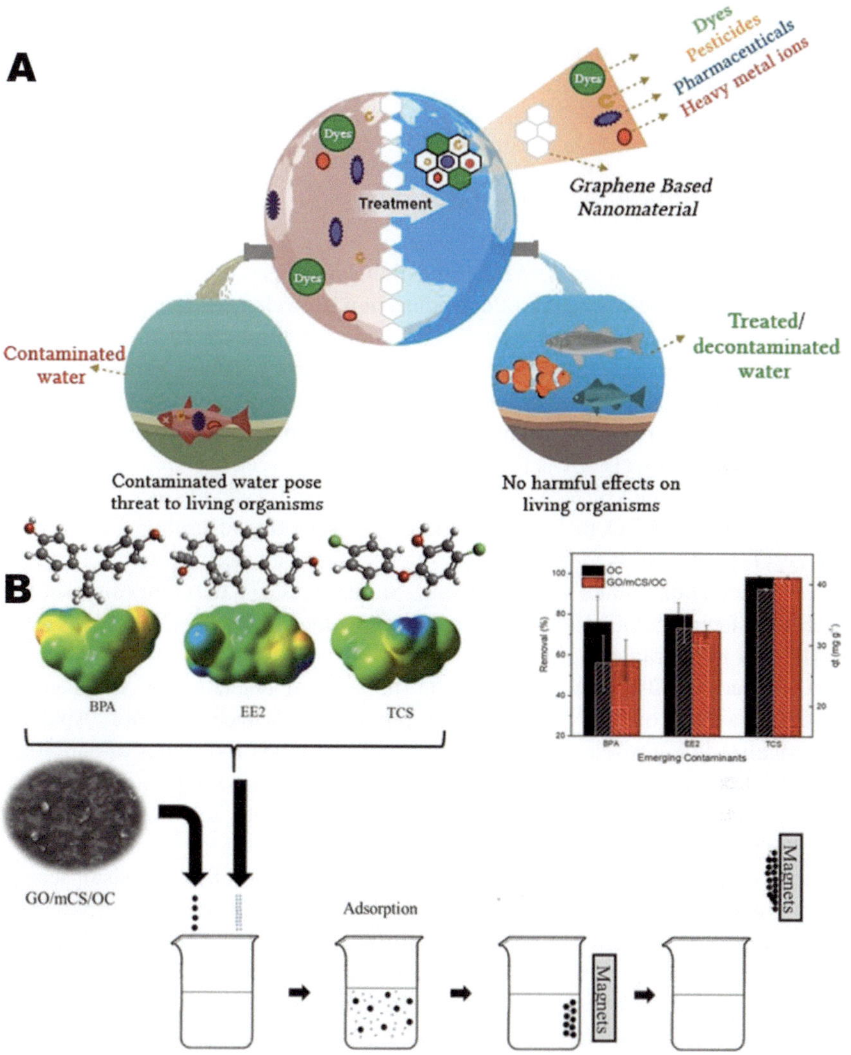

Fig. 8.3 **a** Illustration of removal of various emerging contaminants (ECs) in water using graphene NMs (Assad et al. 2024). Open access. **b** Magnetic composite based on GO, chitosan, and organoclay for the removal of ECs like BPA, 17α-ethinylestradiol, and triclosan (de Almeida 2022). Reproduced with permission from Elsevier

of samples and report results within a short time interval, thus being suitable for far-flung and resource-constrained locations (Rai et al. 2022). The recent establishment of the Internet of Things (IoT) enabled graphene sensors to create new avenues for environmental monitoring. IoT-enabled devices can forward real-time data to the central systems for analysis in order to provide proactive management of water quality (Ganguly and Sengupta 2024). This further implies that graphene-based technologies will change the environmental sense from static to dynamic, responsive scenarios. Detection of various emerging pollutants using graphene nanocomposites has been illustrated in Table 8.1.

8.3.1 Detection of Antibiotic Residues

One of them is a critical problem requiring global address is the rise in antimicrobial resistance associated with antibiotic residues in water released through pharmaceutical manufacture, agricultural run-off, and improper disposal. Environmental and public health monitoring requires trace detection of these residues. Various graphene derivatives and nanocomposites are developed to trace them in aquatic systems. GO and rGO-based sensors are highly sensitive to tetracycline, ciprofloxacin, and sulfamethazine. However, it is the π-π interactions between the aromatic planes of antibiotics and the GO surface, coupled with hydrogen bonding assisted by oxygen-functional groups in GO, that result in efficient and selective adsorption. For instance, GO-Fe_3O_4 nanohybrids for fluorescence quenching-based ciprofloxacin detection in water at a detection limit of as low as 10 nM. Similarly, rGO-ZnO composites showed improved electrochemical activity for tetracycline detection with a limit of detection (LOD) of 1 nM (Tran et al. 2020).

Advanced GQDs were also used for fluorescent detection of sulfonamides taking advantage of the size-dependent photoluminescence behavior. Aptamer-functionalized GQDs selectively hybridize with the target molecules of antibiotics to enable trace-level detection in complex matrices of water samples. For example, sulfamethoxazole was recently detected in water samples at a very low LOD of 2 nM using GQD-aptamer hybrids (He et al. 2018).

Another emerging application is the development of field-effect transistor sensors where graphene nanocomposites are integrated into the sensor. This has been found to utilize graphene's sensitivity to changes in surface charge. For example, molecularly imprinted polymers (MIPs) functionalized rGO was used in a field-effect transistor-based sensor for the detection of amoxicillin, achieving a LOD of 1.5 nM (Thakur and Kumar 2022). Such devices have high specificity and real-time sensing capabilities to make them very suitable for on-site monitoring. Hybrid nanocomposites consisting of graphene combined with other NMs have also demonstrated substantial promise, aside from electrochemical and optical methods. Composites of GO and TiO_2 improve photocatalytic activity under UV illumination, allowing for both sensing and degradation of antibiotics in water. These multifunctional systems minimize the number of analytical steps while resolving the problem of contamination (Tayel et al. 2018).

Table 8.1 Graphene nanocomposite-based detection of various emerging contaminants (ECs)

EC	Pollutant	Graphene nanocomposite	Detection techniques	LOD	Linear range	References
Antibiotic	Ciprofloxacin	rGO-Zr (IV)-based metal–organic framework NH_2 –UiO-66	Anodic stripping voltammetry (electrochemical sensing)	10.86 μA μM^{-1}	0.02 to 1 μM	Fang et al. (2019)
	Lomefloxacin	PANI-GOx-MIP-QDs	Fluorescence spectroscopy (turn-on)	0.07 μg L^{-1}	0.10 to 50.0 μg L^{-1}	Orachorn and Bunkoed (2019)
	Levofloxacin	Poly (l-Cysteine)@AuNPs @rGO film	Differential pulse voltammetry (electrochemical sensing)	3.0×10^{-12} M	1.0×10^{-4} M to 1.0×10^{-11} M	Ghanbari et al. (2019a, b)
	Levofloxacin	rGO-electropolymerized β-cyclodextrin	Voltametric determination	30 pmol L^{-1}	100 pmol L^{-1} to 100 nmol L^{-1}	Ghanbari et al. (2019a, b)
	Acetaminophen and Levofloxacin	Ultrathin graphitic carbon nitride nanosheet-doped GO (CNNS/GO)	Electrochemical sensing	1.7×10^{-8} and 7.9×10^{-8} mol L^{-1}	5×10^{-7} to 3.0×10^{-5}; 5×10^{-7} to 1.5×10^{-5} mol L^{-1}	Yi et al. (2021)
	Norfloxacin	Hydroxyapatite-GQDs-MIP nanoprobes	Fluorescence spectroscopy (turn-off)	0.35 μg L^{-1}	1.0 to 100.0 μg L^{-1}	Bunkoed et al. (2020)
	Norfloxacin	β-Cyclodextrin modified-rGO	Electrochemical detection	0.01 μM	0.4 to 80 μM	Cui et al. 2019
PCP	Methylparaben	rGO and AuNP-chitosan-coated glassy carbon electrode (GCE)	Voltametric determination	13.77 nmol L^{-1}	0.03 to 1.30 μmol L^{-1}	Piovesan et al. (2018)

(continued)

Table 8.1 (continued)

EC	Pollutant	Graphene nanocomposite	Detection techniques	LOD	Linear range	References
	Methyl paraben (MP), Ethyl paraben (EP), Propyl paraben (PP), Isobutyl paraben (IBP), and Butyl paraben (BP)	Graphene-polyvinylpyrrolidone (PVP)-polyaniline (PANI) modified onto the screen-printed carbon electrode	High-performance liquid chromatography (HPLC)	0.01 μg mL^{-1} (MP, EP, PP), 0.02 μg mL^{-1} (IBP), 0.03 μg mL^{-1} (BP)	0.1 and 30 μg mL^{-1}	Kajornkavinkul et al. (2016)
	Benzotriazole ultraviolet filters (PCP)	Graphene sponge (GS)-based solid-phase extraction (SPE)	High-performance liquid chromatography with ultraviolet detection (HPLC–UV)	0.02 to 0.08 μg L^{-1}	20.0 to 1000 μg L^{-1}	Wang et al. (2018)
	Acid green 25 of brilliant blue in hair dye	rGO-$CoFe_{12}O_{19}$-ionic liquid nanocomposite modified carbon paste electrode (CPE)	Cyclic and differential pulse voltammetry	0.067 μmol L^{-1}	0.08 to 900 μmol L^{-1}	Darabi and Shabani-Nooshabadi (2020)
	Caffeine	GO-reduced glutathione-modified carbon paste electrode (CPE)	Electrochemical sensing	0.153 μM	8 to 800 μM	Shehata et al. (2019)
Pharmaceuticals	Naproxen	Au@f-CNT/GO on the glassy carbon electrode	Electrochemical sensing	1.7484 μA $μM^{-1}$ cm^2	100 nM to 15.6 μM	Alagumalai et al. (2021)
	Diclofenac	MWCNTs-GO-AuNP nanocomposite film	Electrochemical sensing	0.09 μM	0.4 to 80 μM	Nasiri et al. (2021)
	Ibuprofen and naproxen	HBP/GO – HF-SLPME	High performance liquid chromatography	2.95 ng mL^{-1} (ibuprofen) 1.51 ng mL^{-1} (naproxen)	5 to 30,000 ng mL^{-1} (ibuprofen), 2 to 10,000 ng mL^{-1} (naproxen)	Rezaeifar et al. (2016)
	Carbamazepine	Ce-ZnO/rGO	Electrochemical sensing	1.2 nM	0.05 to 100 μM	Dhanalakshmi et al. (2018)

(continued)

Table 8.1 (continued)

EC	Pollutant	Graphene nanocomposite	Detection techniques	LOD	Linear range	References
Microplastics	Polystyrene (PS)	Graphene modified electrode	Electrochemical impedance spectroscopy		0.01 to 25 mg L^{-1} (particle size of 1 µm)	Du et al. (2023)
	Bisphenol A in microplastics (MPs)	Graphene-AuNP on the glassy carbon electrode	Electrochemical sensing	15.0 nmol L^{-1}		Baumgarten et al. (2023)
	Polystyrene nano plastics	GO/MWCNT composite membranes	SERS detection	5×10^{-5} mg/mL (50 nm)		Jiang et al. (2024)
Hormone	Progesterone	Magnetic GO	Cyclic voltammetry (CV) and differential pulse voltammetry (DPV)	0.15 pM (DPV) and 0.17 pM (CV)	0.01 pM to 1000 nM	Kumari et al. (2021)
	17β-estradiol	rGO-porphyrin	Electrochemical sensing	5.3 nmol L^{-1}	0.1 to 1.0 µmol L^{-1}	Moraes et al. (2015)
	17α-ethinylestradiol (EE2),	PAN/rGO composite membrane	Electrochemical sensing	100 pmol L^{-1}	10^{-5} to 10^{-11} mol L^{-1}	Facure et al. (2022)
	17-estradiol (E2)	MIP-silanized GO	Electrochemical sensing	0.037 ppm	0.1 to 0.81 mg L^{-1}	Mughal et al. (2024)

[a]NPs: Nanoparticles

8.3.2 Detection of Personal Care Products (PCPs)

PCPs, including parabens, synthetic fragrances, and UV filters are being found as ECs regularly detected in water systems due to their extensive use and inadequate removal during conventional wastewater treatment processes. They are often considered EDs, hence, an appropriate detection of these is an essential step for preserving health and ecosystems. As evidenced by the examples of triclosan and fragrance compounds, PCPs remain in the environment for very long periods and bioaccumulate, bypassing conventional treatments and thereby entering aquatic ecosystems at significantly high concentrations (Abiodun and Ayeleru 2025).

GQDs are promising materials in that they possess fluorescence properties which make them effective in residue detection for PCPs. Functionalized GQDs may selectively bind to compounds like parabens, exhibiting quenching or enhancing fluorescence that has already been used for developing optical sensors with high specificity and extremely low detection limits (Wang et al. 2024). Electrochemical sensors based on GO can also be used for the detection of UV filters by applying the properties of GO-high conductivity and large surface area for rapid and very precise measurements (Xu et al. 2017). GO functionalized with Au nanoparticles has been utilized in the detection of parabens with a very good LOD of 10 nM for methylparaben in complex water matrices. Similarly, rGO-based electrodes with MIPs were applied for the detection of triclosan. This method achieved a high LOD of 5 nM with excellent selectivity (Malarat et al. 2023). GQD-polymer hybrids also demonstrated high sensitivity to UV filters such as oxybenzone and octinoxate which are commonly used in sunscreens (Hopkins et al. 2017).

Hybrid materials, which consist of graphene with metal oxides or other NMs, enhance the detection capabilities. For example, GO-TiO_2 composites have been used to detect and degrade fragrances such as musk ketone through photocatalytic oxidation, thus providing dual functionality in detecting and mitigating pollutants. Another example is the rGO-ZnO hybrid, which can effectively sense UV filters and provide a LOD of 50 nM due to their enhanced electron transfer properties (Pargoletti et al. 2019). Electrochemical methods using graphene nanocomposites have been used extensively for PCP detection, GO-functionalized electrodes with aptamers have been reported for BPA detection (Zhu et al. 2015), a widely used plasticizer in PCPs, with an LOD of 0.1 nM. Moreover, graphene-metal nanocomposites can also be used to detect avobenzone, one of the main sunscreen components, with good response time and selectivity.

8.3.3 Detection of Pharmaceutical Compounds

The persistent nature of pharmaceutical residues and their potential impacts on the ecosystem makes it obligatory to trace down pharma residues appearing in water. Pharmaceuticals include various active molecules, such as steroids, antidepressants, antipyretics, stimulants, cosmetics, antimicrobials, disinfectants along with antibiotics. In most cases, the pharmaceutical residues excreted from living organisms remain largely unaltered metabolites in excreta and urine and are contaminated through untreated sludge or sewage. Graphene nanocomposites have shown major steps forward in terms of sensitivity and selectivity toward detecting pharmaceuticals, enabling the detection of residues even at the nanomolar level.

Functionalized GO and rGO were used for electrochemical sensing, either with aptamers or metal nanoparticles. For example, GO-TiO_2 composites have shown good promise in detecting and photo-catalytically degrading pharmaceuticals (Sengupta and Patel 2024), with the great surface area and electron-transfer ability of graphene. The second innovative approach is with GQDs, a class of fluorescent nanocomposites used for drug analysis like carbamazepine in fluorescence-based sensors (Hazarika et al. 2021) and reported LODs as low as 5 nM. These nanocomposites make use of the GQD's optical properties for sensing purposes in complex water matrices. The versatility of graphene nanocomposites extends to various pharmaceutical classes. Research by Qian et al. (2020) successfully applied a graphene-based sensor to detect non-steroidal anti-inflammatory drugs (NSAIDs) like naproxen in wastewater samples.

Real-time monitoring solutions through advances in smart sensor technologies have recently gained the limelight to detect and treat pollutants at the source. GQD-metal nanoparticle hybrids were also applied for monitoring fragrances in wastewater treatment plants (WWTPs), ensuring precise quantification and removal efficiency. These developments underscore the use of graphene nanocomposites in pharmaceutical detection. They have scalable, efficient, and environmentally sustainable applications. Recent research studies indicate that graphene-based materials are effective for the removal of different pharmaceutical compounds in water.

The incorporation of metal nanoparticles into graphene composites has further improved their performance. Ahmad et al. (2021) developed a GO nanocomposite decorated polymer that showed improved adsorption capabilities for multiple pharmaceuticals, including ibuprofen and diclofenac, with LODs in the low nanomolar range. Furthermore, research by Cao et al. (2016) demonstrated the development of functionalized graphene field-effect transistor sensors that can be used for the detection of trace amounts of atrazine with high sensitivity and specificity.

8.3.4 Detection of Microplastics (MPs)

MPs that are produced by the breakdown of plastics present in the environment promote various health hazards and pose significant risks to the aquatic ecosystem (Bhardwaj 2024; Bhardwaj et al. 2024). Graphene nanocomposite-based materials for identifying MPs in water are a thriving sector of research to combat the worsening problem of plastic pollution. Graphene nanocomposites represent an innovative alternative in contaminant detection and quantification in trace levels, providing properties unique to their high surface area and exceptional conductivity combined with tunable chemistry. Some of these properties make it highly favorable for sensing MPs specifically. GO membranes functionalized with MIPs may bind selectively to MP particles (Das et al. 2024), thus presenting a possible pathway to detect trace contaminants with high sensitivity and specificity. GO-MIP sensors have been known to detect MPs such as polyethylene terephthalate (PET) and polystyrene (PS) at parts per billion levels in complex water matrices.

GQDs are used widely in fluorescence-based sensors to detect MPs like polypropylene (PP) and polyethylene (PE) in real time. Composites of GQD with polymers have proved to be highly sensitive toward these materials, exploiting fluorescence quenching mechanisms to obtain nanomolar LODs (Praveena et al. 2024). Hybrid nanocomposites, such as GO-TiO_2, not only sense MPs but also degrade the organic pollutants attached to them through photocatalytic activity under UV irradiation, thus imparting dual functionality. In addition to their detection capabilities, graphene nanocomposites exhibit antimicrobial properties that can help reduce microbial contamination associated with MP pollution. Functionalized graphene sheets can disrupt bacterial membranes due to their sharp edges and oxidative stress effects (Xie et al. 2020). This dual functionality enhances the overall effectiveness of graphene-based materials in addressing water quality issues related to MPs.

Despite these advances, synthesis at the nanocomposite level still lags behind due to potential environmental concerns raised by derivatives of graphene. This future study should involve environmentally friendly methods of synthesizing green nanocomposites that improve the reusability of graphene-based sensors to assure sustainable usage in MP detection. As such, the detection of MPs within water samples using graphene nanocomposites offers a novel revolution in monitoring technologies toward the detection of this contaminant using sensitive techniques.

8.3.5 Detection of Plasticizers

Some of the most widespread organic water pollutants come from plasticizers used widely in plastics, mainly to package consumer and construction items, and other products. Of these, several, like BPA and phthalates, are already known to cause hormonal dysfunction, potentially impairing reproductive and developing health, among others, increasing their concern (Cimmino et al. 2020; Bhardwaj and Sharma 2021; Bhardwaj 2022). Graphene-based materials have found their applications as potential tools in the trace detection of plasticizers by taking advantage of properties such as high surface area, excellent electrical conductivity, and tunable chemical functionalities.

GO and rGO functionalized with specific antibodies or aptamers were applied for the detection of BPA at very low concentration levels down to 0.1 nM by using hydrogen bonding and π-π stacking interactions (Zahra et al. 2023). Electrochemical sensors that utilize the conductivity of graphene can measure BPA at concentrations much lower than the regulatory thresholds. For example, rGO doped with Au-nanoparticles increases the electrochemical sensitivity of sensors, which allows them to detect phthalates (Acharya et al. 2024) with LODs of 0.2 nM in real water samples. The hybrid sensors made up of graphene combined with metal nanoparticles and conductive polymers have the ability for signal amplification when certain plasticizers are present in water.

Recent studies have demonstrated that graphene-NMs are also effective in the selective detection of different plasticizers such as phthalates and non-phthalate alternatives. A GO-based sensor would be able to detect the concentration of di(2-ethylhexyl) phthalate (DEHP) as low as 0.1 ng/mL, showing the sensitivity of graphene nanocomposites for trace analysis in water samples. Similarly, rGO composite is functionalized with Ag-nanoparticles that have superior electrochemical properties for sensing dibutyl phthalate (DBP) by achieving a very low LOD of 0.05 ng/mL from differential pulse voltammetry (Azzouz et al. 2019).

The incorporation of different functional groups into graphene nanocomposites has further enhanced their selectivity toward specific plasticizers. For instance, the GO-based composite functionalized with amino groups that showed high affinity toward the detection of butyl benzyl phthalate (BBP) in complex water matrices, and the LOD was found to be 0.02 ng/mL (Guo et al. 2019). This targeted approach allows for the development of selective sensors capable of distinguishing between different plasticizers in environmental samples. Magnetic graphene nanocomposites have also been applied for the extraction and analysis of plasticizers from water samples. For example, it was reported that magnetic GO composites might significantly recover phthalates at recovery rates of over 95% from polluted water (Qinghua et al. 2024). It is a sensitive method yet decreases sample preparation before analysis thus it is well suited to practical applications. Labra-Vázquez et al. (2023) showed promise for field-effect transmitter sensors functionalized for the detection of low concentrations of diethyl phthalate (DEP) having an outstanding LOD of 0.1 ppb. Such research would be really important to achieve on-site monitoring devices, which offer rapid outputs during environmental diagnostics.

8.3.6 Detection of Perfluorinated Organic Compounds (PFOCs)

Perfluorinated organic compounds (PFOCs), which consist of perfluorooctanoic acid (PFOA) and perfluorooctane sulfonate (PFOS), are the most known persistent environmental pollutants that pose serious implications both for health and ecology owing to their bio-accumulative nature and resistance to degradation. The extensive use of these compounds in industrial processes and consumer products has made detection in water systems a challenge. Various graphene nanocomposites have been identified as promising materials for the detection of PFOCs with high sensitivity, selectivity, and scalability. GO composites are widely used, and studies have shown that GO-quaternary ammonium hybrids can reach LODs below 10 nM for PFOS because of increased electrostatic and hydrophobic interactions (Caroleo et al. 2022).

Another promising material for electrochemical sensing is rGO integrated with metal nanoparticles, such as rGO-Fe_3O_4. These composites allow for the simultaneous detection and magnetic separation of PFOCs, with LODs as low as 5 nM in real-world water samples (Safajou et al. 2024). The rGO-ZnO nanohybrids have been found to exhibit improved photocatalytic activity, such that PFOS is degraded into less toxic intermediates while offering high detection accuracy (Shanaah et al. 2023). GQDs are widely applied in fluorescence-based detection. GQD-metal organic framework composites can have high specificity toward PFOA and could rely on fluorescence quenching mechanisms for accurate quantification down to 1 nM. Graphene field-effect transmitter sensors functionalized with aptamers provide for real-time, label-free detection of PFOCs at ultra-low concentrations with rapid response time and excellent sensitivity, including LODs below regulatory limits for such substances (Sun et al. 2023).

Graphene nanocomposites, therefore, have huge potential to solve the issues that ECs cause in water systems. Their properties make sensitive detection methods critical to challenges posed by the contamination in aquatic systems. As research continues to evolve, these technologies hold promise for enhancing public health outcomes and shielding ecosystems from the adverse impacts associated with exposure to such pollutants. Advanced materials and novel sensor designs will be instrumental in developing effective strategies to manage this critical environmental concern.

8.4 Removal of Emerging Contaminants (ECs) from Water Using Graphene Nanocomposites

Graphene nanocomposites, which have become popular candidates in the research field of the purification of water, show excellent properties due to their large surface area, multifunctionality, stability, and scope for surface modifications. These advantages of graphene nanocomposites help in the effective extraction and removal of a variety of pollutants. Graphene polymer composites are also efficient in the adsorption of organic pollutants such as dyes, pharmaceutical compounds, and EDCs (Fig. 8.3a). Emerging pollutants, as discussed, have advanced structures that remain undetected and difficult to remove through conventional methods and materials. To combat the issues, various nanocomposites of graphene are developed to cope with the upcoming ordeals and overcome the limitations of various traditional materials. Moreover, in the past few years, various modifications have been performed in the nanocomposite structures for better results and improved functionality. Herein, various methods and mechanisms of removal of emerging pollutants using graphene nanocomposites are being discussed.

8.4.1 Adsorption

Removal of emerging pollutants through adsorption is one of the most predominant techniques adopted with graphene nanocomposites due to their exceptionally high surface area of graphene derivatives that provide strong interaction with contaminants through van der Waals forces, hydrogen bonding, π-π interactions, and electrostatic attraction (Fig. 8.3b). Moreover, the ability to functionalize with desirable groups effectively captures a variety of contaminants, such as heavy metals, pharmaceuticals, pesticides, and PCPs. For instance, GO effectively adsorbs heavy metals like Pb^{2+} and Cd^{2+} through stable complexation with oxygen-functional groups of GO (Peng et al. 2017). Similarly, π-π interactions between graphene and aromatic pollutants, like benzene derivatives, increase the capacity to adsorb. GO functionalized with iron oxide nanoparticles efficiently removes As and chromium (Cr) from contaminated samples, achieving adsorption capacities of 99.5%.

Another variant is the rGO composites with metal nanoparticles, which also showed improved electrochemical detection. These composites utilize the conductivity of graphene to capture and quantify MPs through size exclusion and surface interactions. For instance, Che Nan et al. (2022) found that GO functionalized with magnetic nanoparticles selectively captured PS MPs from water samples with removal efficiencies of over 90%. Similarly, Yuan et al. (2020) prepared an rGO composite that showed improved adsorption ability for PE MPs with LOD as low as 0.1 ng/mL. rGO-TiO_2 composites exhibit dual functionality by adsorbing pesticides like glyphosate while facilitating photocatalytic degradation under UV light.

GQDs are photoluminescent and have been used for the removal of pharmaceuticals like ibuprofen and diclofenac through enhanced adsorption mechanisms by surface functionalization. GO-polymer hybrids, such as GO-chitosan composites, have also shown excellent performance in capturing EDCs and pharmaceuticals through electrostatic interactions (Alves et al. 2021). GO-MOFs also show high adsorption efficiency for PFOCs, allowing for targeted removal of PFOS and PFOA at sub-nanogram levels (Lath et al. 2018). Advances in graphene-based membranes provide more leeway for the specific removal of MPs and heavy metals by size exclusion and chemical affinity.

8.4.2 Photocatalysis

Photocatalytic degradation is one of the emerging methods based on graphene composites, especially by rGO and metal oxide hybrids, for organic pollutant removal from water. High electron transfer properties of the graphene significantly reduce charge recombination in the photocatalysts, like TiO_2, and ZnO, thus achieving better efficiency under UV or even visible light. For instance, rGO-TiO_2 composites have already been noted to convert pesticides and pharmaceuticals into non-toxic byproducts through reactive oxygen species (ROS) generation (Dotti et al. 2024).

The already high pollution of water bodies with such pharmaceuticals, pesticides, and industrial chemicals has even led to the conception of incorporating graphene-based materials for photocatalysis in wastewater treatment. Despite some flaws, the efficiency of graphene-TiO_2 and graphene-ZnO nanocomposites in degrading organic pollutants is proving to be effective in the latest research.

Researchers have considered the applicability of the graphene-TiO_2 composites in wastewater treatment and analyzed the capability of these materials to outperform traditional dyes and pharmaceuticals by means of better light absorption and charge carrier dynamics. More than 90% of the methylene blue dye was degraded with UV light for 60 min within the GO-TiO_2 composite, which led to an effective, fast elimination of the pollutants, as demonstrated by Fu et al. (2019). In their study, however, Moslan et al. (2023) also included the GO-ZnO nanocomposite, but due to the activation of photocatalysis, the BPA degradation was enhanced, so a maximum of 95% removal efficiency for 120 min is maintained under visible light.

Another innovative technique for the improvement of the treatment process is the development of a photocatalytic membrane loaded with graphene nanocomposites. GO-TiO_2 membrane can show superior permeability and rejection performances of organic pollutants; even 98% removal of various dyes was achieved upon solar light exposure. Sevil Akçağlar (2024) demonstrated 100% photodegradation of aldrin and 99% of dieldrin using TiO_2/GO/$CuFe_2O_4$ nanocomposite at a sunlight intensity of 9 W m^{-2} (photodegradation time of 25 min) using the dosage of 1.3 mg L^{-1}. The integration of membrane filtration with photocatalysis will enhance operational efficiency and the removal of pollutants.

In addition, improvement in the synthesis of graphene-based nanocomposites has provided the way for alternative materials and methods. Nitrogen-doped TiO_2 combined with graphene is focused upon by Zhao et al. (2020), shifting the absorption spectrum toward visible light thus enabling it to be applied under the natural sunlight conditions. This alteration helped degrade biological contaminants such as bacteria and viruses, so the capability of these materials was depicted to address various issues with water quality. Very recently, magnetic graphene nanocomposites have been designed for easy separation after treatment. Barakat et al. (2020) reported that the heavy metals in contaminated water could be removed, and organic pollutants degraded with the presence of UV light by a magnetic GO-TiO_2 composite, with removal efficiencies above 90% for both types of contaminants. Thus, the application of graphene nanocomposites in photocatalytic processes is a step forward in the removal of emerging pollutants from water.

8.4.3 Membrane Filtration (MF)

Graphene nanocomposites-based MF has been shown as the new frontier in the removal of emerging pollutants with highly efficient and sustainable efficiencies from water. Graphene-based membranes have excellent mechanical strengths and surface tunabilities to be more feasible towards the selective removal of contaminating substances. GO membranes are primarily found to be used broadly on filtering micropollutants such as pharmaceuticals and pesticides by size exclusion or chemical interaction. GO-bridged polyamide membranes can also remove antibiotics including tetracycline and sulfamethazine efficiently (Xia et al. 2023). Here, rejection rates reach greater than 90% by way of mechanisms, electrostatic repulsion, and π-π interactions. Membrane performance may increase manifold by adding graphene derivatives associated with metal oxides and/or polymers. A noteworthy example is GO-TiO_2, a membrane wherein filtrating and photocatalytic degradation of contaminants made it possible for them efficient against dye contamination and POPs (Xu et al. 2013). Similarly, rGO-Ag nanoparticle composites incorporated into ultrafiltration membranes have antimicrobial properties to ensure the removal of chemical pollutants as well as biological contaminants like bacteria.

Hybrid materials that involve graphene with MOFs show additional adsorption sites to the pollutants BPA and PFOS and have improved strength and selectivity in complex water matrices handling. Many studies have emphasized the removal of emerging pollutants through the use of nanocomposite membranes based on graphene. For instance, Yadav et al. (2022) synthesized a GO-polysulfone composite membrane which showed more than 94% rejection of sodium sulfate at a high-water flux of 14.2 L/m^2/h. In addition, Dadvar and Heidari (2018) recently reported that GO-based nanocomposite membranes can effectively remove greater than 97% of methylene blue dye from aqueous solutions. In addition to that, the incorporation of other additives to graphene composites also improved filtration performance. For example, Zhang et al. (2022) prepared a hybrid membrane of GO and carbon nanotubes (CNTs), which showed improved mechanical strength and antimicrobial activity. It remained over 99% organic dyes in the filtration process.

Furthermore, magnetic graphene nanocomposites have been used to make separation easier after treatment. Even magnetic GO composite could effectively remove Pb and As from the polluted water containing an organic pollutant under UV light irradiation with removal efficiency above 90%. Graphene nanocomposite-based photocatalytic membrane reactors (PMRs) are some of the innovative methods developed (de Oliveira et al. 2022). Environmental sustainability is also another area of interest regarding graphene nanocomposite membranes; numerous studies report green synthesis of these materials by low energy consuming and environmentally benign processes. Thus, the application of graphene nanocomposites in MF technologies marked yet another new milestone toward eliminating emerging pollutants from the waters.

8.4.4 Electrochemical Degradation

Electrochemical degradation with graphene nanocomposites has recently been one of the most effective techniques to remove new pollutants from water, showing scalability, high efficiency, and environmental sustainability. Graphene-based materials, which include rGO and GO, possess excellent conductivity and surface reactivity; therefore, they are very compatible with electrochemical systems. Thus, electron transfer is enhanced in these nanocomposites as they contain pharmaceuticals, pesticides, or EDCs, among others; therefore, efficient pollutant degradation is facilitated. For example, degradation of antibiotics like ciprofloxacin and tetracycline at efficiencies greater than 90% was achieved utilizing rGO-functionalized electrodes (Ma et al. 2022). Graphene's superior conductivity increases ROS generation during electrochemical oxidation, breaking down the complex organic molecules into their nontoxic byproducts.

Graphene-metal hybrids, like rGO-Ag or rGO-copper (Cu) composites, improve the catalytic activity for electrochemical degradation, especially on PFOCs, such as PFOS and PFOA by cleaving their robust carbon–fluorine bonds (Duinslaeger et al. 2022). GQDs, incorporated in electrochemical systems, show high electrocatalytic performance for trace-level removals of PCPs, such as triclosan. Electrochemical degradation technologies have further advanced through hybrid graphene nanocomposites with metal–organic frameworks, which guarantee that MOF-rGO electrodes permit the effective removal of BPA through electrochemical oxidation and under stable harsh water matrices (Báez et al. 2024).

Furthermore, functionalized graphene nanocomposites embedded in field-effect transmitter sensors facilitate the monitoring and degradation of pollutants in real time with sensing and remediation functions combined together. Krishnan et al. (2023) demonstrated that trace PFOCs could be detectable using functionalized graphene field-effect transmitter sensors capable of electrochemical degradation of the said compounds simultaneously.

Research continues to optimize the scalability and environmental impact of graphene-based electrochemical systems. Developing green synthesis methods and improving electrode reusability are key steps toward sustainable applications. For example, mechanical exfoliation techniques can produce high-quality graphene without harmful chemicals, making these membranes suitable for sustainable water treatment applications. Generally, electrochemical degradation in water using graphene nanocomposites is a great technology for removing emerging pollutants. These properties are responsible for efficient degradation mechanisms; therefore, these are necessary to fight modern water quality issues. This technology, therefore, is expected to improve public health and minimize the negative impacts associated with certain contaminants in water resources.

8.5 Recommendations

Therefore, it is evident that graphene nanocomposites have multiple applications, such as multifunctional systems that can both detect and degrade contaminants because of their excellent structural and physic-chemical properties. However, the challenges include synthesis scalability, cost, and environmental safety, which need to be addressed for widespread implementation. Improvements in synthesis techniques will be directed toward scalable and eco-friendly approaches, for example, green synthesis, using plant extracts, and low-energy techniques such as microwave-assisted techniques. The standardized protocols are essential for large-scale applications. Strategies of functionalization are the key, and innovations like aptamers and MIPs have enhanced selectivity for pollutants such as pharmaceuticals, pesticides, and EDs. Combination with metal oxides or polymers has further optimized adsorption and catalytic efficiency in systems like rGO-TiO_2 composites. Graphene nanocomposites are also loaded into filtration membranes, photocatalytic reactors, and electrochemical systems to enhance water purification, aside from incorporation into a wide range of new technologies for real-time pollutant monitoring. For example, enhanced membranes result in increased permeability and contaminant rejection rates. The rGO-metal hybrids accelerate photocatalytic processes for the degradation of organic pollutants. The AOPs utilize electrocatalytic properties in cost-effective and energy-efficient pollutant removal. Besides, safety and environmental factors are also significant. If long-term ecological impact studies of graphene along with sustainable disposal methods are implemented, then the risk factors would be reduced. Future directions include three-dimensional architectures of graphene for greater efficiency in pollutant capture as a bioinspired filtration system mimicked to purify water.

8.6 Conclusion and Future Perspectives

Pollutants, such as MPs, pharmaceuticals, pesticides, and industrial chemicals are growing in prevalence; the demand for new and efficient remediation technologies is, therefore, turning increasingly acute. Graphene nanocomposites have proven to be one of the most efficient and versatile solutions in the area of water purification for both the detection and removal of ECs and critical environmental challenges. Graphene-based materials, mainly GO and rGO, have emerged as promising candidates for nanocomposite synthesis because of their excellent properties in regard to very high surface area, mechanical strength, and high electrical conductivity, such properties help graphene nanocomposites effectively adsorb and degrade various contaminants through van der Waals forces, hydrogen bonding, and electrostatic interactions.

Recent research has shown the potential of graphene nanocomposites to remove specific contaminants. In this regard, scalability remains a critical area for future research. Developing cost-effective synthesis methods and enhancing the reusability of these materials will be essential for their widespread adoption in real-world applications. Additionally, integrating graphene-based materials into hybrid systems that combine adsorption with photocatalytic degradation can provide dual functionality for pollutant removal. It is further possible to design MF technologies that utilize graphene nanocomposites to attain more efficient separation processes without fouling and longer operating lives than previously possible for such technology.

Addressing the environmental concerns of sustainability in the synthesis and disposal of graphene derivatives is a must. Studies should be directed toward environmentally friendly green synthesis methods of graphene nanocomposites that still retain their potency. This not only saves the ecological footprint but is also in line with worldwide efforts toward sustainable development. The future scope of graphene nanocomposites in the advanced detection and removal of emerging water contaminants is bright. Its unique properties will position them as vital components in next-generation water treatment technologies that will efficiently handle complex pollution challenges. It will, therefore, depend on continued research and innovation to unlock its full potential in public health protection and conservation of aquatic ecosystems from the adverse effects of various contaminants in water resources.

Acknowledgements The authors are grateful to the Department of Physics, Jadavpur University, India, for providing them with support to do this study. Dr. Souravi Bardhan sincerely expresses her gratitude to the Department of Environmental Science, Netaji Nagar College for Women, India, and Dr. Dipak Kr Chanda is thankful to Ferro Refratech, India for providing the necessary resources and support for this work.

References

Abiodun OA, Ayeleru OO (2025) Transport, fate, and detection of personal care products (PCPs) in the environment. In: Smart nanomaterials for environmental applications. Elsevier, pp 487–525

Acharya PB, George A, Shrivastav PS (2024) A status update on the development of polymer and metal-based graphene electrochemical sensors for detection and quantitation of bisphenol A. Crit Rev Anal Chem 54(4):669–690

Ahmad AL, Ebenezer OI, Shoparwe NF, Ismail S (2021) Graphene oxide-doped polymer inclusion membrane for remediation of pharmaceutical contaminant of emerging concerns: Ibuprofen. Membranes 12(1):24

Ahmed MB, Zhou JL, Ngo HH, Guo W, Thomaidis NS, Xu J (2017) Progress in the biological and chemical treatment technologies for emerging contaminant removal from wastewater: a critical review. J Hazard Mater 323:274–298

Akçağlar S (2024) Photocatalytic removal of aldrin and dieldrin using graphene oxide and TiO 2-doped CuFe 2O4. J Chem Technol & Biotechnol

Alagumalai K, Babulal SM, Chen SM, Shanmugam R, Yesuraj J (2021) Electrochemical evaluation of naproxen through Au@ f-CNT/GO nanocomposite in environmental water and biological samples. J Ind Eng Chem 104:32–42

Alam S, Bhardwaj LK, Mallick R, Rai S (2023) Estimation of heavy metals and fluoride ion in vegetables grown nearby the stretch of River Yamuna, Delhi (NCR), India. Indian J Environ Prot 43(1):64–73

Alves DCDS, Healy B, Yu T, Breslin CB (2021) Graphene-based materials immobilized within chitosan: applications as adsorbents for the removal of aquatic pollutants. Materials 14(13):3655

Anh NTN, Chowdhury AD, Doong RA (2017) Highly sensitive and selective detection of mercury ions using N, S-codoped graphene quantum dots and its paper strip based sensing application in wastewater. Sens Actuators, B Chem 252:1169–1178

Anis SF, Hashaikeh R, Hilal N (2019) Reverse osmosis pretreatment technologies and future trends: a comprehensive review. Desalination 452:159–195

Assad H, Lone IA, Kumar A, Kumar A (2024) Unveiling the contemporary progress of graphene-based nanomaterials with a particular focus on the removal of contaminants from water: a comprehensive review. Front Chem 12:1347129

Azzouz A, Kailasa SK, Kumar P, Ballesteros E, Kim KH (2019) Advances in functional nanomaterial-based electrochemical techniques for screening of endocrine disrupting chemicals in various sample matrices. TrAC, Trends Anal Chem 113:256–279

Báez ML, García A, Martínez I, González C, Gómez M, Rodríguez B (2024) Electrochemical detection of 17β-estradiol and bisphenol a using graphene oxide and reduced graphene oxide modified electrodes: a review. Int J Electrochem Sci 100538

Bahrdwaj LK, Upadhyay D, Jain H, Rath P, Kumar D, Bajpai S (2025a) Recent trends of phytoremediation for the wastewater treatment and resource recovery. In: Biotechnologies for wastewater treatment and resource recovery. Elsevier, pp 135–146

Barakat MA, Anjum M, Kumar R, Alafif ZO, Oves M, Ansari MO (2020) Design of ternary Ni (OH) 2/graphene oxide/TiO2 nanocomposite for enhanced photocatalytic degradation of organic, microbial contaminants, and aerobic digestion of dairy wastewater. J Clean Prod 258:120588

Baumgarten LG, Freitas AA, Santana ER, Winiarski JP, Dreyer JP, Vieira IC (2023) Graphene and gold nanoparticle-based bionanocomposite for the voltammetric determination of bisphenol A in (micro) plastics. Chemosphere 334:139016

Bhardwaj LK (2022) Evaluation of bis (2-ethylhexyl) phthalate (DEHP) in the PET bottled mineral water of different brands and impact of heat by GC–MS/MS. Chemistry Africa 5(4):929–942

Bhardwaj LK, Sharma A (2021) Estimation of physico-chemical, trace metals, microbiological and phthalate in PET bottled water. Chemistry Africa 4(4):981–991

Bhardwaj LK, Rath P, Choudhury M (2023) A comprehensive review on the classification, uses, sources of nanoparticles (NPs) and their toxicity on health. Aerosol Sci Eng 7(1):69–86

Bhardwaj LK, Rath P, Yadav P, Gupta U (2024) Microplastic contamination, an emerging threat to the freshwater environment: a systematic review. Environ Syst Res 13(1):8

Bhardwaj LK, Rath P, Rehman F, Yadav P, Massahi T, Mittal R, Jain H (2025b) Status of micro/nano-plastic (MNP) pollution in the aquatic system. In: Societal and environmental ramifications of plastic pollution. IGI Global, pp 43–62

Bhardwaj LK (2024) Occurrence of microplastics (MPs) in Antarctica and its impact on the health of organisms. Maritime Technol Res 6(2):265418 [online]

Bunkoed O, Donkhampa P, Nurerk P (2020) A nanocomposite optosensor of hydroxyapatite and graphene quantum dots embedded within highly specific polymer for norfloxacin detection. Microchem J 158:105127

Calderón-Preciado D, Matamoros V, Bayona JM (2011) Occurrence and potential crop uptake of emerging contaminants and related compounds in an agricultural irrigation network. Sci Total Environ 412:14–19

Cao TT, Nguyen HB, Bui HT, Vu TT, Phan NH, Phan BT, Hoang L, Bayle M, Paillet M, Sauvajol JL, Phan NM (2016) Fabrication of few-layer graphene film based field effect transistor and its application for trace-detection of herbicide atrazine. Adv Nat Sci: Nanosci Nanotechnol 7(3):035007

Caroleo F, Magna G, Naitana ML, Di Zazzo L, Martini R, Pizzoli F, Muduganti M, Lvova L, Mandoj F, Nardis S, Stefanelli M (2022) Advances in optical sensors for persistent organic pollutant environmental monitoring. Sens 22(7):2649

Chamoli P, Banerjee S, Raina KK, Kar KK (2020) Characteristics of graphene/reduced graphene oxide. In: Handbook of nanocomposite supercapacitor materials I: characteristics. pp 155–177

Chang H, Wu H (2013) Graphene-based nanocomposites: preparation, functionalization, and energy and environmental applications. Energy Environ Sci 6(12):3483–3507

Che Nan SNB, Wan Hazman D, Miskon MF, Abd Hamid S, Mohd. Salim R, Razali A (2022) Reduced graphene oxide functionalized magnetic nanocomposites for environmental pollutant removal. In: Materials science forum, vol. 1076. Trans Tech Publications Ltd., pp 109–117

Cimmino I, Fiory F, Perruolo G, Miele C, Beguinot F, Formisano P, Oriente F (2020) Potential mechanisms of bisphenol A (BPA) contributing to human disease. Int J Mol Sci 21(16):5761

Cui X, Cao D, Djellabi R, Qiao M, Wang Y, Zhao S, Mao R, Gong Y, Zhao X, Yang B (2019) Enhancement of Ni/NiO/graphitized carbon and β-Cyclodextrin/reduced graphene oxide for the electrochemical detection of norfloxacin in water sample. J Electroanal Chem 851:113407

Dadvar E, Heidari A (2018) A review on separation techniques of graphene oxide (GO)/base on hybrid polymer membranes for eradication of dyes and oil compounds: recent progress in graphene oxide (GO)/base on polymer membranes-related nanotechnologies. Clin Med Rev Case Rep 5(8):228–247

Darabi R, Shabani-Nooshabadi M (2020) Determination of acid green 25 in the presence of Brilliant Blue in hair dye and wastewater using voltammetric sensor amplified with reduced graphene oxide/CoFe12O19-ionic liquid nanocomposite. Environ Technol Innov 19:101020

Das TK, Basak S, Ganguly S (2024) 2D nanomaterial for microplastic removal: a critical review. Chem Eng J 152451

de Almeida ADSV, de Figueiredo Neves T, da Silva MGC, Prediger P, Vieira MGA (2022) Synthesis of a novel magnetic composite based on graphene oxide, chitosan and organoclay and its application in the removal of bisphenol A, 17α-ethinylestradiol and triclosan. J Environ Chem Eng 10(1):107071

de Oliveira CPM, Farah IF, Koch K, Drewes JE, Viana MM, Amaral MCS (2022) TiO2-Graphene oxide nanocomposite membranes: a review. Sep Purif Technol 280:119836

Devi S, Pal V, Yadav N, Jaiswal R, Roy D, Tripathi UK (2024) Graphene and its derivatives: various routes of synthesis. In: Electrochemical exfoliation of graphene and its derivatives: commercial applications. Springer Nature Singapore, Singapore, pp 61–112

Dhanalakshmi N, Priya T, Thinakaran N (2018) Highly electroactive Ce-ZnO/rGO nanocomposite: ultra-sensitive electrochemical sensing platform for carbamazepine determination. J Electroanal Chem 826:150–156

Dong L, Yang J, Chhowalla M, Loh KP (2017) Synthesis and reduction of large sized graphene oxide sheets. Chem Soc Rev 46(23):7306–7316

Dotti A, Basso Peressut AS, Matarrese R, Latorrata S (2024) Reduced graphene oxide and TiO_2 coated floating foams for photocatalytic wastewater treatment. In: 4th European school on environmental applications of advanced oxidation processes

Dreyer DR, Park S, Bielawski CW, Ruoff RS (2010) The chemistry of graphene oxide. Chem Soc Rev 39(1):228–240

Du H, Chen G, Wang J (2023) Highly selective electrochemical impedance spectroscopy-based graphene electrode for rapid detection of microplastics. Sci Total Environ 862:160873

Duinslaeger N, Radjenovic J (2022) Electrochemical degradation of per-and polyfluoroalkyl substances (PFAS) using low-cost graphene sponge electrodes. Water Res 213:118148

Durodola SS, Adekunle AS, Olasunkanmi LO, Oyekunle JA, Ore OT, Oluwafemi SO (2022) A review on graphene quantum dots for electrochemical detection of emerging pollutants. J Fluoresc 32(6):2223–2236

Facure MH, Mercante LA, Correa DS (2022) Polyacrylonitrile/reduced graphene oxide free-standing nanofibrous membranes for detecting endocrine disruptors. ACS Appl Nano Mater 5(5):6376–6384

Fang X, Chen X, Liu Y, Li Q, Zeng Z, Maiyalagan T, Mao S (2019) Nanocomposites of Zr (IV)-based metal–organic frameworks and reduced graphene oxide for electrochemically sensing ciprofloxacin in water. ACS Appl Nano Mater 2(4):2367–2376

Fu Z, Zhang S, Fu Z (2019) Preparation of multicycle GO/TiO2 composite photocatalyst and study on degradation of methylene blue synthetic wastewater. Appl Sci 9(16):3282

Fu H, Gray KA (2024) Graphene-encapsulated nanocomposites: synthesis, environmental applications, and future prospects. Sci Total Environ 176753

Ganguly S, Sengupta J (2024) Graphene-based nanotechnology in the internet of things: a mini review. Discover Nano 19(1):110

Ganguly S (2022) Preparation/processing of polymer-graphene composites by different techniques. In: Polymer nanocomposites containing graphene. Woodhead Publishing, pp 45–74

Geim AK (2009) Graphene: status and prospects. Sci 324(5934), 1530–1534

Ghanbari MH, Khoshroo A, Sobati H, Ganjali MR, Rahimi-Nasrabadi M, Ahmadi F (2019a) An electrochemical sensor based on poly (L-Cysteine)@ AuNPs@ reduced graphene oxide nanocomposite for determination of levofloxacin. Microchem J 147:198–206

Ghanbari MH, Shahdost-Fard F, Khoshroo A, Rahimi-Nasrabadi M, Ganjali MR, Wysokowski M, Rębiś T, Żółtowska-Aksamitowska S, Jesionowski T, Rahimi P, Joseph Y (2019b) A nanocomposite consisting of reduced graphene oxide and electropolymerized β-cyclodextrin for voltammetric sensing of levofloxacin. Microchim Acta 186:1–11

Gogoi A, Mazumder P, Tyagi VK, Chaminda GT, An AK, Kumar M (2018) Occurrence and fate of emerging contaminants in water environment: a review. Groundw Sustain Dev 6:169–180

Gomes IB, Maillard JY, Simões LC, Simões M (2020) Emerging contaminants affect the microbiome of water systems—strategies for their mitigation. NPJ Clean Water 3(1):39

Guo L, Ma X, Xie X, Huang R, Zhang M, Li J, Zeng G, Fan Y (2019) Preparation of dual-dummy-template molecularly imprinted polymers coated magnetic graphene oxide for separation and enrichment of phthalate esters in water. Chem Eng J 361:245–255

Hanvey JS, Lewis PJ, Lavers JL, Crosbie ND, Pozo K, Clarke BO (2017) A review of analytical techniques for quantifying microplastics in sediments. Anal Methods 9(9):1369–1383

Hazarika I, Kalita T, Deka P, Gogoi SK, Althubeiti K, Thakuria R, Gogoi B (2021) Soot-based reduced graphene quantum dot/hemin conjugate for favipiravir sensing. ACS Appl Nano Mater 4(12):13927–13937

He Y, Zhang B, Fan Z (2018) Aptamer based fluorometric sulfamethazine assay based on the use of graphene oxide quantum dots. Microchim Acta 185:1–8

Hopkins ZR, Snowberger S, Blaney L (2017) Ozonation of the oxybenzone, octinoxate, and octocrylene UV-filters: Reaction kinetics, absorbance characteristics, and transformation products. J Hazard Mater 338:23–32

Jasrotia R, Langer S, Dhar M (2021) Endocrine disrupting chemicals in aquatic ecosystem: an emerging threat to wildlife and human health. In: Proceedings of the zoological society, vol 74, no (4), pp 634–647. Springer India, New Delhi

Jiang Y, Wang X, Zhao G, Shi Y, Wu Y, Yang H, Zhao F (2024) Silver nanostars arrayed on GO/MWCNT composite membranes for enrichment and SERS detection of polystyrene nanoplastics in water. Water Res 255:121444

Kajornkavinkul S, Punrat E, Siangproh W, Rodthongkum N, Praphairaksit N, Chailapakul O (2016) Graphene/polyvinylpyrrolidone/polyaniline nanocomposite-modified electrode for simultaneous determination of parabens by high performance liquid chromatography. Talanta 148:655–660

Krishnan SK, Nataraj N, Meyyappan M, Pal U (2023) Graphene-based field-effect transistors in biosensing and neural interfacing applications: recent advances and prospects. Anal Chem 95(5):2590–2622

Kumar R, Qureshi M, Vishwakarma DK, Al-Ansari N, Kuriqi A, Elbeltagi A, Saraswat A (2022) A review on emerging water contaminants and the application of sustainable removal technologies. Case Stud Chem Environ Eng 6:100219

Kumari P, Nayak MK, Kumar P (2021) An electrochemical biosensing platform for progesterone hormone detection using magnetic graphene oxide. J Mater Chem B 9(26):5264–5271

Labra-Vázquez P, Gressier M, Rioland G, Menu MJ (2023) A review on solution-and vapor-responsive sensors for the detection of phthalates. Analytica Chimica Acta 341828

Lath S, Navarro DA, Losic D, Kumar A, McLaughlin MJ (2018) Sorptive remediation of perfluorooctanoic acid (PFOA) using mixed mineral and graphene/carbon-based materials. Environ Chem 15(8):472–480

Lawal AT (2015) Synthesis and utilisation of graphene for fabrication of electrochemical sensors. Talanta 131:424–443

Li X, Shen X, Jiang W, Xi Y, Li S (2024) Comprehensive review of emerging contaminants: detection technologies, environmental impact, and management strategies. Ecotoxicol Environ Saf 278:116420

Ma S, Zhang Q, Zhu J, Shi H, Zhang K, Shen Y (2021) Rational engineering of Ag-doped reduced graphene oxide as electrochemical sensor for trace mercury ions monitoring. Sens Actuators, B Chem 345:130383

Ma R, Xue Y, Ma Q, Chen Y, Yuan S, Fan J (2022) Recent advances in carbon-based materials for adsorptive and photocatalytic antibiotic removal. Nanomater 12(22):4045

Mahajan MR, Patil PO (2022) Design of zero-dimensional graphene quantum dots based nanostructures for the detection of organophosphorus pesticides in food and water: a review. Inorg Chem Commun 144:109883

Majdinasab M, Mishra RK, Tang X, Marty JL (2020) Detection of antibiotics in food: New achievements in the development of biosensors. TrAC, Trends Anal Chem 127:115883

Malarat N, Oin W, Kanjana K, Makkliang F, Siaj M, Poorahong S (2023) Electrochemical platform based on activated carbon/graphene oxide-gold nanoparticle composites for the electrochemical sensing of methylparaben in cosmetic samples. Microchem J 188:108473

Malm L, Liigand J, Aalizadeh R, Alygizakis N, Ng K, Fro Kjær EE, Nanusha MY, Hansen M, Plassmann M, Bieber S, Letzel T (2024) Quantification approaches in non-target LC/ESI/HRMS analysis: an interlaboratory comparison. Anal Chem 96(41):16215−16226

Manivannan B, Nallathambi G, Devasena T (2022) Alternative methods of monitoring emerging contaminants in water: a review. Environ Sci Process Impacts 24(11):2009–2031

Mao HY, Lu YH, Lin JD, Zhong S, Wee ATS, Chen W (2013) Manipulating the electronic and chemical properties of graphene via molecular functionalization. Prog Surf Sci 88(2):132–159

Marques PAAP, Gonçalves G, Cruz S, Almeida N, Singh M, Grácio J, Sousa AACM (2011) Functionalized graphene nanocomposites. Adv Nanocomposite Technol 11:247–272

Memisoglu G, Murugesan RC, Zubia J, Rozhin AG (2023) Graphene nanocomposite membranes: fabrication and water treatment applications. Membranes 13(2):145

Monjezi R, Azadi R, Hamoule T (2024) Design, synthesis, and characterization of thiol-decorated cross-linked graphene oxide framework for high-capacity Hg2+ ion adsorption. J Saudi Chem Soc 28(1):101786

Moraes FC, Rossi B, Donatoni MC, de Oliveira KT, Pereira EC (2015) Sensitive determination of 17β-estradiol in river water using a graphene based electrochemical sensor. Anal Chim Acta 881:37–43

Moslan MS, Othman MHD, Zakria HS, Ismail NJ, Borhamdin S, Rahman MA, Jaafar J, Puteh MH, Hashim N, Kerisnan NDA, Yahaya NKE (2023) Photocatalytic degradation of bisphenol A and energy storage capability of novel MoO3/ZnO/GO under visible light irradiation. J Mater Sci 58(22):9233–9250

Mughal ZUN, Aylaz G, Shaikh H, Memon S, Andac M (2024) Development of a molecularly imprinted polymer on silanized graphene oxide for the detection of 17-estradiol in wastewater. Water Environ Res 96(3):e11006

Nasiri F, Rounaghi GH, Ashraf N, Deiminiat B (2021) A new electrochemical sensing platform for quantitative determination of diclofenac based on gold nanoparticles decorated multiwalled carbon nanotubes/graphene oxide nanocomposite film. Int J Environ Anal Chem 101(2):153–166

Nurrohman DT, Chiu NF (2021) A review of graphene-based surface plasmon resonance and surface-enhanced raman scattering biosensors: current status and future prospects. Nanomaterials 11(1):216

Orachorn N, Bunkoed O (2019) A nanocomposite fluorescent probe of polyaniline, graphene oxide and quantum dots incorporated into highly selective polymer for lomefloxacin detection. Talanta 203:261–268

Pan D, Jiao J, Li Z, Guo Y, Feng C, Liu Y, Wang L, Wu M (2015) Efficient separation of electron–hole pairs in graphene quantum dots by TiO_2 heterojunctions for dye degradation. ACS Sustain Chem & Eng 3(10):2405–2413

Pan J, Liu P, Yu X, Zhang Z, Liu J (2024) The adverse role of endocrine disrupting chemicals in the reproductive system. Front Endocrinol 14:1324993

Pargoletti E, Hossain UH, Di Bernardo I, Chen H, Tran-Phu T, Lipton-Duffin J, Cappelletti G, Tricoli A (2019) Room-temperature photodetectors and VOC sensors based on graphene oxide–ZnO nano-heterojunctions. Nanoscale 11(47):22932–22945

Pena-Pereira F, Bendicho C, Pavlović DM, Martín-Esteban A, Díaz-Álvarez M, Pan Y, Cooper J, Yang Z, Safarik I, Pospiskova K, Segundo MA (2021) Miniaturized analytical methods for determination of environmental contaminants of emerging concern–a review. Anal Chim Acta 1158:238108

Peng C, Zhang X (2021) Chemical functionalization of graphene nanoplatelets with hydroxyl, amino, and carboxylic terminal groups. Chemistry 3(3):873–888

Peng W, Li H, Liu Y, Song S (2017) A review on heavy metal ions adsorption from water by graphene oxide and its composites. J Mol Liq 230:496–504

Perreault F, De Faria AF, Elimelech M (2015) Environmental applications of graphene-based nanomaterials. Chem Soc Rev 44(16):5861–5896

Piovesan JV, Santana ER, Spinelli A (2018) Reduced graphene oxide/gold nanoparticles nanocomposite-modified glassy carbon electrode for determination of endocrine disruptor methylparaben. J Electroanal Chem 813:163–170

Pooresmaeil M, Namazi H (2024) Surface modification of magnetic nanoparticles for biomaterials. In: Surface engineering of biomaterials. CRC Press, pp 430–482

Praveena U, Raja V, Ragavan KV, Anandharamakrishnan C (2024) Optical detection probes and sensors for micro-/nano-plastics. Rev Environ Sci Bio/technol 23(3):569–599

Qamar S, Ramzan N, Aleem W (2024) Graphene dispersion, functionalization techniques and applications: a review. Synthetic Metals 117697

Qian L, Thiruppathi AR, Elmahdy R, van der Zalm J, Chen A (2020) Graphene-oxide-based electrochemical sensors for the sensitive detection of pharmaceutical drug naproxen. Sens 20(5):1252

Qinghua YANG, Zhimin HE, Yilan SHI, Jinxin WANG, Qianying QIU, Honghai DENG, Wenhui PING (2024) Determination of 17 phthalate esters in water using GC–MS combined with magnetic solid phase extraction with Fe3O4@ GO. Chin J Anal Chem 52(4):100378

Rai PK, Islam M, Gupta A (2022) Microfluidic devices for the detection of contamination in water samples: a review. Sens Actuators, A 347:113926

Pratap Singh Raman A, Thakur G, Pandey G, Kumari K, Singh P (2024) An updated review on functionalized graphene as sensitive materials in sensing of pesticides. Chem & Biodivers e202302080

Ramírez C, Belmonte M, Miranzo P, Osendi MI (2021) Applications of ceramic/graphene composites and hybrids. Materials 14(8):2071

Rathi BS, Kumar PS (2021) Application of adsorption process for effective removal of emerging contaminants from water and wastewater. Environ Pollut 280:116995

Rezaeifar Z, Es' haghi Z, Rounaghi GH, Chamsaz M (2016) Hyperbranched polyglycerol/graphene oxide nanocomposite reinforced hollow fiber solid/liquid phase microextraction for measurement of ibuprofen and naproxen in hair and waste water samples. J Chromatogr B 1029:81−87

Rowley-Neale SJ, Randviir EP, Dena ASA, Banks CE (2018) An overview of recent applications of reduced graphene oxide as a basis of electroanalytical sensing platforms. Appl Mater Today 10:218–226

Safajou H, Mizwari ZM, Rostaminia A, Khojasteh H, Aspoukeh P, Mazhari MP (2024) Green synthesis and enhanced photocatalytic performance of rGO/ZnO/Fe3O4 nanocomposites: a sustainable approach to environmental remediation. J Fluoresc 1–20

Sengupta P, Patel SB (2024) Multifaceted applications of nanoparticle-functionalized graphene. 2D Mater: Chem Appl (Part 2) 17

Shahid MK, Kashif A, Fuwad A, Choi Y (2021) Current advances in treatment technologies for removal of emerging contaminants from water–a critical review. Coord Chem Rev 442:213993

Shanaah HH, Alzaimoor EF, Rashdan S, Abdalhafith AA, Kamel AH (2023) Photocatalytic degradation and adsorptive removal of emerging organic pesticides using metal oxide and their composites: recent trends and future perspectives. Sustain 15(9):7336

Shehata M, Azab SM, Fekry AM (2019) May glutathione and graphene oxide enhance the electrochemical detection of caffeine on carbon paste sensor in aqueous and surfactant media for beverages analysis? Synth Met 256:116122

Skaltsas T, Tagmatarchis N, Pispas S (2015) Non-covalent graphene/polymer functional materials. Curr Org Chem 19(18):1800–1818

Sun J, Shao B (2023) Aptamer-based analysis of persistent organic pollutants and endocrine disruptor compounds. In: Aptamers for food applications. Academic Press, pp 227–240

Tayel A, Ramadan AR, El Seoud OA (2018) Titanium dioxide/graphene and titanium dioxide/graphene oxide nanocomposites: synthesis, characterization and photocatalytic applications for water decontamination. Catalysts 8(11):491

Thakur K, Kandasubramanian B (2019) Graphene and graphene oxide-based composites for removal of organic pollutants: a review. J Chem Eng Data 64(3):833–867

Thakur A, Kumar A (2022) Recent advances on rapid detection and remediation of environmental pollutants utilizing nanomaterials-based (bio) sensors. Sci Total Environ 834:155219

Tran HL, Darmanto W, Doong RA (2020) Ultrasensitive detection of tetracycline using boron and nitrogen co-doped graphene quantum dots from natural carbon source as the paper-based nanosensing probe in difference matrices. Nanomater 10(9):1883

Urban-Chmiel R, Marek A, Stępień-Pyśniak D, Wieczorek K, Dec M, Nowaczek A, Osek J (2022) Antibiotic resistance in bacteria—a review. Antibiotics 11(8):1079

Venkatesan M, Vusa CSR, Koyappayil A, Chavan SG, Lee MH (2024) Biological recognition element anchored 2D graphene materials for the electrochemical detection of hazardous pollutants. Electrochimica Acta 144413

Wang X, Wang J, Du T, Kou H, Du X, Lu X (2018) Determination of six benzotriazole ultraviolet filters in water and cosmetic samples by graphene sponge-based solid-phase extraction followed by high-performance liquid chromatography. Anal Bioanal Chem 410:6955–6962

Wang L, Gu C, Wu L, Tan W, Shang Z, Tian Y, Ma J (2024) Recent advances in carbon dots for electrochemical sensing and biosensing: a systematic review. Microchem J 111687

Wille K, De Brabander HF, Vanhaecke L, De Wulf E, Van Caeter P, Janssen CR (2012) Coupled chromatographic and mass-spectrometric techniques for the analysis of emerging pollutants in the aquatic environment. TrAC, Trends Anal Chem 35:87–108

Xia X, Yan Y, Luo J, Liu T, Wu B, Qian F (2023) Effects of peroxide types on the removal performance and mechanism of sulfonamide antibiotics using graphene-based catalytic membranes. Process Saf Environ Prot 179:362–372

Xie C, Zhang P, Guo Z, Li X, Pang Q, Zheng K, He X, Ma Y, Zhang Z, Lynch I (2020) Elucidating the origin of the surface functionalization-dependent bacterial toxicity of graphene nanomaterials: oxidative damage, physical disruption, and cell autolysis. Sci Total Environ 747:141546

Xu C, Cui A, Xu Y, Fu X (2013) Graphene oxide–TiO2 composite filtration membranes and their potential application for water purification. Carbon 62:465–471

Xu J, Wang Y, Hu S (2017) Nanocomposites of graphene and graphene oxides: synthesis, molecular functionalization and application in electrochemical sensors and biosensors. A review. Microchim Acta 184:1–44

Yadav S, Ibrar I, Samal AK, Altaee A, Déon S, Zhou J, Ghaffour N (2022) Preparation of fouling resistant and highly perm-selective novel PSf/GO-vanillin nanofiltration membrane for efficient water purification. J Hazard Mater 421:126744

Yi W, Han C, Li Z, Guo Y, Liu M, Dong C (2021) A strategy of electrochemical simultaneous detection of acetaminophen and levofloxacin in water based on gC_3 N_4 nanosheet-doped graphene oxide. Environ Sci Nano 8(1):258–268

Yuan F, Yue L, Zhao H, Wu H (2020) Study on the adsorption of polystyrene microplastics by three-dimensional reduced graphene oxide. Water Sci Technol 81(10):2163–2175

Zahra QUA, Fang X, Luo Z, Ullah S, Fatima S, Batool S, Qiu B, Shahzad F (2023) Graphene based nanohybrid aptasensors in environmental monitoring: concepts, design and future outlook. Crit Rev Anal Chem 53(7):1433–1454

Zhang X, Kong H, Yang G, Zhu D, Luan X, He P, Wei G (2022) Graphene-based functional hybrid membranes for antimicrobial applications: a review. Appl Sci 12(10):4834

Zhao Y, Wang G, Li L, Dong X, Zhang X (2020) Enhanced activation of peroxymonosulfate by nitrogen-doped graphene/TiO_2 under photo-assistance for organic pollutants degradation: Insight into N doping mechanism. Chemosphere 244:125526

Zhu Y, Cai Y, Xu L, Zheng L, Wang L, Qi B, Xu C (2015) Building an aptamer/graphene oxide FRET biosensor for one-step detection of bisphenol A. ACS Appl Mater Interfaces 7(14):7492–7496

Zubair M, Syamaladevi R, Ullah A (2024) Challenges and prospects: graphene oxide-based materials for water remediation including metal ions and organic pollutants. Environ Sci: Nano

Chapter 9
Emerging Trends in Challenges in Technological Innovation for Sustainable Water Management

Qudrat Ullah, Muhammad Waqar, Muhammad Sajad, Muhammad Qasim, and Thanet Khomphet

Abstract Over 2.2 billion people experience water scarcity, with expected deterioration due to an increase in the demand for freshwater by 55% in the year 2050 because of increased population, urbanization, and industrialization. Innovative technologies, such as precision irrigation and rainwater collection, utilize only 10% water yet convey 90% efficiency and reduce wastage by 30 to 50%. Integrated Water Resource Management (IWRM) optimizes and improves water quality, pooling the required resources and harmonizing relations between stakeholders, leading to a 20 to 30% increase in efficiency. These challenges are intensified by climate change and water scarcity: The world's arid regions are projected to experience a 40% reduction in their water availability by 2050 and therefore smart water and wastewater reuse. This chapter highlights the importance of water management and efficient practices to cope with water insecurity and encourage production from the agriculture and livestock industry.

Keywords Sustainable water management (SWM) · Emerging contaminants (ECs) · Sustainability · Technology · Water scarcity · Integrated resource management

Q. Ullah (✉) · M. Waqar · T. Khomphet
Food Technology and Innovation Research Center of Excellence, School of Agricultural Technology and Food Industry, Walailak University, Nakhon Si Thammarat, Thailand
e-mail: qudratullahmpur2@gmail.com

M. Sajad
Department of Plant Breeding and Genetics, Faculty of Agriculture and Environment, The Islamia University of Bahawalpur, Bahawalpur, Punjab, Pakistan

M. Qasim
Department of Environmental Science, Government College University Faisalabad, Punjab, Pakistan

H. Jain et al. (eds.), *Emerging Contaminants in the Aquatic Environment*, Advances in Wastewater Research, https://doi.org/10.1007/978-981-96-8548-6_9

List of Abbreviations

AI	Artificial Intelligence
AOPs	Advanced oxidation processes
BMPs	Best management practices
CRE	Contaminant removal efficiency
ECs	Emerging contaminants
GIS	Geographic Information Systems
IoT	Internet of Things
IPCC	Intergovernmental Panel on Climate Change
IWRM	Integrated Water Resource Management
MBR	Membrane bioreactor
MF	Membrane filtration
ML	Machine learning
NGOs	Non-governmental organizations
PPCPs	Pharmaceuticals and personal care products
PPPs	Public-private partnerships
R&D	Research and development
RO	Reverse osmosis
SWM	Sustainable Water Management
SWMM	Storm Water Management Model
TSS	Total suspended solids
UN	United Nations
US	United States
UV	Ultraviolet
WER	Water efficiency ratio
WQM	Water quality management

9.1 Introduction to Sustainable Water Management (SWM)

Sustainable Water Management (SWM) is a wider systematic approach to meeting competing water demands while making sure that water resources are available in the future. This involves measures that enhance the efficient use of water, the conservation of the quality of water supplies as well as the conservation of the biological quality of water resources (Alsaeed et al. 2022). The appropriateness of SWM is further enhanced by the fact that more than 2 billion people live in areas with high water stress today and this number is expected to increase in the future due to climate change and population growth. The problem of water management is closely connected with the economic aspects: It was revealed that a lack of water resources can cost certain areas up to 6% of the GDP by 2050. In addition, the availability and usage of water, energy, and food are mutually dependent, and calls for synergy in resource management because world demand for fresh water is estimated to rise by 55% by year 2050 mainly for agricultural and industry usage (Smith et al. 2024). Thus, SWM is not just a technical problem; it is a core aspect of sustainable development, involving cross-stakeholder interventions from governments, private entities, and intensive public participation.

Water management practices as a discipline have their origin/initiation with the primitive societies/agricultural civilizations where societies wanted to add value and productivity to agriculture through the construction of irrigation systems. For instance, the Minoans and Egyptians required complex approaches to gain mastery in terms of the available water control with the integration of the hydrological regimes of the two areas (Lieberman and Gordon 2021). In bigger societies and with increased urbanization, especially during the industrialization process, more complex forms of management have evolved into central administration. This change promoted the design of extensive water supply and wastewater treatment, which, although it met the needs of the time, failed to take long-term concerns about water into account. During the past few decades, the world has shifted to a new approach to water resource management known by its acronym IWRM abbreviated as Integrated Water Resource Management for Water Resources (Hughes et al. 2021).

Currently, the global community is experiencing several water issues that threaten the sustainability of water supplies in the world. Climate change is attributed to all these challenges because it affects the precipitation regimes, increases the occurrence of severe weather, and increases temperatures that worsen the scarcity of water (Yildiz et al. 2022). For instance, sub-Saharan Africa is expected to likely have 20% less water per capita by 2050 because of climate change. Thirdly, rapid urbanization is exerting unrelieved pressure on water supply infrastructures, while more than half of the world population lives in urban areas, the figure is projected to reach two-thirds by 2050 (Letwin 2021). This is often accompanied by assimilation of infrastructural development hence millions of people within these urban areas are denied basic privileges of access to clean water or even sanitation. This is made worse by pollution whereby industrial waste, chemically fertilized effluent from agricultural practices, and untreated sewage find their way into the freshwater sources. The regularization of only 20% of wastewater produced worldwide creates environmental problems that threaten human health and wildlife (Freitas and Radis-Baptista 2021). To tackle them, new technology, sound policies, and engagements from the community to ensure that sustainable, equitable, and effective water management practices prevail because of these changes.

New strategies are crucial in enhancing the systems used in the management of sustainable water resources. Technological progress underway in water treatment solutions has integrated recent technologies like membrane filtration (MF) and advanced oxidation processes (AOPs) to enhance effective water reuse and recycling to ease competition at the water user end (Yu et al. 2022). In addition, by using elements of big data and real-time monitoring, a smart water management system offers solutions for proper and timely management of water supply so that the stakeholders have the chance to react to the change in water supply and quality situations on time. The concepts of rainwater harvesting structures and desalinization structures and types of equipment have come into light as the variation to the traditional supply of water, in areas where water is rare. These innovations also benefit economic development by decreasing the costs associated with water scarcity, as well as increasing the drought and climate change shock coping ability of communities.

This chapter aims to present the concept, the development, the current issues facing SWM, and the new technologies. Through these aspects, it draws attention to the need to take a systems approach to water management with a view to water-energy-food nexus. In addition, it establishes the lessons learned and measures that can be taken to improve water management due to increasing pressures and changes in the weather.

9.2 Technological Innovations in Water Resource Management

Technological advancement has played a major role in shifting the management of water resources to the provision of adequate and quality water that meets global demand. Another measure that has been proven to be very effective in the development of this field is smart water management systems (Giri 2021). Such systems use high-technology sensors, data analysis, and automation of the water supply and usage. For instance, smart meters have been found to decrease water use by as much as 15% in urban dwellings, thanks to conductivity feedback information offered to consumers (Syrmos et al. 2023). This availability saves a lot of water and contributes to the identification of leakage or wastage in the water distribution networks. Furthermore, smart water management fosters improved cross-linkage with other systems such as Geographic Information System (GIS); this makes it easier for water utilities to view data regarding installations, investment, distribution, and maintenance spatially (Nižetić et al. 2020).

Apart from smart water management systems, Internet of Things (IoT) application in water management is another notable innovation known in the arena. IoT devices which include smart sensors and individual devices enable constant assessment of water quality and water quantity data can be useful in decision-making. For instance, smart sensors in the IoT application integrate the monitoring of water quality parameters including pH, total suspended solids (TSS), and contamination. This capability is specifically useful in handling water in urban settings as this is characterized by pollution and contamination. Research has shown that IoT technologies can result in efficient operations and achieve potential annual cost savings of 20 to 30% in water utility organizations because the use of these systems makes it easier to identify problematic situations and prevent them in advance (Boyle et al. 2022). Moreover, through the data generated by IoT applications, a detailed analysis of predictive models that would predict water demand and supply would reach increased efficiencies of water resource utilization.

New communication technologies are also useful in water resources management to the extent that modern remote sensing systems are essential in the assessment of available supplies. These technologies apply satellite images and aerial data to observe and evaluate water bodies because of availability, quality, and usage. For instance, remote sensing can be used in monitoring variations in surface water, mapping areas with water deficits, and the effects caused by climate change on water. Remote sensing has been proven to enhance the reliability of the water resource analysis and quantification by up to 40% to ensure proper nutrition of the water resource (Sanders and Masri 2016). Also, remote sensing can contribute to the identification of unauthorized extractions from water sources thus complementing its regulation to ensure that sustainable water practices are followed. When available remotely sensed data is incorporated with the ground measurements, this leads to a better understanding of the hydrological process and facilitates the planning for water resources management which is adaptive for climate change scenarios (Acharya et al. 2021).

The challenge of incorporating data analytics into decision-making related to water resources issues has successfully evolved into an innovative process to improve the management and distribution of resources. Data analysis is the process of making sense of the data using computational techniques to make findings or patterns useful in water management. With the specific help of the methods outlined in this chapter, water officials and consumers can anticipate problems, solve them with high efficiency, and increase productivity. For example, predictive analytics can make supply since they could predict water consumption due to history, seasonal changes, or demographic shifts. Current statistics prove that, when optimized, the use of data analytics can cause a decline in operations expense that ranges between 20 to 25%, due to better resource management and minimization of loss (Pasupuleti et al. 2024). Moreover, solutions based on the use of big data can help in improving the indicators of water quality based on reliable findings of contaminant sources, which will help in early prevention thus preserving the health of the people and the environment.

Some technological options in the specific application of water resource management can effectively be delivered and are well illustrated by the following examples. A recent case is the application of smart water metros in different urban cities that has led to considerable provision of water usage and leakage reduction. An empirical study carried out in a large city showed that the use of smart meters brought relative savings of 20% in water usage by people because users can track their usage in real time (Daminato et al. 2021). Another successful application is the application of IoT in monitoring water quality in farms remotely. With IoT sensors in the field, farmers can watch over moisture and water quality in the soil in real time and use it to irrigate crops with efficiency and cut water usage by 30% (Ayaz et al. 2019). It also saves water resources, increases crop productivity, and decreases operating expenses. These case studies show the role of technology in decision-making on water management and the desire for innovation in solving these problems for increased benefits.

Water Efficiency Ratio (WER):

$$WER = \frac{Effective\ Water\ Utilization\ (E)}{Total\ Water\ Input\ (T)}$$

where

E = Water effectively utilized for productive purposes (liters).

T = Total water supplied to the system (liters).

This ratio evaluates the efficiency of water utilization in various systems, with values closer to 1.0 indicating higher efficiency.

WER quantifies the effectiveness of water-use systems, serving as a critical metric in sustainable resource management. For example, precision irrigation systems can achieve a WER of up to 0.9, reflecting their capability to direct up to 90% of water input toward productive ends. In contrast, conventional systems often exhibit a WER below 0.7 due to higher losses from evaporation and runoff. Such evaluations are vital for comparing technologies, optimizing water allocations, and reducing wastage in agriculture and urban applications. Integrating WER into decision-making frameworks supports the transition toward more SWM practices.

Table 9.1 provides an overview of different technological advancements in water management and treatment; description, application, benefits/advantages, limitations/disadvantages, and references are presented. Such developments as the IoT in the water sector and smart water grid are revolutionizing urban water systems through increasing water systems efficiency and monitoring the systems in real time, although come with some other burdens such as high capital costs and security threats. For water scarcity, there are rainwater harvesting systems for places other than the coastal region, and for the coastal regions desalination technologies but with the following challenges: Cost of installing the systems and the effect they have on the environment. They are vital components of water treatment because they provide, among other qualities, high removal efficiency as well as environmentally friendly solutions for wastewater treatment but present problems in their operations. Moreover, the connection of water quality sensors and smart irrigation systems is one of the examples of how the use of data in agriculture and the environment is becoming more and more popular. In total, these novelties are important improvements in coping with international water problems; however, they prove the necessity of further research and development (R&D) to solve current obstacles.

9.3 Water Quality Management (WQM)

Water Quality Management (WQM) represents an important component of sustainable water resource use because its quality and condition are closely associated with people's health, the condition of ecosystems, and the overall environment. Failed water quality was not just a risk to public health but fundamental for drinking, farming, and supporting aquatic life. That is why water quality is of great importance; diseases such as waterborne diseases affect millions of people around the world. For example, about 2 billion people face a problem with affordable access to

Table 9.1 Summary of technological innovations

Innovation	Description	Application area	Benefits	Challenges	References
IoT in water management	Integration of IoT devices for real-time monitoring	Water supply and distribution	Enhanced efficiency, reduced wastage	High initial costs, data security concerns	Maroli et al. (2021)
Smart water grids	Advanced grid systems for optimized water distribution	Urban water systems	Improved resource management, leak detection	Infrastructure upgrades required	Nižetić et al. (2019)
Rainwater harvesting systems	Systems designed to collect and utilize rainwater	Urban and rural areas	Sustainable water source, reduced runoff	Initial setup costs, maintenance needs	Rahman et al. (2020)
Desalination technologies	Methods for converting seawater to freshwater	Coastal regions	Provides fresh water in arid areas	High energy consumption, environmental impact	Ayaz et al. (2022)
Advanced filtration systems	Innovative filtration methods for water purification	Drinking water treatment	The high removal efficiency of contaminants	Membrane fouling, operational costs	Li et al. (2022)
Bioremediation techniques	Use of microorganisms to clean contaminated water	Polluted water bodies	Eco-friendly, cost-effective	Time-consuming, effectiveness varies	Kuppan et al. (2024)
Water quality sensors	Sensors for real-time water quality monitoring	Environmental monitoring	Immediate data for decision-making	Calibration and maintenance requirements	Garrido-Momparler and Peris (2022)
Smart irrigation systems	Automated irrigation systems using sensor data	Agriculture	Water conservation, optimized crop yield	Initial investment, technology adoption	Zambrano-Prado et al. (2021)
Wastewater treatment innovations	New technologies for treating wastewater	Municipal and industrial sectors	Enhanced treatment efficiency, resource recovery	Regulatory compliance, public acceptance	Kehrein et al. (2020)

clean water for drinking which causes a high rate of morbidity and mortality due to waterborne diseases (Qamar et al. 2022). Additionally, this study reveals that deterioration of water quality in the region leads to severe economic consequences as they involve costs of healthcare, water treatment, and lost productivity that run to billions of United States (US) dollars. These risks mean that WQM measures must be implemented to reduce such impacts and ensure provisions of the needed quality of water in the future.

On this background, emerging contaminants (ECs) present a new challenge in WQM since they cannot be dealt with by the most common forms of water treatment. These pollutants comprise drugs, cosmetics, and chemical products that find their way into water systems through releases from sewage and irrigation practices. According to the literature, a lot of groups of pharmaceuticals have been found at views up to 1,000 ng/L in water and it is suspected that they may cause some harm both to animals and to people (Ghai et al. 2024). Such pollutants can interfere with the endocrine function in lives which in turn affects reproductive behaviors and what you might call death. Additionally, despite reduced concentrations, the chronic impact of such chemicals on the overall quality of drinking water is not yet fully established and requires more investigation. The economic factor of dealing with such new pollutants is also costly, since ozone-based AOPs may be necessary to remove such compounds from water sources, which may add up to 30% of the overall cost of managing the water supply (Tripathi and Hussain 2022).

Hence, to cater to ECs and supply elaborate tools for WQM, the incorporation of modern sensing techniques is inevitable. Several technologies, including real-time water quality sensors, remote sensing, and data analytics can become tools that can be used to enhance the understanding of water quality parameters to make early and effective preventive measures. For instance, the institution of a real-time monitoring system may avert the response time of contamination by up to 50%, which means faster remediation (Fan et al. 2020). Besides, the application of data analysis can improve the knowledge of long-term changes in water quality and improve the modeling for better management strategies of related occasions in the future. Sophisticated technologies bring important and positive economic effects; it has been estimated that implementation of advanced monitoring systems yields 20% cost savings in water treatment operations because further treatment intensifications are not as often required (Ganiyu et al. 2022). On the whole, integration of the best practices of WQM, understanding of new threats concerning water quality, and application of the most efficient means of monitoring are the major factors in securing water supplies and serving the population's needs.

Water quality, therefore, can be described as the general processes, methods, and apparatus used to maintain the quality of water to acceptable standards. This management involves several factors, among them being the management of polluted water which is important in conserving public health as well as the environment. Siloam treatment technologies have also developed and advanced where there are different technologies to eliminate the different contaminants. For example, coagulation, sedimentation, and filtration are the routine steps in water treatment to eliminate suspended materials and pathogens. However, these methods may not be sufficient for the removal of new and complex compounds present in water such as pharmaceuticals and personal care products (PPCPs) which have been found to have adverse impacts on human health and aquatic life. New treatment processes that can remove these contaminants have, however, been invented, these include MF, AOPs, and biological treatment. Outcomes show that the efficiency of removing pollutants in MF can reach 99% for some of them, thus enhancing the quality of water (Titchou et al. 2021). These are sophisticated technologies capable of sparking off a raft of potential economic repercussions which include initial capital investment that costs from as low as 1.0 million dollars and as high as 5.0 million dollars for small to medium-sized treatment facilities, but the advantages include reduction in morbidity and mortality rates among the populace occasioned by water-borne diseases, or pollution of the physical surroundings by poisonous wastes (Sanders 2017).

Contaminant Removal Efficiency (CRE):

$$CRE(\%) = \left(\frac{C_i - C_f}{C_i} \times 100 \right)$$

where

C_i = Initial contaminant concentration (mg/L).

C_f = Final contaminant concentration after treatment (mg/L).

This equation calculates the percentage efficiency of a water treatment process in removing specific contaminants.

Assessing the efficiency of water treatment systems cannot be fully completed without the determination of the CRE. For example, regarding the Organic Pollutant Removal Efficiency, AOPs can yield a CRE of up to 99%, thus eradicating contaminants from 100.0 to 1.0 mg/L. Efficiencies of such high levels are required for pollutant treatment, especially the new-generation pollutants like pharmaceuticals that pose threats to humans and fish. Through establishing measurable indicators for the treatment effectiveness of wastewater, CRE contributes toward better decision-making in developing and implementing the WQM strategies, meeting the legal requirements on environmental pollution, and conserving public health.

Table 9.2 is a comprehensive comparison between a variety of water treatment technologies; the removal efficiency, operating cost, and the type of contaminant suitable have been summarized. Previous methods of treatment, whether suspended carrying or pathogen elimination, entail removal efficiencies of only 70 to 90% and have set costs of operations ranging between 0.5 million to 2.0 million. As for the low technology like sedimentation and coagulation, they exhibit a low removal efficiency of between 40 to 70% while the high technologies including the MF and AOPs show high efficiencies ranging from 90 to 99% at higher costs of operation, which can cost between 1.0 million to 5.0 million. The use of biological treatment methods also has an estimated removal efficiency of 80 to 95% and is best suited for nutrients and biodegradable organics. UV disinfection and RO are clearly shown in the table where UV disinfection reached the removal efficiency of over 95% for paths but has a higher operating cost than RO which similarly attained over 95% removal efficiency for salts.

Table 9.2 Comparison of water treatment technologies

Treatment technology	Removal efficiency (%)	Operational cost (USD)	Suitability for contaminants	References
Conventional treatment	70 to 90	0.5 to 2.0 million	Suspended solids, pathogens	Tzanakakis et al. (2020)
Membrane filtration (MF)	90 to 99	1.0 to 5.0 million	Microorganisms, organic pollutants	
Advanced oxidation processes (AOPs)	90 to 99	1.0 to 3.0 million	Organic pollutants, pharmaceuticals	Jha et al. (2021)
Biological treatment	80 to 95	0.3 to 1.0 million	Nutrients, biodegradable organics	Kumar et al. (2024)
Activated carbon filtration	70 to 90	0.5 to 2.0 million	Organic compounds, taste, and odor	Mikulčić et al. (2022)
Ultraviolet (UV) disinfection	99+	0.2 to 1.0 million	Pathogens, viruses	Chen et al. (2021)
Reverse osmosis (RO)	95 to 99	1.0 to 4.0 million	Salts, heavy metals, organic pollutants	Giannakoudakis et al. (2024)
Rainwater harvesting	50 to 80	1.0 to 0.5 million	Non-potable uses, stormwater management	Tzanakakis et al. (2020)
Desalination	95 to 99	3.0 to 10.0 million	Saline water, brackish water	

Moreover, rainwater management is another cheapest method of water conservation for non-potable use whereas water desalination is another important tool in managing saline water problems but at a higher cost. This comparison indicates that the choice of proper treatment technologies should be underlined by certain water quality issues and cost estimations.

9.4 Integrated Water Resource Management (IWRM)

Integrated Water Resource Management (IWRM) on the other hand is a communal approach to the perpetration of water resources in a sustainable manner, for purposes of needful, communal, and environmental benefits. IWRM principles, as recognized now, stress systemic thinking on water management and show that water is a limited and susceptible commodity without which life of all forms cannot be sustained. Another of the starting tenets of IWRM is the acknowledgment of the linkages between water resources and sectors of the economy such as agriculture, energy, and urban development. This has the effect of making many water management decisions inter-sectoral, which means that the effects on other sectors must be considered. For example, over-exploitation of water sources through agricultural practices impacts negatively the water supply to urban society, these problems, therefore, call for the implementation of integrated planning and management solutions (Talat 2021). In addition, IWRM seeks to balance the equitable use of water resources, challenging the cultures that lead to the wastage of this important natural resource. Research confirms that the use of IWRM principles requires that water usage efficiency increases with some efficiency gains ranging from 20 to 30% within the regions because of effective water management practices (Koech and Langat 2018).

Although there is disagreement on the core components of IWRM, different authors agree that stakeholder involvement is an important aspect of IWRM since it involves representatives of society in the decision-making process. Stakeholder management enables the coordination of many entities such as government departments, communities, business entities, and non-governmental organizations (NGOs). This teamwork is useful, especially in considering the multiple and often antithetical demands of various stakeholders encountered in this process. For instance, in water resource allocation, farmer's interest will be irrigation while consumers' interest will be drinking water supply. When more stakeholders are involved in planning and executing the plans on IWRM, then the chances of arriving at the best solutions will be more proper and more sustainable. Some research also shows that the involvement of active stakeholders in water management enhances success and improvement in the broad water sustainability goals in the region where water quality and availability increase in more than 60% of the cases studied (Chuenchum et al. 2024). Furthermore, the engagement of its stakeholders can also improve the legitimacy of decisions made concerning water management and escalate support and thus compliance levels on issues regarding the same.

Integrative approaches are central to IWRM strategies as the latter cannot be isolated from other sectors. Water resources management at its central part intersects with policies and practices in agriculture, energy, health, and urban planning. For example, the application of smart irrigation systems in agriculture can cut the use of water by half while improving production, this is one clear example of the government sector extending its benefits to other sectors (Panel 2018). Also, centralization of water and energy infrastructure finally fosters improved water management, for instance, treated wastewater can be used for cooling power plants, thus cutting water consumption. The multi-sectoral approach can promote a variety of benefits, and some studies reveal the potential for up to 20% in operational cost savings for utilities for integrated water-energy management (Mongird et al. 2023). In addition, the application of IWRM principles embraced through intersectoral cooperation can improve the coping capacity to climate change effects because it incorporates timely management measures that address the variation and unpredictability associated with weather issues. Therefore, to address water challenges sustainably; principles of IWRM, stakeholder involvement, and multi-sectionalism are crucial.

9.5 Problems in Practicing Integrated Water Resources Management (IWRM)

Integrated Water Resources Management (IWRM) involves many challenges that make it challenging to produce the needed results sustainably. Among the more compelling factors for consideration is the weakness of institutional and governance frameworks. In many regions, current water management arrangements are largely sectoral and bureaucratic with poorly coordinated institutions. This fragmentation can create duplicating issues of jobs, misunderstanding of roles by different stakeholders as well as conflicting roles. For instance, a study revealed that due to the absence of integrated policies, the issue led to a 25% rise in water supply costs due to mismanagement and poor utilization of resources in various urban zones (Koop et al. 2022). Also, inadequate legal frameworks slow the execution of water management laws and can greatly affect IWRM implementation.

A fourth activity is that another factor is socio-economic cleavage that exists in different societies. In many instances, women, children, the elderly, minorities, and people of color, as well as low-income and Indigenous populations are excluded from water control and access. This exclusion can result in unfair distribution of water and can worsen other inequalities that are already there. According to research, regions that show higher socio-economic differentiation are approximately 30% more likely to develop water-based conflicts, as groups work to control scarce water resources (Döring et al. 2024). In addition, there is little awareness and knowledge of the principles of resource management and water within communities, which is a major factor that strongly affects participation and is important for the implementation of IWRM initiatives. Whenever there is no informed stakeholder, collaborative management or the ability to practice sustainability reduces considerably.

Another formidable threat that fuels climate change is also a threat to IWRM implementation. As climate change continues to play its part and extreme weather events like floods and droughts become more frequent, intense water resource management is a flexibility exercise (Bhardwaj 2023). For instance, areas that depend on annual precipitation for water requirements are currently characterized by considerable fluctuations, and the water deficit and competition with other sectors are growing. Reports suggest that climate change might decrease the availability of water by 40% by 2050 as it continues to exert additional pressure on water-stressed systems (du Plessis 2023). Such stability means requiring adaptations to management operations that are challenging when there are strict frameworks in place.

9.6 Learned and Best Practices from Integrated Water Resource Management (IWRM) Implementation Cases

However, it also elaborated on how the principles of IWRM can be successfully implemented through examples of the best practices. Among them, it is possible to list the experience in the organization of IWRM in the metropolitan area of Rome, Italy where a stormwater management approach was applied. The study that was conducted in the peri-urban watershed established that the application of best management practices (BMPs) reduced water runoff from the area by 20% and enhanced the condition of water in water bodies in the area (Tota-Maharaj et al. 2024). This case presents an example of how incorporating various management practices can produce substantial environmental returns apart from handling issues of urbanization.

A similar success story has been established in Tehran, Iran where the application of spatial–temporal planning was applied to increase urban resilience to climate change. The study showed that if the principles of IWRM were integrated into the urban planning of the city, water resources management would be enhanced and hence the overall efficiency would be 15% (Koop et al. 2022). This approach was instrumental in increasing the availability of water and attained the other milestones of IWRM, such as community participation and stakeholder involvement.

Of all the IWRM sub-sectors, the best developmental outcomes have been realized in urban stormwater management in Australia. IWRM principles when implemented reduced flood occurrences by 30% and enhanced the water quality of urban streams according to comparative analysis with other cities in the world (Pokhrel et al. 2022). The above case establishes that integrating various sectors in water management increases resilience, hence creating sustainable approaches. Therefore, these case studies reveal the adaptive utility of IWRM in dealing with compounded water management issues and in the context of establishing closer links between water management and development. Customers should therefore emulate these regions to figure out the best solutions that would fit into the constituent's socio-economic and environmental setting.

Equation 1: IWRM Effectiveness Model

To quantify the effectiveness of IWRM, we can develop a model that incorporates various factors influencing water management outcomes. The IWRM effectiveness model can be expressed as follows:

$$\textbf{IWRM Effectiveness (E)} = (\mathbf{S} + \mathbf{C} + \mathbf{Q} + \mathbf{R})/\mathbf{T}$$

where

E = IWRM Effectiveness.

S = Stakeholder Engagement Index (measures the level of participation and collaboration among stakeholders).

C = Climate Resilience Score (assesses the adaptability of water management practices to climate change impacts).

Q = Water Quality Improvement (quantifies the enhancement in water quality due to implemented practices).

R = Resource Efficiency Ratio (evaluates the efficiency of water resource utilization).

T = Total Management Costs (includes all costs associated with implementing IWRM practices).

This capable equation represents an overall evaluation of IWRM productivity alongside the social, environmental, as well as economic scopes. Each of them can be quantified based on its indicators resulting from the analyses of case studies and theoretical premises. For example, where stakeholder engagement is the criterion for evaluation, surveys on the level of community involvement could be used; where climate resilience is being assessed, the levels of vulnerability to acts of climatic violence (extreme weather incidences) may be used (Radović and Iglesias 2019). It is possible to find the improvement in water quality calculated before and after the adoption of IWRM (Duc et al. 2021) and calculate the water resource use efficiency pre-and post the adoption of practices from IWRM (Timotewos and Barjenbruch 2024). Last of all, total management costs may be obtained from expenses connected to the implementation of IWRM programs.

The use of this model helps in the assessment of its performance in the manner that the weaknesses in the formulated strategies on IWRM are easily revealed, hence the formulation of a way forward on the strengths that are incorporated into the IWRM. This equation helps the stakeholders make appropriate decisions that may improve the sustainability and resiliency of water resources in his/her region.

9.7 Climate Change and Water Resources

Transformations in the earth's climate have been singled out as a critical threat to water supply and quality at the global level. According to the Intergovernmental Panel on Climate Change (IPCC), by 2050 absence or insufficient quality quantity of fresh water could affect 2.3 billion people mainly in arid and semi-arid areas (Gaaloul and Eslamian 2022). This scarcity results mainly from changes in precipitation, evaporation arising from rising temperatures, and increased occurrences of extreme weather conditions. The study pointed out that in the Middle East, there will be an increased consumption of water resources due to climate change and a decrease in their supply in the same region. In the agricultural sector, irrigation with excess nutrients and pest control chemicals combined with unsustainable use and management practices lead to high levels of water pollution, which in turn contributes more to difficulties of water availability (Mateo-Sagasta et al. 2018). The consequences of these changes thus affect human health and global biodiversity and threaten the stability of all economies with an impact on sectors that require a stable and fixed water supply, such as agriculture and industry.

Reacting to the impacts of climate change, adaptation measures of water are crucial. These strategies must be complex and contain the best practices as well as innovative practices. For instance, the adoption of IWRM frameworks improves the chances of problem-solving by addressing all the circles because it takes physical, economic, environmental, and social considerations into account. Like previous work, Roy et al. (2023) underlined the significance of stakeholder participation in building sound IWRM frameworks and approaches pointing to the fact that collective work rendered sustainable results. On the same note, harvesting of rainwater systems has also been recommended as part of adaptation measures, especially in urban settings. A cross-sectional field study by Kuntz Maykot and Ghisi (2020) and Baptista et al. (2023) in Brazil concluded that rainwater harvesting systems could potentially save 50% of potable water consumption, with shower usage being the maximum at 54.2% with washing machines next at 21.3%. It not only relieves the demand for current water sources but also brings about the other advantages of a smaller amount to be paid on water bills together with the extra option of subsidy for having it fitted. Additionally, increasing the degree of climate cop management about agricultural practices by implementing a climate-smart agriculture approach may reduce the adverse effects of climate variability on yields thereby addressing food and water security for vulnerable groups.

It can therefore be concluded that there is no aspect to improving climate resilience where the application of technology is not important. Climate change impacts challenge us to enhance the means for observing systems, control mechanisms, and systems for allocating water resources enhanced by technology. For instance, that option, knowledge nourished by remote sensing and GIS, helps inform initiatives related to water supply and quality with much-needed data. Further, stormwater management and modeling tools like the Storm Water Management Model (SWMM) enable to modeling of different management practices for their hydrological and water quality impacts (Ramovha et al. 2024). It can put a finger on the choices, which is useful in lessening the impact of climate change on water resources. Furthermore, the implementation of smart water management systems—IoT technologies applied to water networks—can also augment the effectiveness of water supply networks and lessen leakage levels and delays in service delivery. According to Giri (2021), balanced schemes of reservoir management can enable the ecological quality of the sections of the river passing through urban areas proving that the use of technology can help in addressing both quantity and quality of water problems. With these technologies being put in apposition, new stakeholders can design improved water management frameworks capable of protecting water supplies against climatic volatilities.

9.8 Policy Responses to Climate Change

The more recent experiences of climate change impacts on water resources have meant that policy interventions need to be strong and able to address both current and future challenges. Policymakers should come to realize that water resources management issues are complex and are worsened by climate fluctuations, population pressure, and urbanization. That is why policy frameworks should come up with policies for improving water resilience based on scientific research and the experience of local people. For example, the use of IWRM is important as it utilizes a sector-wide and scale approach. It has then helped enhance the available water quality and availability and at the same time has had a minimal impact on water users' conflicts. And where demographic changes are most pronounced, as in the Mediterranean Basin where warming has been even more intense than the averages globally, IWRM can assist with the emerging battles for water scarcity (Mueller et al. 2021).

Also, policies which also promote the use of water from other sources need to be put into operation. The use of non-conventional sources of water that have been treated domestically or otherwise, and rainwater can go a long way in easing the burden on fresh water. Investment in improved wastewater treatment technologies has made it possible for water to be reused in developed countries, hence balancing usage and cutting the competition (Obaideen et al. 2022). For instance, rainwater harvesting in buildings cuts potable water use by 50% in urban centers, creating both cost and environmental savings. Also, Information Technology for Agriculture (ITA) which the promotion of Climate-Smart Agriculture (CSA) can boost water use efficiency in farming is important because globally, agriculture is the leading consumer of freshwater at 70% (Michel-Villarreal et al. 2019). There is, therefore, the need to maximize funding in basic infrastructure for the practice of these measures in a manner that makes them easily available to as many players as possible, especially in needy areas.

In addition, mitigation measures for climate change require that policies to reverse the deterioration of water quality should form part of the response strategies. This pollution caused by agricultural drainages, industrial effluents, and increased urbanization is generally a danger to human beings and aquatic organisms. For example, analysis of the nutrient density of soil has revealed that the average nitrate status in control fields where conventional agriculture practice is used is almost double the levels found in organic farming (Montgomery and Biklé 2021). They urged the policymakers to strengthen the policy regarding nutrient application and encourage practicable farming techniques without adverse effects on the environment. In the same regard, public awareness campaigns are very important since they inform the needed changes and give out information on how communities can protect the quality of water.

Table 9.3 presents a detailed consideration of the effects of climate change on different aspects of the water resource system and shows that climate change is a complex issue for water resource systems and different regions. Water resources, water quality, agriculture, and ecosystems are highlighted as important areas of focus with the table showing projections for change and the impacts. For example, it is expected that water accessibility will be lowered significantly and most severely in the Mediterranean Basin where the future population of 2.3 billion people may have no access to freshwater resources by 2050. Furthermore, there is still much scope for better practice of water use inside urban areas; rainwater harvesting and wastewater reuse can greatly reduce per capita use of potable water. There is, therefore, a need for IWRM as one of the important approaches to address vulnerabilities resulting from climate change impacts, improve water quality, and avoid conflict among users. In summary, the assessment draws attention to the particular importance of developing climate change-responsive approaches and raising public consciousness on water conservation.

9.9 The Nexus Between Water, Food, and Animal Sciences

Irrigation is a vital component of agriculture and animal husbandry, and world consumption of fresh water is estimated to be 70% used by agriculture (Singh 2021). With water-saving techniques such as drip irrigation, the water usage rate is enhanced to 90%, and water wastage is cut by 30 to 40% (Lakhiar et al. 2024). Animal husbandry is also vulnerable since it uses 15,400 L of water per kilogram of beef; water quality affects production; contamination leads to the reduction of milk output by 40% (Pexas and Kyriazakis 2023). The directed use of rainwater and treated wastewater in agriculture reduces the water demand by 25 to 50% and helps to provide food and animal feed (Al-Hazmi et al. 2023).

- **Ensuring Sustainability in Agriculture and Livestock Through Water Management**

Water resource management has become central in determining both food production for human consumption, and overall animal health. Irrigation alone uses about 70% of freshwater globally while agriculture utilizes about 40% of the global food production (Wang et al. 2021; Khondoker et al. 2023). Sophisticated methods of irrigation have been developed and practiced including drips irrigation and sprinkler systems that may reduce water wastage from irrigation to its minimum of between 8 to 10% (Chauhdary et al. 2023). For instance, smart usage that entails the installation of precision irrigation technology can cut water use by 30 to 40% but raise crop yields by up to 20% in arid and semi-arid climates (Bwambale et al. 2022). Likewise, on crops, traits are available that can potentially reduce water needs by a quarter while not compromising on yield but are represented by drought-resistant varieties of maize. Water quality also affects agricultural yield directly because heavy metals or excessive salt concentrations can decrease germination by 15 to 50% according to crop type. SWM, therefore, is essential to feed the growing population of the world, estimated to be 9.7 billion in 2050 (Issaoui et al. 2022).

Other sectors such as farming and live stocks which form a base of food systems also wholly rely on both water availability and quality. Often, at least 15,400 L

Table 9.3 Climate change impact assessment

Impact area	Description of impact	Affected regions	Projected changes	References
Water availability	Decrease in streamflow and changes in water balance components due to climate change	Mediterranean Basin	Significant reductions in water availability by 2050, affecting 2.3 billion people	Gaaloul et al. (2021)
Water quality	Increased nutrient runoff leads to higher nitrate levels in conventional agriculture	Global, particularly in agricultural regions	Nitrate levels are nearly double in conventional fields compared to organic fields	Harrison et al. (2019)
Agricultural water use	Increased water use efficiency in crops due to elevated CO_2 levels	Global	Potential increases in water use efficiency may be negated by severe drought conditions	Lenka et al. (2021)
Drought frequency	Increased frequency and intensity of droughts affecting water supply and agriculture	Mediterranean Basin, arid regions	Higher frequency of droughts projected, leading to significant water scarcity	Tanarhte et al. (2024)
Urban water management	Need for improved urban water management practices to cope with climate impacts	Urban areas globally	Adoption of rainwater harvesting and wastewater reuse systems to reduce potable water consumption by up to 50%	Canteiro et al. (2024)
Ecosystem health	Degradation of aquatic ecosystems due to altered hydrological cycles	Global	Increased ecological degradation risks, impacting biodiversity, and ecosystem services	Ekka et al. (2023)
Policy and governance	Need for IWRM to address climate challenges	Global	Implementation of IWRM can improve water quality and availability while reducing conflicts among users	Nagata et al. (2022)

(continued)

Table 9.3 (continued)

Impact area	Description of impact	Affected regions	Projected changes	References
Climate adaptation strategies	Adoption of climate-smart agricultural practices to enhance resilience	Agricultural regions globally	Shift toward more water-efficient irrigation methods, reducing water use by up to 50%	Kumar et al. (2023)
Public awareness	The importance of educating communities on water conservation and quality protection	Global	Increased public engagement in water management practices to protect water resources	Siddique (2021)

of water are used per kilogram of beef, and many of them are spent on cultivating feedstock such as maize and alfalfa (Rodrigues Junior and Dziedzic 2021). Pathogens, pesticides, or heavy metals in water decrease the productiveness of livestock by 20 to 30% of cases of severe diseases such as bovine tuberculosis or reproductive disorders (Scialabba 2021; Bhardwaj et al. 2019, 2023; Bhardwaj and Jindal 2020, 2022; Alam et al. 2023). Additionally, the lack of clean water supply affects the production of milk from dairy cows by an average of 40% as a result showing the relationship between water and animal husbandry (Golher et al. 2021). Modern approaches to intermediate technologies and rainwater harvesting systems in combination with livestock farms save 30 to 50% of the water and drive young people to overuse groundwater resources (Korlakunta 2022). From the latest examples of Indian states, it has been observed that blending treated wastewater in animal feed crop aquifers, saves 25% of fresh water for agricultural use, both environmental and food security purposes (Lahlou et al. 2021). Such integrated approaches therefore stress the need for synchronizing water food and animal sciences to realize a sustainable agricultural future.

9.10 Water Scarcity and Demand Management

Available literature shows that water scarcity is a growing concern in the world today owing to the phenomenon of supply demand equality, which is preconditioned by various causes including population increase, climatic conditions, as well as poor water resource use. Projections on water scarcity show that by 2025, 1.8 billion people could be living in areas characterized by absolute water scarcity and 2/3 of the world population shall be under water-stressed conditions (Tinazzi 2024). This is more evident in areas that are characterized by low water availability and in areas where this water is fully exploited. For instance, and more closely related to our topic, the Mediterranean Basin has seen a rate of warming that is much higher than the global means and this has affected the availability of water and the frequency of droughts. Climate change expects to increase hydroclimatic cycles of water evaporation together with changes in precipitation that may also complicate the global water shortage challenges.

Water shortage not only implies the absence of resources but also has economic, social, and even environmental effects. Other factors made worse by this include water pollution and nutrient runoff that also causes difficulties especially when it comes to agriculture because of conventional methods that enhance high nitrate. There is a clear suggestion of a nitrate risk to drinking water and public health when conventional farming techniques are adopted since it was revealed in the research that the nitrate concentrations can be as high as double in conventional fields as compared to organic fields. Additionally, the channels of the fast pace of urbanization and industrialization influence rivers and their ecological services, reducing them in the process. For instance, Chinese urban reaches of river systems have suffered elevated levels of ecological deterioration and require appropriate reservoir management to improve ecological conditions (Ai et al. 2022). Therefore, it is important to establish a more effective and broader definition of water scarcity as a condition that refers not only to the amount of water that people can get but also to the quality of that water as well as the social and economic conditions of its use.

- **Demand Management Strategies**

This chapter concludes that demand management is critically important to adequately deal with water scarcity issues and achieve rational use of water. These strategies include measures that seek to realize high efficiency in water usage and improvements to values in different sectors like agriculture, industries, and homes. The most prospective of them is IWRM whose urgent implementation is aimed at the integrated development and management of water, land, and related resources. This situation is likely to be addressed by IWRM since conflicts between water users and overall water quality and availability can be improved. For example, the implementation of rainwater harvesting and wastewater reuse systems reduceable potable water demands the latter with some research work pointing to 50% of urban usage as achievable (Ritter et al. 2002).

Apart from IWRM, water use efficiency requires the use of improved water technologies and other practices in the management of water resources. Since the farming system is a primary contributor to water losses, there is a need to adapt to climate-smart systems to enhance the use of irrigation systems. Despite the use of water-efficient crops and planting time, the surveys suggest that the adoption of a more efficient irrigation system can save water by up to 30% while at the same time improving yields (Lakhiar et al. 2024). In addition, general advertising and promotional measures for the province suggest that practical steps to ensure water conservation are being taken with a specific consumer focus. The outcomes that resulted from carrying out the interventions and achieving the above goals are the establishment of a water stewardship culture whereby through their effort and support the communities can continue to minimize water use and strengthen water resources in managing climate change challenges.

- **The Use of Pricing and Economic Instruments**

Pricing and other economic instruments have been found to play a central role in the demand management of water. The use of proper price stimuli should therefore lead to efficient usage of water and the general promotion of preservation by consumers. For example, flexible water tariffs, in which water usage costs more with every increase in units consumed, can bring about behavior changes that lead to rational use of water. Water trading and other market-based mechanisms can also be used in determining how water should be used to achieve the biggest results in terms of efficiency. In areas with scarce water supply, it becomes convenient to manage the water supply through market-based approaches due to the flexibility offered to reorganize the supply based on certain demands if they alter.

Moreover, there are requirements in water infrastructure and technology that help to solve the problem of water supply and quality. Several studies in the economics of water supply and sanitation have estimated an economic rate of return of about $4 for each $1 invested. This has further stressed the need to devote adequate attention to water investments in the continuity of sustainable human development and economic growth (Daloğlu Çetinkaya et al. 2023). At the same time, the consideration of environmental factors in water pricing will ensure the protection of the environment and SWM services. When decision-makers consider water beyond a mere product but as a social asset important for human and other living organisms' use, an improved framework for managing water is possible.

- **Technological Advances for Water Saving**

In the following research, technology has been identified as central to dealing with water shortage and improving water utilization. Better water use efficiency can be achieved through adopting new Information and Communication Technology in the water management systems because the agricultural sector is the largest user of freshwater resources. For example, the use of efficient water application methods like drip and sprinkler has been found to use water by as little as 30% while keeping yields on the same level. Such systems enable precise or uniform water delivery thus reducing evaporation and runoff losses, something that is useful where water is scarce. Moreover, the level idea is to incorporate smart water management technologies, which include sensors and data analysis, to improve water supply and consumption in a city implying still greater water use efficiency. Researchers suggest that it is possible to save about 20% of water supply through innovative technologies in smart cities in specific areas such as leak detection and usage pattern monitoring (Adedeji et al. 2022).

In addition, the adoption of water resources other than those from conventional sources like rainwater harvesting and wastewater reuse is seemingly undergoing the process of finding a permanent solution to the scarcity of water. Rainwater harvesting systems provide storage for rainwater for applications other than drinking; hence, reducing the amount of water that would have been drawn from the ordinary source. Studies have shown that rainwater harvesting when practiced together with other practices can reduce the use of potentially potable water in urban areas by up to

50%. In the same way, the use of wastewater to produce crops and for industrial purposes takes a lot of pressure off fresh water and can contribute to environmental sustainability. In many cases, a new wastewater treatment technology was found to improve the quality of the reclaimed water thereby making it useful in different areas. Where water scarcity remains severely manifested, for example, in the Middle East and North Africa, wastewater recycling needs to be stepped up to satisfy the demand for water and support economic growth.

These technological solutions have a large economic impact. Efficiencies in water investment remain promising, as each dollar channel led toward water-efficiency solutions is regarded to provide economic returns of approximately $4 (Koseoglu-Imer et al. 2023). Such a return on investment is more apparent in agricultural industries where the application of proper methods of irrigation brings more productivity of crops as well as cuts contrary expenses. In addition, water-efficient technologies can help to increase the climate change vulnerability of communities by making sure there is a constant demand for water even as there is fluctuating and unpredictable rainfall. Therefore, the encouragement of technology for application in water demand management is a strategic tool for effective demand management but on the same note, enables economic growth and sustainability of the environment.

Table 9.4 summarizes numerous water demand management strategies based on the description, advantages, and supporting literature. They include precision irrigation and rainwater harvesting, wastewater reuse, and smart water management systems which all come in handy in improving water conservation and efficiency. The advantages include the effectiveness of these practices, which include up to 50% savings in water use in urban areas through harvested rainwater and 30% savings in agricultural water use, mostly through precise irrigation. Other relevant strategies that are also called for in enhancing the community attitude needed for the efficient preservation of water include community participation and public education. Incorporation of monetary rewards increases people's interest in taking part in water conservation measures; the co-management systems improve the functioning of the measures that are put in place. Thus, altogether these techniques comprise a complex approach toward solving the topical problems of water deficit and stimulation of rational water utilization.

9.11 Innovative Financing for Water Projects

Fresh water has become a scarce resource in the current world, especially due to concerns like population pressure, increase in urban populations, and climate change. Due to population growth the global population is expected to reach 9.7 billion by 2050, with a major chunk of the population living in urban areas, there will be great demand for freshwater resources (Falcon et al. 2022). This position calls for creative methods of financing to promote sustainable management and the development of water systems. Different kinds of financing mechanisms such as public–private partnerships (PPPs), crowdfunding, and international funds play a crucial role in helping

Table 9.4 Water demand management techniques

Technique	Description	Benefits	References
Precision irrigation	Utilizes advanced technologies to deliver water directly to plant roots	Reduces water consumption by up to 30% while maintaining crop yields	Lakhiar et al. (2024)
Rainwater harvesting	Captures and stores rainwater for non-potable uses	Can reduce potable water consumption by up to 50% in urban settings	Lepcha et al. (2024)
Wastewater reuse	Treats and reuses wastewater for agricultural and industrial purposes	Alleviates pressure on freshwater sources and promotes sustainability	Popoola et al. (2024)
Smart water management systems	Employs sensors and data analytics for real-time monitoring of water use	Achieves water savings of approximately 20% through leak detection	Joseph et al. (2022)
Community engagement programs	Involves local stakeholders in water conservation initiatives	This can lead to reductions in water use of up to 25% through behavioral change	WARD et al. (2021)
Economic incentives	Provides subsidies and tax breaks for water-efficient technologies	Motivates community participation in water conservation efforts	Sohu et al. (2024)
Technological innovations	Introducing new technologies to improve water use efficiency	Enhances resilience to climate change impacts and reduces operational costs	Ahmed et al. (2023)
Public awareness campaigns	Educates the public on the importance of water conservation	Encourages simple actions that collectively lead to substantial water savings	Emile et al. (2022)
Collaborative water management	Engage community members in decision-making processes	Improves governance and compliance with conservation measures	Egbemhenghe et al. (2023)

arrangements of funds for water projects. For example, the World Bank has identified an investment need of roughly $114 billion every year to address the water and sanitation needs of the developing world only (Medina et al. 2021).

As a result of these challenges, PPPs have been seen as an option for financing water projects due to harnessing the best features of both the public and private sectors for improved service delivery and infrastructure development. This kind of private participation can harness the efficiency and innovation of the private sector while PPPs can maintain the public value of accountability and equity. A good example is a contractual relationship with the government of Ghana to facilitate the development of better water provision in urban areas; the use of water kiosks created by a private firm led to the percentage rise in the population that has access to clean water in Ghana from 32 to 62% within a space of two years (Akelo 2023). Nevertheless, for PPPs to be effective, there is a need for an efficient legal and accountable framework that will safeguard all investors. The task relations within the PPPs that cause substantial discussion such as the risks, contract management, and necessity to achieve the optimal results in water financing must be considered and managed. Moreover, risk management techniques must be incorporated to manage the vagueness of content on the part of project implementation up to the achievement of the targeted financial outcomes as well as to guarantee that both the public and private sectors will be able to reap advantages out of these partnerships.

Besides PPPs, there is crowdfunding and community financing that have lately been fashionable to support water projects, especially in unserved areas. These models allow local communities to assume full ownership of water management programs and projects. For example, a campaign in India recently received more than $1 million to finance rainwater harvesting projects at more than 10 thousand rural households (Aayog 2019). Besides funding, this approach also improves the level of community participation and sensitization on the use of water. However, the problems of crowdfunding are still about the scalability of the initiatives as most labor is based on necessary funding. Thus, for cooperation with development policy and credit-based financing models of sustainable development, the most crucial aspect is the framework that supports people's participation and rewards their efforts.

Foreign institutions including multilateral development banks and NGOs are the most significant source of financing for water projects, especially in developing countries with emerging economies. These organizations offer funds in the form of grants and loans, as well as technical assistance toward the achievement of their amended goals of "Clean Water for the World." For example, the Asian Development Bank presently supports more than $2 billion in water supply and sanitation projects in the Asia–Pacific region to supply most of the population with safe drinking water. Nonetheless, the utility of international funding is not optimal due to issues; formalities and all the cash do not meet the local requirements. Hence, there is a need to encourage partnerships between international organizations and communities to ensure that funds are utilized to social challenges following the needs of the communities. Also, there is virtue in strengthening the role of stakeholders in the formulation of policies that relate to the management of water, to ensure that the existing governance structures and regulations aid in the sustainable achievement of effective water management goals.

9.12 Community-Based Water Management

In community-based water management, the community manages the water resources of a specific area or locality. Engaging local communities can provide richer insights, knowledge, and governance into the problems facing the water sector. Hence, the relevance of community-based water management is in reinforcing the capacity to manage water resources within the context and customarily by the concerned communities. Because water is becoming scarce due to urbanization and climate change, community-based water management is a progressive strategy to alter the distribution of water. There is hesitancy to change some historical management systems although countries do require improved water management in 2015, the United Nations (UN) reported that over 2 billion people around the world were living in areas with high water stress (Vanham et al. 2021). Therefore, adopting local knowledge and practices of community-based water management empowers people to own the program and strengthens the systems to resist outside influence.

9.12.1 Examples of Community Mobilization Programmed

Many examples can be listed to prove that community-oriented approaches work well in water management. For instance, in rural areas in India, the program called Jal Samvardhan has turned out effective in mobilizing communities for effective management of water resources in terms of rainwater harvesting and watershed management (Maiorano 2024). They have contributed through this initiative to a 30% replenishment of the water tables and access to drinking water for over 10,000 households (Bharti 2023). Likewise in Brazil, the "Agua para Todos" has also created a forum for communities to engage in the planning and delivery of water supply systems and has curbed 25% of water-borne diseases in the stipulated areas (Paredes and Woolley 2022). The examples given above illustrate how safe, sustainable, and equitable exploitation of water resources might be realized by actualizing the capacity of community-based water management to tackle local water concerns proactively and in supporting health and wellbeing. In addition, by combining traditional approaches with contemporary ones, there was an improvement in water resource reliability among the indigenous people whose water systems have been developed for several years now.

9.12.2 *Difficulties Experiences by Communes*

However, several challenges constrain the effectiveness of community-based water management although improving access to water through community-based water management has potential benefits. A more critical concern pertains to inadequate financial capital and technological skills in terms of the selection, installation, and management of water systems among communities. A cross-sectional survey carried out in Africa's rural region indicated that more than 60% of all the water point systems run by the communities were defunct due to poor funding, and absolutely no backing from technical personnel (Bishoge 2021). Further, there are aggravating factors that are outside the control of the local authorities but still impose pressure on water resources, which contributes to the scarcity of water and adds to the difficulties of their management. Water resource issues and conflicts for water usage by farm, industrial, and urban users add to the equation, and in many cases, disempower communities. To successfully meet these challenges, there is a need to enhance the capacities of local communities by providing them with requisite training, and products, and enabling appropriate policies that support community management initiatives.

9.13 Trends in New Water Management

With the continuing deterioration of water supply worldwide, new technologies can bring major changes to water resource management. Technologies like smart water meters, remote sensing, and GIS have found their way into water monitoring and management in most organizations. For instance, smart water meters would help the community get an instant view of how the quantities of water are being utilized such that everyone would see where there is wastage or leakage. Smart water meters will be implemented in homes in England starting with a pilot project in California that proved the usage of smart meters reduced consumption by 15% (Saleh 2021). In addition, remote sensing can also be used to detect the quality and availability of water hence assisting decision-making makers and managers in the communities. Some of these technologies improve the utilization of water resources and provide the communities with the right measures that should be taken for their water resources.

9.13.1 *Opportunities Regarding Artificial Intelligence (AI) and Machine Learning (ML)*

Machine Learning (ML) and Artificial Intelligence (AI) innovations are rapidly gaining a spot in the water management domain. These technologies can also entail analyzing large data sets to make patterns and predictions and therefore, water demand as well as the resources to be allocated. For instance, the use of AI algorithms can predict water usage based on previous history hence communities will be able to prepare for future uses adequately. According to a study conducted in Singapore, the use of AI-driven predictive models enhances Water Supply Management by 20% (Adedeji et al. 2022). Furthermore, it can improve the assessment of water quality using the data from sensors in real time; it makes it possible to respond to different episodes of contamination. The application of these technologies in community-based water management systems shows good potential in enhancing sustainability and resilience as these technologies advance.

9.13.2 New and Creative Ways of Reusing Water

Technological advancements in water recycling and reuse remain important in the management of water-scarce situations, especially for urban users. Modern methods such as membrane bioreactor (MBR) and RO make it safe to use the wastewater for different purposes including watering crops or for industrial uses. For instance, Los Angeles has developed a wastewater recycling program where over 100 million gallons of wastewater are recycled daily and used instead of imported water (Alley and Alley 2022). In addition, there has been a slow rise in the adoption of greywater recycling by community-based organizations for water conservation and use for irrigation because it saves potable water while tending parts or full lawns. Like many innovations, these problems do not solely solve the stress laid on the freshwater sources but also support the sustainable management of water resources.

9.14 Recommendations

Regarding the complex problem of SWM, the promotion of new approaches and technology, as well as the application of a systemic approach, is imperative. It is suggested that stakeholders invest more in better water treatment technology and smart sensors to monitor and manage water resources better and more effectively. Both the governments and the private sectors need to work hand in hand in policy formulation for the utilization of rainwater, re-use of wastewater, and coming up with climate-smart farming techniques. The following are also important in the public domain: Awareness and enlightenment programs also play an important role in getting the public involved in conservation programs and ultimately encouraging sustainable use of water resources. Moreover, it will be pertinent to enhance collaboration with the international community and share experiences because the fast adoption of the utmost effective strategies and innovative approaches that match various regional difficulties will be possible then. By so doing, there will be strong evidence of water management systems that are sustainable both in the natural environment and in other socio-economic aspects.

9.15 Conclusions

The future of water management can be best understood concerning modern technologies such as smart sensors, AI, complication water reuse technologies, and systems that augment the modicum and optimize water supply more efficiently. The use of these advanced technologies, which enables control, monitoring, and even predictive analysis, for water leakage and the quality of water supplied to the communities, is greatly minimized. As the trend of IWRM advances globally, it is possible to focus on the involvement of stakeholders to overcome the complex problem associated with water deficiency and climate change. Expanding corporate investments in R&D with affordable products for water treatment and increasing public awareness of water conservation can go a long way toward the development of a secure and sustainable future for water, which would be capable of meeting future demands for water among a rising population, and at the same time protecting extensive necessary ecosystems resources for future generations.

Acknowledgements The author, Qudrat Ullah, sincerely acknowledges my co-authors Muhammad Waqar, Nimra Sajjad, Muhammad Qasim, and Thanet Khomphet for their valuable contributions and collaboration. I appreciate the support from Walailak University, Deakin University, and Government College University Faisalabad in facilitating this Book Chapter (Emerging Trends and Challenges in Technological Innovation for Sustainable Water Management, which is to be published in the book Emerging Contaminants in the Aquatic Environment: Global Overview). Special thanks to the editors for their guidance and to Springer Nature Singapore Pvt. Ltd. for this publication opportunity. This work reflects our collective effort toward Sustainable Water Management innovations.

References

Aayog NITI (2019) Composite water management index. Government of India, New Delhi, p 21. .http://social.niti.gov.in/uploads/sample/water_index_report2.pdf

Acharya BS, Bhandari M, Bandini F, Pizarro A, Perks M, Joshi DR, Wang S, Dogwiler T, Ray RL, Kharel G, Sharma S (2021) Unmanned aerial vehicles in hydrology and water management: applications, challenges, and perspectives. Water Resour Res 57(11):e2021WR029925

Adedeji KB, Ponnle AA, Abu-Mahfouz AM, Kurien AM (2022) Towards digitalization of water supply systems for sustainable smart city development—Water 4.0. Appl Sci 12(18):9174

Ahmed Z, Gui D, Murtaza G, Yunfei L, Ali S (2023) An overview of smart irrigation management for improving water productivity under climate change in drylands. Agron 13(8):2113

Ai Y, Ma Z, Xie X, Huang T, Cheng H (2022) Optimization of ecological reservoir operation rules for a northern river in China: balancing ecological and socio-economic water use. Ecol Ind 138:108822

Akelo JO (2023) Influence of household water use characteristics, technical and governance factors on financial sustainability of community-managed borehole water kiosk service in Kisumu County, Kenya. Doctoral dissertation, St. Paul's University

Alam S, Bhardwaj LK, Mallick R, Rai S (2023) Estimation of heavy metals and fluoride ion in vegetables grown nearby the stretch of River Yamuna, Delhi (NCR), India. Indian J Environ Prot 43(1):64–73

Al-Hazmi HE, Mohammadi A, Hejna A, Majtacz J, Esmaeili A, Habibzadeh S, Saeb MR, Badawi M, Lima EC, Mąkinia J (2023) Wastewater treatment for reuse in agriculture: prospects and challenges. Environ Res 116711

Alley WM, Alley R (2022) The water recycling revolution: tapping into the future. Rowman & Littlefield

Alsaeed BS, Hunt DV, Sharifi S (2022) Sustainable water resources management assessment frameworks (SWRM-AF) for arid and semi-arid regions: a systematic review. Sustain 14(22):15293

Ayaz M, Ammad-Uddin M, Sharif Z, Mansour A, Aggoune EHM (2019) Internet-of-things (IoT)-based smart agriculture: Toward making the fields talk. IEEE Access 7:129551–129583

Ayaz M, Namazi MA, ud Din MA, Ershath MM, Mansour A (2022) Sustainable seawater desalination: Current status, environmental implications and future expectations. Desalination 540:116022

Baptista VS, Coelho VHR, Bertrand GF, da Silva GB, Caicedo NO, Montenegro SMG, Stefan C, Glass J, Heim R, Conrad A, Almeida CDN (2023) Rooftop water harvesting for managed aquifer recharge and flood mitigation in tropical cities: towards a strategy of co-benefit evaluations in João Pessoa, northeast Brazil. J Environ Manage 342:118034

Bhardwaj LK, Jindal T (2020) Persistent organic pollutants in lakes of Grovnes Peninsula at Larsemann Hill area, East Antarctica. Earth Syst Environ 4(2):349–358

Bhardwaj L, Sharma S, Ranjan A, Jindal T (2019) Persistent organic pollutants in lakes of Broknes peninsula at Larsemann Hills area, East Antarctica. Ecotoxicol 28:589–596

Bhardwaj LK, Sharma S, Jindal T (2023) Estimation of physico-chemical and heavy metals in the lakes of Grovnes & Broknes Peninsula, Larsemann Hill, East Antarctica. Chemistry Africa 6(5):2677–2694

Bhardwaj LK, Jindal T (2022) Evaluation of coliform and Faecal coliform bacteria in the lakes of Broknes and Grovnes Peninsula, East Antarctica. Nat Environ & Pollut Technol 21

Bhardwaj LK (2023) A comprehensive review on the climate change and its impact on health

Bharti DDDS (2023) Global consequences of ground water scarcity and replenishment prospects at the end of current millennium by 3000 AD. Remit Rev 8(4)

Bishoge OK (2021) Challenges facing sustainable water supply, sanitation and hygiene achievement in urban areas in sub-Saharan Africa. Local Environ 26(7):893–907

Boyle C, Ryan G, Bhandari P, Law KM, Gong J, Creighton D (2022) Digital transformation in water organizations. J Water Resour Plan Manag 148(7):03122001

Bwambale E, Abagale FK (2022) Smart irrigation monitoring and control. Encyclopedia of smart agriculture technologies. Springer International Publishing, Cham, pp 1–7

Canteiro M, Cotler H, Mazari-Hiriart M, Babinet N, Maass M (2024) Natural dynamics and watershed approach incorporation in urban water management: a scoping review. PLoS ONE 19(8):e0309239

Chauhdary JN, Li H, Jiang Y, Pan X, Hussain Z, Javaid M, Rizwan M (2023) Advances in Sprinkler irrigation: a review in the context of precision irrigation for crop production. Agron 14(1):47

Chen C, Guo L, Yang Y, Oguma K, Hou LA (2021) Comparative effectiveness of membrane technologies and disinfection methods for virus elimination in water: a review. Sci Total Environ 801:149678

Chuenchum P, Meneesrikum C, Teerapanuchaikul C, Sriariyawat A (2024) Community participation and effective water management: a study on water user organizations (WUOs) in Thailand. World Dev Perspect 34:100589

Daloğlu Çetinkaya I, Yazar M, Kılınç S, Güven B (2023) Urban climate resilience and water insecurity: future scenarios of water supply and demand in Istanbul. Urban Water J 20(10):1336–1347

Daminato C, Diaz-Farina E, Filippini M, Padrón-Fumero N (2021) The impact of smart meters on residential water consumption: evidence from a natural experiment in the Canary Islands. Resour Energy Econ 64:101221

Döring S, Kim K, Swain A (2024) Integrating socio-hydrology and peace and conflict research. J Hydrol 131000 du Plessis A (2023) Water resources from a global perspective. South Africa's Water predicament: Freshwater's unceasing decline. Springer International Publishing, Cham, pp 1–25

Duc NH, Avtar R, Kumar P, Lan PP (2021) Scenario-based numerical simulation to predict future water quality for developing robust water management plan: a case study from the Hau river, Vietnam. Mitig Adapt Strat Glob Chang 26:1–38

Egbemhenghe AU, Ojeyemi T, Iwuozor KO, Emenike EC, Ogunsanya TI, Anidiobi SU, Adeniyi AG (2023) Revolutionizing water treatment, conservation, and management: harnessing the power of AI-driven ChatGPT solutions. Environ Chall 13:100782

Ekka P, Patra S, Upreti M, Kumar G, Kumar A, Saikia P (2023) Land degradation and its impacts on biodiversity and ecosystem services. Land Environ Manag For 77–101

Emile R, Clammer JR, Jayaswal P, Sharma P (2022) Addressing water scarcity in developing country contexts: a socio-cultural approach. Humlties Soc Sci Commun 9(1):1–10

Falcon WP, Naylor RL, Shankar ND (2022) Rethinking global food demand for 2050. Popul Dev Rev 48(4):921–957

Fan Y, Xu Z, Huang Y, Wang T, Zheng S, DePasquale A, Brüeckner C, Lei Y, Li B (2020) Long-term continuous and real-time in situ monitoring of Pb (II) toxic contaminants in wastewater using solid-state ion selective membrane (S-ISM) Pb and pH auto-correction assembly. J Hazard Mater 400:123299

Freitas LDAA, Radis-Baptista G (2021) Pharmaceutical pollution and disposal of expired, unused, and unwanted medicines in the Brazilian context. J Xenobiotics 11(2):61–76

Gaaloul N, Eslamian SAEID, Katlance R (2021) Impacts of climate change and water resources management in the southern mediterranean countries. Water Prod J 1(1):51–72

Gaaloul N, Eslamian S (2022) Groundwater quality in arid environments. Clean Water Sanit 260–272

Ganiyu SO, Sable S, El-Din MG (2022) Advanced oxidation processes for the degradation of dissolved organics in produced water: a review of process performance, degradation kinetics and pathway. Chem Eng J 429:132492

Garrido-Momparler V, Peris M (2022) Smart sensors in environmental/water quality monitoring using IoT and cloud services. Trends Environ Anal Chem 35:e00173

Ghai MK, Khatri A, Kumar K, Thakur IS (2024) Multi-omics and advance technologies in biodegradation of emerging contaminants and eco-estrogens in environmental waste. Total Environ Adv 200113

Giannakoudakis DA, Das P, Pal P, Nayak J, Chakrabortty S (eds) (2024) Metal value recovery from industrial waste using advanced physicochemical treatment technologies. Elsevier

Giri S (2021) Water quality prospective in twenty first century: status of water quality in major river basins, contemporary strategies and impediments: a review. Environ Pollut 271:116332

Golher DM, Patel BHM, Bhoite SH, Syed MI, Panchbhai GJ, Thirumurugan P (2021) Factors influencing water intake in dairy cows: a review. Int J Biometeorol 65:617–625

Harrison S, McAree C, Mulville W, Sullivan T (2019) The problem of agricultural 'diffuse' pollution: getting to the point. Sci Total Environ 677:700–717

Hughes J, Cowper-Heays K, Olesson E, Bell R, Stroombergen A (2021) Impacts and implications of climate change on wastewater systems: a new Zealand perspective. Clim Risk Manag 31:100262

Issaoui M, Jellali S, Zorpas AA, Dutournie P (2022) Membrane technology for sustainable water resources management: Challenges and future projections. Sustain Chem Pharm 25:100590

Jha G, Kankarla V, McLennon E, Pal S, Sihi D, Dari B, Diaz D, Nocco M (2021) Per-and polyfluoroalkyl substances (PFAS) in integrated crop–livestock systems: environmental exposure and human health risks. Int J Environ Res Public Health 18(23):12550

Joseph K, Sharma AK, Van Staden R (2022) Development of an intelligent urban water network system. Water 14(9):1320

Kehrein P, Van Loosdrecht M, Osseweijer P, Garfí M, Dewulf J, Posada J (2020) A critical review of resource recovery from municipal wastewater treatment plants–market supply potentials, technologies and bottlenecks. Environ Sci: Water Res Technol 6(4):877–910

Khondoker M, Mandal S, Gurav R, Hwang S (2023) Freshwater shortage, salinity increase, and global food production: a need for sustainable irrigation water desalination—a scoping review. Earth 4(2):223–240

Koech R, Langat P (2018) Improving irrigation water use efficiency: a review of advances, challenges and opportunities in the Australian context. Water 10(12):1771

Koop SH, Grison C, Eisenreich SJ, Hofman J, van Leeuwen K (2022) Integrated water resources management in cities in the world: global solutions. Sustain Cities Soc 86:104137

Korlakunta RS (2022) Groundwater crisis in Southern Rural India: understanding farmers' perspectives and local participatory mitigation strategies

Koseoglu-Imer DY, Oral HV, Calheiros CSC, Krzeminski P, Güçlü S, Pereira SA, Surmacz-Górska J, Plaza E, Samaras P, Binder PM, Van Hullebusch ED (2023) Current challenges and future perspectives for the full circular economy of water in European countries. J Environ Manage 345:118627

Kumar S, Yadav A, Kumar A, Hasanain M, Shankar K, Karan S, Rawat S, Sinha A, Kumar V, Gairola A, Prajapati SK (2023) Climate smart irrigation practices for improving water productivity in India: a comprehensive review. Int J Environ Clim Chang 13(12):333–348

Kumar R, Mishra A, Goyal MK (2024) Wastewater management policies: a perspective and roadmap for India. Curr Sci (00113891) (7)

Kuntz Maykot J, Ghisi E (2020) Assessment of a rainwater harvesting system in a multi-storey residential building in Brazil. Water 12(2):546

Kuppan N, Padman M, Mahadeva M, Srinivasan S, Devarajan R (2024) A comprehensive review of sustainable bioremediation techniques: eco friendly solutions for waste and pollution management. Waste Manag Bull

Lahlou FZ, Mackey HR, Al-Ansari T (2021) Wastewater reuse for livestock feed irrigation as a sustainable practice: a socio-environmental-economic review. J Clean Prod 294:126331

Lakhiar IA, Yan H, Zhang C, Wang G, He B, Hao B, Han Y, Wang B, Bao R, Syed TN, Chauhdary JN (2024) A review of precision irrigation water-saving technology under changing climate for enhancing water use efficiency, crop yield, and environmental footprints. Agriculture 14(7):1141

Lenka NK, Lenka S, Yashona DS, Jat D (2021) Elevated temperature and low nitrogen partially offset the yield, evapotranspiration, and water use efficiency of winter wheat under carbon dioxide enrichment. Agric Water Manag 250:106856

Lepcha R, Patra SK, Ray R, Thapa S, Baral D, Saha S (2024) Rooftop rainwater harvesting a solution to water scarcity: a review. Groundw Sustain Dev 101305

Letwin O (2021) China versus America: a warning. Biteback Publishing

Li S, Wang X, Guo Y, Hu J, Lin S, Tu Y, Chen L, Ni Y, Huang L (2022) Recent advances on cellulose-based nanofiltration membranes and their applications in drinking water purification: a review. J Clean Prod 333:130171

Lieberman B, Gordon E (2021) Climate change in human history: prehistory to the present. Bloomsbury Publishing

Maiorano D (2024) India 2023: towards the general elections amid rising social tension. Asia Maior 34(1):259–300

Maroli AA, Narwane VS, Raut RD, Narkhede BE (2021) Framework for the implementation of an Internet of things (IoT)-based water distribution and management system. Clean Technol Environ Policy 23:271–283

Mateo-Sagasta J, Zadeh SM, Turral H (eds) (2018) More people, more food, worse water?: a global review of water pollution from agriculture

Medina D, Kullmann C, Felter G (2021) Revisiting resilience in the Caribbean: water supply and sanitation. World Bank, Washington, DC

Michel-Villarreal R, Hingley M, Canavari M, Bregoli I (2019) Sustainability in alternative food networks: a systematic literature review. Sustain 11(3):859

Mikulčić H, Baleta J, Klemeš JJ (2022) Cleaner technologies for sustainable development. Clean Eng Technol 7:100445

Mongird K, Rice JS, Oikonomou K, Homer J (2023) Energy-water interdependencies across the three major United States electric grids: a multi-sectoral analysis. Utilities Policy 85:101673

Montgomery DR, Biklé A (2021) Soil health and nutrient density: beyond organic versus conventional farming. Front Sustain Food Syst 5:699147

Mueller A, Detges A, Pohl B, Reuter MH, Rochowski L, Volkholz J, Woertz E (2021) Climate change, water and future cooperation and development in the Euphrates-Tigris basin. ResearchGate/Geosci/Rep 38–53

Nagata K, Shoji I, Arima T, Otsuka T, Kato K, Matsubayashi M, Omura M (2022) Practicality of integrated water resources management (IWRM) in different contexts. Int J Water Resour Dev 38(5):897–919

Nižetić S, Djilali N, Papadopoulos A, Rodrigues JJ (2019) Smart technologies for promotion of energy efficiency, utilization of sustainable resources and waste management. J Clean Prod 231:565–591

Nižetić S, Šolić P, Gonzalez-De DLDI, Patrono L (2020) Internet of Things (IoT): Opportunities, issues and challenges towards a smart and sustainable future. J Clean Prod 274:122877

Obaideen K, Shehata N, Sayed ET, Abdelkareem MA, Mahmoud MS, Olabi AG (2022) The role of wastewater treatment in achieving sustainable development goals (SDGs) and sustainability guideline. Energy Nexus 7:100112

Panel MM (2018) Water-wise: smart irrigation strategies for Africa. Intl Food Policy Res Inst

Paredes CL, Woolley K (2022) ¿ Agua para todos? differences in access to clean water and a bathroom at home by ethno-racial characteristics in contemporary Peru. Sociol Dev 8(4):450–484

Pasupuleti V, Thuraka B, Kodete CS, Malisetty S (2024) Enhancing supply chain agility and sustainability through machine learning: optimization techniques for logistics and inventory management. Logistics 8(3):73

Pexas G, Kyriazakis I (2023) Hotspots and bottlenecks for the enhancement of the environmental sustainability of pig systems, with emphasis on European pig systems. Porc Health Manag 9(1):53

Pokhrel SR, Chhipi-Shrestha G, Hewage K, Sadiq R (2022) Sustainable, resilient, and reliable urban water systems: Making the case for a "one water" approach. Environ Rev 30(1):10–29

Popoola OJ, Ogundele OD, Ladapo EA, Senbore S (2024) The impact of gold mining on soil biogeochemistry and environmental health. In: Soil microbiome in green technology sustainability. Springer Nature Switzerland, Cham, pp 483–509

Qamar K, Nchasi G, Mirha HT, Siddiqui JA, Jahangir K, Shaeen SK, Islam Z, Essar MY (2022) Water sanitation problem in Pakistan: a review on disease prevalence, strategies for treatment and prevention. Ann Med Surg 82

Radović V, Iglesias I (2019) Extreme weather events: definition, classification, and guidelines towards vulnerability reduction and adaptation management. Climate Action. Springer International Publishing, Cham, pp 464–476

Rahman A, Snook C, Haque MM, Hajani E (2020) Use of design curves in the implementation of a rainwater harvesting system. J Clean Prod 261:121292

Ramovha N, Chadyiwa M, Ntuli F, Sithole T (2024) The potential of stormwater management strategies and artificial intelligence modeling tools to improve water quality: a review. Water Resour Manag 1–34

Ritter KS, Sibley P, Hall K, Keen P, Mattu G, Beth Linton L (2002) Sources, pathways, and relative risks of contaminants in surface water and groundwater: a perspective prepared for the Walkerton inquiry. J Toxicol Environ Health A 65(1):1–142

Rodrigues Junior UJ, Dziedzic M (2021) The water footprint of beef cattle in the amazon region, Brazil. Ciência Rural 51:20190294

Roy R, Monju MH, Tan ML, Rahman MS, Kundu S, Rahman MS, Talukder B, Bhuyan MS (2023) Determining synergies and trade-offs between adaptation, mitigation and development in coastal socio-ecological systems in Bangladesh. Environ Dev 48:100936

Saleh A (2021) Prepaid digital water meters and the challenges of sustainable innovation

Sanders KT, Masri SF (2016) The energy-water agriculture nexus: the past, present and future of holistic resource management via remote sensing technologies. J Clean Prod 117:73–88
Sanders RR (2017) The rise of Africa'S small & medium size enterprises: spurring development & growing the middle class. Xlibris Corporation
Scialabba NEH (ed) (2021) Managing healthy livestock production and consumption. Academic Press
Siddique I (2021) Sustainable water management in urban areas: integrating innovative technologies and practices to address water scarcity and pollution. Pharm Chem J 8(1):172–178
Singh A (2021) A review of wastewater irrigation: environmental implications. Resour Conserv Recycl 168:105454
Smith MD, Sikka A, Taguta C, Dirwai TL, Mabhaudhi T (2024) Embracing complexities in agricultural water management through nexus planning. Irrig Drain
Sohu JM, Tian H, Kherazi FZ, Junejo I, Ejaz F, Ejaz S, Hossain MB (2024) Leveraging loss aversion and self-efficacy: the role of water pricing and risk in driving individual innovation for sustainable water consumption. Water 16(23):3510
Syrmos E, Sidiropoulos V, Bechtsis D, Stergiopoulos F, Aivazidou E, Vrakas D, Vezinias P, Vlahavas I (2023) An intelligent modular water monitoring iot system for real-time quantitative and qualitative measurements. Sustain 15(3):2127
Talat N (2021) Urban water-supply management: indirect issues of climate change leading to water scarcity scenarios in developing and underdeveloped nations. In: Water conservation in the era of global climate change. Elsevier, pp 47–71
Tanarhte M, De Vries AJ, Zittis G, Chfadi T (2024) Severe droughts in North Africa: a review of drivers, impacts and management. Earth-Sci Rev 104701
Timotewos MT, Barjenbruch M (2024) Examining the prospects of residential water demand management policy regulations in Ethiopia: implications for sustainable water resource management. Sustain 16(13):5625
Tinazzi I (2024) Water scarcity, migrations and climate change: an assessement of their Nexus
Titchou FE, Zazou H, Afanga H, El Gaayda J, Akbour RA, Nidheesh PV, Hamdani M (2021) Removal of organic pollutants from wastewater by advanced oxidation processes and its combination with membrane processes. Chem Eng Process-Process Intensif 169:108631
Tota-Maharaj K, Karunanayake C, Kunwar K, Chadee AA, Azamathulla HM, Rathnayake U (2024) Evaluation of permeable pavement systems (PPS) as best management practices for stormwater runoff control: a review. Water Conserv Sci Eng 9(1):32
Tripathi S, Hussain T (2022) Water and wastewater treatment through ozone-based technologies. In: Development in wastewater treatment research and processes. Elsevier, pp 139–172
Tzanakakis VA, Paranychianakis NV, Angelakis AN (2020) Water supply and water scarcity. Water 12(9):2347
Vanham D, Alfieri L, Flörke M, Grimaldi S, Lorini V, de Roo A, Feyen L (2021) The number of people exposed to water stress in relation to how much water is reserved for the environment: a global modelling study. Lancet Planet Health 5(11):e766–e774
Wang R, Wang Q, Dong L, Zhang J (2021) Cleaner agricultural production in drinking-water source areas for the control of non-point source pollution in China. J Environ Manage 285:112096
Ward PIWU, County RSCIS, Aloo FO (2021) Environmental and social impact assessment (summary project report) for proposed rehabilitation and expansion of Panyako water
Yildiz V, Hatipoglu MA, Kumcu SY (2022) Climate change impacts on water resources. Water and wastewater management: global problems and measures. Springer International Publishing, Cham, pp 17–25
Yu C, Xiong Z, Zhou H, Zhou P, Zhang H, Huang R, Yao G, Lai B (2022) Marriage of membrane filtration and sulfate radical-advanced oxidation processes (SR-AOPs) for water purification: current developments, challenges and prospects. Chem Eng J 433:133802
Zambrano-Prado P, Muñoz-Liesa J, Josa A, Rieradevall J, Alamús R, Gasso-Domingo S, Gabarrell X (2021) Assessment of the food-water-energy nexus suitability of rooftops. A methodological remote sensing approach in an urban Mediterranean area. Sustain Cities Soc 75:103287

Chapter 10
Trends of Emerging Contaminants in the Aquatic Environment—Risk Assessment and Management Strategies

Preeti Mehta Kakkar, Ayushi Aggarwal, and Natasha Nair

Abstract Emerging contaminants (ECs) are a burgeoning class of pollutants which include pharmaceutical drugs, personal care products (PCPs), home care products, pesticides, and industrial chemicals that are finding their way into water bodies because of improved technologies and population growth. Such pollution would not be considered "recently developed" since ECs are not a spillage or emission from a singular source but rather a concoction of micropollutants that enter the environment through various trajectories but can be extremely damaging due to their characteristics and distributions in the ecosystem. ECs pose a threat to ecosystems or human health due to their permanence, potential factories or bioaccumulation, and ability to disrupt the ecological niche. ECs can leach through the food chain caused by environmental contamination which leads to long exposure effects. It is important that global management strategies be adopted. Numerous studies have pointed out that conventional wastewater treatment methods such as filtration, sedimentation, and activated sludge systems do not treat nor eliminate all ECs and their residue levels. Advanced removal methods, including advanced oxidation processes (AOPs), adsorption, hybrid membrane bioreactors (MBRs), and integrated bioremediation systems, have shown a remarkable impact. By integrating technological innovations with appropriate policies and regulatory frameworks, it is feasible to mitigate the risks linked to ECs and safeguard water resources and ecosystems.

Keywords Emerging contaminants (ECs) · Micropollutants · Microplastics (MPs) · Personal care products (PCPs) · Human health · Environment

List of Abbreviations

Ag	Silver
Ag-NPs	Silver nanoparticles

P. M. Kakkar (✉) · A. Aggarwal · N. Nair
Amity Institute of Biotechnology, Amity University, Noida, Uttar Pradesh, India
e-mail: pmkakkar@amity.edu

H. Jain et al. (eds.), *Emerging Contaminants in the Aquatic Environment*, Advances in Wastewater Research, https://doi.org/10.1007/978-981-96-8548-6_10

AI	Artificial intelligence
AOP	Advanced oxidation process
APCI	Atmospheric pressure chemical ionization
APPI	Atmospheric pressure photoionization
As	Arsenic
Au	Gold
BOD	Biological oxygen demand
BP	Buckypaper
BPA	Bisphenol A
BTEX	Benzene, toluene, ethylbenzene, xylenes
Cd	Cadmium
CNTs	Carbon nanotubes
COD	Chemical oxygen demand
Co	Cobalt
Cr	Chromium
Cu	Copper
ECs	Emerging contaminants
ECOTOx	ECOTOXicology
EDCs	Endocrine-disrupting chemicals
EQSs	Environmental Quality Standards
EU	European Union
Fe	Iron
FO	Forward osmosis
GC-MS	Gas chromatography-mass spectrometry
GO	Graphene oxide
GPS	Global positioning systems
H_2O_2	Hydrogen peroxide
Hg	Mercury
HQ	Hazard quotient
HR-MS/MS	High-resolution tandem mass spectrometry
LC-HRMS	Liquid chromatography-high resolution mass spectrometry
LC-MS	Liquid chromatography-mass spectrometry
MBRs	Membrane bioreactors
MCNPs	Magnetic cellulose nanoparticles
MEPS	Microextraction by packed sorbent
MF	Microfiltration
Mg	Magnesium
ML	Machine learning
MPs	Microplastics
MSPE	Magnetic solid phase extraction
MTBE	Methyl tertiary-butyl ether
NF	Nano-filtration
Ni	Nickel
NPs	Nanoparticles
NRs	Nuclear receptors

O&G	Oil and grease
PA	Polyamide
Pb	Lead
PCAs	Polychlorinated alkanes
PCBs	Polychlorinated biphenyls
PCDDs	Polychlorinated dibenzo-P-dioxins
PCPs	Personal care products
PE	Polyethylene
PEPs	Polyethylene plastics
PET	Polyethylene terephthalate
PFAS	Per and poly-fluoroalkyl substances
PFECAs	Perfluoroether carboxylic acids
PFOA	Perfluorooctanoic acid
POCIS	Polar organic chemical integrative samplers
POPs	Persistent organic pollutants
PP	Polypropylene
PS	Polystyrene
PVA	Polyvinyl alcohol
PVC	Polychloride
QNs	Quality norms
RO	Reverse osmosis
ROS	Reactive oxygen species
RQ	Risk quotient
SPE	Solid phase extraction
m-SPE	Micro-solid phase extraction
SSO	Submarine sewage outfalls
TiO_2	Titanium dioxide
TSS	Total suspended solids
UF	Ultra-filtration
US	United States
UV	Ultraviolet
USEPA	United States Environmental Protection Agency
WFD	Water Framework Directive
Zn	Zinc

10.1 Introduction

Water plays a crucial role in sustaining all forms of life, and its significance is well-recognized in the operation of ecosystems and the well-being of humans (Bhardwaj et al. 2025a). Despite the abundance of water found on the Earth, huge populations do not have access to freshwater sources. Only 2.5% of the water on the planet can be used by people and over 97.5% is ocean or polar ice-locked. Such scarcity has pronounced the need to efficiently utilize water resources, especially in areas that already are under droughts or scarcity of freshwater resources. The threat that water demand poses is, however, compounded due to water pollution, which directly affects water availability. Contaminated water sources due to industrial, agricultural, or household sources of pollution make water unsuitable for use without extensive, costly, and energy-demanding treatment.

Water pollution is caused by the influx of toxic substances in water that are toxic to humans and organisms in the environment, as well as the entire ecosystem (Zhang et al. 2024). There is also a notable risk from untreated wastewater that is dumped into rivers and lakes as this has been shown to release pathogens and chemicals that are detrimental to aquatic organisms and can destroy an ecosystem. In recent years, a particular focus has been on ECs. These pollutants, most notably sourced from pharmaceuticals, PCPs, and industrial chemicals, can enter aquatic ecosystems, and threaten biodiversity (Table 10.1). ECs can be described as pollutants, but the major difference is that these ECs are not yet fully regulated as conventional pollutants which in turn complicates their detection as well as their removal.

Many pharmaceutical substances such as antibiotics and anti-inflammatory drugs are present to enter the water system through human feces and poor disposal of those drugs. There are so many therapeutic drugs such as sulfamethoxazole and diclofenac which have been reported in different water bodies all around the world (Hussain et al. 2023). Such drugs are never entirely metabolized by the human body, and their waste materials enter water bodies. These residue pollutants are biodegradable over time and cause severe environmental harm (Ahmed et al. 2024). Such pollutants have a deep ecological effect on phytoplankton, which are key components of the food web. These tiny creatures reside within water bodies. Aquatic ecosystems containing various species from fish to microscopic organisms are impacted by low levels of contamination.

The pharmaceutical compounds present in the water have been shown to influence the diversity of species in an ecosystem and even cause reproductive problems in some fish together with halting the growth phases in phytoplankton which are very essential within the food structure. Alterations in the water cycle and increased numbers and intensities of flooding and droughts can be mentioned as factors that raise the most concern in global categories and define the symptoms of climatic changes. The particularity of flooding is that they wash off the pollutants from urban, industrial, and agricultural areas into the water systems. Droughts, conversely, tend to reduce water volumes in rivers and lakes resulting in higher concentrations of these contaminants.

Given that climate change and ECs are interconnected problems, their interactions introduce severe levels of risks to the aquatic systems and water quality since the elimination of ECs such as antibiotics and industrial chemicals has the potential to cause debilitating changes in food webs and ecosystems. Elimination of ECs such as pharmaceuticals and agricultural chemicals has severe consequences for the entire biodiversity even in trace levels (Gangopadhyay et al. 2024). To tackle the problems of ECs and water contamination, responsive strategies are necessary. ECs have been successfully tackled by many technical notions, some being advanced appliances

Table 10.1 Various emerging contaminants (ECs) and their certain characteristics

Contaminant	Source/origin	Persistence and bioaccumulation	Occurrence in the environment	Emerging contaminants (ECs) group
Nonylphenol ethoxylates, sodium lauryl sulfate	Detergents, industrial cleaners, emulsifiers	Moderately persistent, bioaccumulative	Soil, sediment, sludge, water	Surfactant and surfactant metabolism
Bisphenol A (BPA), phthalates, nonylphenol, dioxins	Plastics, detergents, pesticides, industrial chemicals	Not bioaccumulative	Surface water, groundwater	Endocrine-disrupting chemicals (EDCs)
Human and veterinary medicine residues, hospital and household waste	Ibuprofen, carbamazepine, sulfamethoxazole, diclofenac	Persistent, bioaccumulative	Water, soil, sediment	Pharmaceuticals
PBDEs	Methyl tertbutyl ether (MTBE), ethanol	Highly persistent, bioaccumulative, atmospheric deposition	Fuel additives, automotive emissions, petroleum refineries	Gasoline additives
Nanomaterials	Agricultural runoff, urban pest control, and forestry practices. E.g. DDT, glyphosate	Persistent, bioaccumulative	Surface water, soil, sludge	Pesticides

and materials that led to advanced water treatment like adsorption, filtration, and excrement oxidation.

However, technical advancements will have little impact without furthering responsible practices concerning pharmaceuticals and water usage. It is imperative for developed countries to adopt policies that will enable them to come up with clean water management systems, enhance recycling and re-use practices, and carry out further studies into the long-term health effects due to exposure to ECs. Minimum chemical standards and ecological benchmarks for surface waters have been set by policy frameworks (e.g., the European Union's Water Framework Directive (WFD)). However, these frameworks require revision to recognize ECs as priority hazardous substances. It is also essential to develop effect-based monitoring tools tailored to chemical and ecological properties in water systems and to standardize exposure and effects-related data to improve the assessment of regulatory compliance.

It would also be advantageous to expand the horizon of Environmental Quality Standards (EQSs) and Quality Norms (QNs) to embrace the impact of ECs on aquatic ecosystem structure and function. It has been shown that the behavior of EC is complex and the transformation processes, such as photodegradation and microbial degradation, may depend on temperature, pH, and interactions with other chemicals (Norvill et al. 2016). Although some ECs degrade relatively quickly, others persist in the environment, where they accumulate in biota and present prolonged ecological and health hazards. In terms of EC management, a multidisciplinary strategy, involving cutting-edge monitoring, enhanced treatment technologies, and strong regulation, is essential to minimize their effects. Adverse outcomes to aquatic ecosystems owned and managed services and industries, organizations, and associations must have at least one coordinating global effort to address water pollution and water scarcity.

10.2 Emerging Contaminants (ECs) Explanation and Signification

ECs are artificial or naturally occurring chemicals that cannot be generally observed in nature but can possibly enter the environment and cause already existing or suspected negative ecological or human health effects (Samal et al. 2022). Tenacious bio-accumulative toxicants, and persistent toxicants, ECs are increasingly recognized to engage in a crucial role as key environmental pollutants. The sources of these pollutants are diverse, incorporating agricultural runoff, industrial discharges, and household waste. These pollutants can enter the environment via various different routes, including leaching into the soil, depositions from the atmosphere, and release through polluted wastewater.

A vast variety of ECs from industrial and household discharge, i.e., pharmaceutical, microplastics (MPs), nanomaterials, PCPs, endocrine-disrupting chemicals (EDCs), pesticides, triclosan, benzophenone and parabens, etc., cannot be separated by wastewater treatments and thus deposits in the water bodies, where they affect aquatic ecosystems and human well-being. EDCs such as bisphenol A (BPA) and phthalates disrupt hormonal systems, resulting in developmental and reproductive problems in humans and wildlife. Their relatively low concentrations in aquatic systems have considerable biological consequences for organisms (Ghosh et al. 2022). Minimum chemical standards and ecological benchmarks for surface waters have been set by policy frameworks (e.g., the European Union's WFD). However, these frameworks require revision to recognize ECs as priority hazardous substances.

It is also essential to develop effect-based monitoring tools tailored to chemical and ecological properties in water systems and to standardize exposure and effects-related data to improve the assessment of regulatory compliance. The behavior of EC is complex, and it has been demonstrated that the efficiency of transformation processes that are involved in its degradation depends on temperature, pH, and interactions with other chemicals (Table 10.1). Although some ECs degrade relatively quickly, others persist in the environment, where they accumulate in biota and present prolonged ecological and health hazards. In terms of EC management, a multidisciplinary strategy, involving cutting-edge monitoring, enhanced treatment technologies, and strong regulation, is essential to minimize their effects. Adverse outcomes to aquatic ecosystems owned and managed services and industries, organizations, and associations must have at least one coordinating global effort to address water pollution and water scarcity.

Substances like atrazine, glyphosate, or neonicotinoids pollute water through agricultural runoff. Such pesticides are damaging to non-target species. They also inflict long-term ecological damage (Pathak et al. 2022). The chemical properties of industrial substances and nanomaterials such as perfluorooctanoic acid (PFOA), silver nanoparticles (Ag-NPs), and titanium dioxide (TiO_2), tend to accumulate in the environment and can have detrimental effects on human health. In freshwater systems, they are increasing. This raises concerns about long-drawn ecological impact (Włodarczyk and Kwarciak-Kozłowska 2021). These are the surfactants, such as sodium lauryl sulfate, found in decontaminating products that break up oils, fats, and grease; they can destroy aquatic ecosystems. They compromise the surface tension of water and impact aquatic species.

Persistent organic pollutants (POPs), e.g., polychlorinated dibenzo-p-dioxins (PCDDs) can expand over large distances in the ecosystem and through food chain build-up in organisms. High-sensitivity methods, such as liquid chromatography and mass spectrometry (LC–MS), detect low levels of ECs. The ng/L to μg/L is the range at which these contaminants exist. EC in surface water is more than in groundwater. Even though natural attenuation decreases EC levels in groundwater, pollution sources in proximity to aquifers raise the risks of contamination. EC behavior is affected by environmental properties like pH, temperature, and organic matter. These results highlight requirements for the latest treatment technologies and stringent regulations (Li et al. 2024).

10.2.1 Personal Care Products (PPCs)

Pharmaceuticals and PCPs are found to be the most common environmental contaminants due to their excessive use and persistence in water. Approximately 3000 pharmaceutical compounds such as antibiotics, analgesics, and anti-inflammatories are manufactured globally and taken worldwide at rates of 50 to 150 gm per capita per year in evolved countries (Ortúzar et al. 2022). These compounds are powerfully built in the environment as they cannot be fully degraded during natural degradation and wastewater treatment processes. These small concentrations of pharmaceuticals and metabolites can be detected by analytical techniques such as LC–MS and gas chromatography-mass spectrometry (GC–MS).

Yet sample heterogeneity hinders data comparability due to a lack of standardization. Pharmaceuticals are subjected to biotic and abiotic transformations (oxidation, hydrolysis, and photolysis) that adjudicate their persistence and bioactivity. Generally detected compounds are ibuprofen, carbamazepine, and triclosan, as well as veterinary drugs such as sulfamethazine and monensin antibiotics. Water contamination from PCPs, such as triclosan, parabens, ultraviolet (UV) filters, and fragrances, is a prevalent concern. Lifestyle-related compounds such as caffeine and nicotine are present as well in surface waters. Such contaminants are often persistent and resist natural degradation, and bioaccumulation has been suspected (Khalikova et al. 2024). Recreational drugs and many veterinary compounds remain under-characterized and understudied, even though they have been detected in several locations, for which there is an urgent need to understand the fate and effects in the environment.

10.2.2 Microplastics (MPs)

Nowadays, plastics are ubiquitous, and the latest estimate shows that 94% of the 320 million tons manufactured per year end up as waste in the environment (Pilapitiya et al. 2024). MPs are fragments estimated between 100 nm to 5 mm in size (Ullah et al. 2024), which can come from a diversity of sources including, but not limited to synthetic textile fibers, cosmetics, pharmaceuticals, and industrial-cleaning machinery. This debris releases particles into the surrounding environment from the direct disposal of plastic waste into water bodies, effluents from wastewater treatment factories, and waste from industries such as fisheries and water management. Secondary MPs also result when larger plastic objects, such as bags and fishing nets, break down over time (Bhardwaj 2024; Bhardwaj et al. 2024).

MPs polymers polypropylene (PP), polyethylene (PE), polyethylene terephthalate (PET), polystyrene (PS), polychloride (PVC), polyamide (PA), polyvinyl alcohol (PVA) are found everywhere in water, soil and air and can interact with living creatures and remain in the ecosystems for a long period of time (Albaseer et al. 2024). MPs can be classified as primary MPs which are produced as intentionally small products, or secondary MPs which are formed through the degradation of larger plastic debris (Ali et al. 2023; Bhardwaj et al. 2025b). Being small makes them a gateway to the food chain, their way to tissues, blocking biological processes. MPs are more buoyant and durable, allowing for their widespread distribution into ocean environments, where they are dispersed with ocean currents, exposing marine life to pollutants throughout the surface and deep waters. Researchers have presented that MPs can stop seed germination, development of roots, and nutrient supply in plants which adversely affect the terrestrial environment.

10.2.3 Nanomaterials

Nanomaterials are in the order 1 to 100 nm and possess high value because of their mechanical strength, thermal stability, electrical conductivity, and tunability and are used in the construction of cosmetics, medical devices, electronics, and sports gear (Bhardwaj et al. 2023a). Common types are nanotubes, nanosilver, TiO_2-NPs, and quantum dots (El-Kalliny et al. 2023). Although these materials provide many opportunities, their chemical reactivity and biological activity can pose environmental and health problems, especially for aquatic ecosystems. Their compact size allows them to persist in the environment and carry out slow degradation, contributing to contamination in the environment and through organisms at a cellular level (El-Kalliny et al. 2023).

In water systems, nanomaterials interfere with microorganisms that drive nutrient cycling and oxygen production, impeding entire aquatic food webs. Bioaccumulation increases risks because these small particles are ingested by smaller organisms, causing these particles to work their way up through the food chain and endanger humans through contaminated seafood. Some inorganic forms, such as Ag-NPs, over time, dissolve into toxic ions that adversely affect aquatic flora and fauna, while TiO_2-NPs under sunlight produce reactive oxygen species (ROS), which instigate oxidative stress in aquatic species (Azimzada et al. 2020). Nanomaterials' fascinating properties are highly favored by industry, but they are also a serious concern to aquatic systems. Their ubiquitous presence, environmental persistence, and potential for toxicity underscore the importance of conducting studies to establish their lasting effects on human health and the environment, as well as the creation of maximum permissible limits to regulate their application in consumer products.

10.2.4 Industrial Chemicals and Unconventional Sources

Industrial chemicals and their byproducts (as closely related to dioxins and organic solvents, various heavy metals) are persistent and toxic substances that affect ecosystems and human health. Contaminants such as heavy metals (lead (Pb) and mercury (Hg)) adulterate soil and water, leading to loss of biodiversity and bioaccumulation in food chains, which presents long-term health hazards (Alam et al. 2023; Li et al. 2024). Regular contaminants suspended in industries are organic solvents, pesticides, and EDCs like BPA and nonylphenol. High transfer efficiency rates of BPA in the wastewater treatment process (up to 99%) cannot be reported for its environmental persistence, and in Europe, it can be found in wastewater and sewage sludge at levels higher than in the United States (US).

Impurities found in groundwater incorporate various other pollutants like polychlorinated alkanes (PCAs), methyl tertiary-butyl ether (MTBE), and benzene, toluene, ethylbenzene, xylenes (BTEX). MTBE is a crucial contaminant to drinking water quality because it is both a potential carcinogen and a contaminant used in gasoline. The lack of knowledge surrounding chemicals such as PCAs, which are known to be present in treated wastewater and sewage sludge, emphasizes the importance of further investigation. Appropriate measures must be taken to reduce the impact on these industries as well as the regulatory bodies for high-precision waste management. Ongoing monitoring, research exploration, and the development of better treatment technologies will be important to reduce the deleterious effects that those contaminants cause on environmental and public health.

10.2.5 Pesticides

Pesticides, synthetic chemicals, or natural compounds have been found in small levels in groundwater sources (Kaur et al. 2024). They are mostly utilized in agricultural practices to eradicate weeds, pests, and other kinds of crop diseases. Over the years, certain pesticides like atrazine, which had been found to pose a high environmental risk, have been banned or restricted. However, newer alternatives, such as diuron, may also carry potential risks to ecosystems and water sources. Metaldehyde, a pesticide of growing concern, has been found in some cases to exceed the European Union's (EU) permissible levels for pesticides in drinking water (Rolph et al. 2020), raising significant concerns about its environmental and health effects. Continuous monitoring and stricter regulation are being emphasized to mitigate these risks and safeguard water quality for human consumption and ecological balance.

10.3 Factors Influencing the Presence and Transport of Contaminants

The behavior and movement of ECs in aquatic environments are complex and largely dependent on their physicochemical characteristics, which can be affected by the surrounding conditions, either in a reinforcing or opposing manner. These interactions are influenced by physical, chemical, and biological elements. Physical factors such as water flow dynamics, temperature, sunlight, and precipitation significantly affect EC transport and transformation. Additionally, suspended particles, including algae, serve as crucial sorbents for ECs in water bodies. Chemical elements like pH levels, nutrient availability, and oxygen levels can alter algae and bacterial populations, indirectly impacting EC distribution.

Biological components, such as bacterial communities, play a role in EC biodegradation and transformation. Interactions between these factors further influence EC degradation rates in aquatic settings. Studying EC degradation through water columns and sediment layers is essential for understanding their distribution, requiring insight into interactions among different phases and the lifecycle of each EC (Tong et al. 2022). Integrated modeling tools that combine hydrodynamic, eutrophication, and multi-phase EC modules with laboratory and field data can help in developing a comprehensive view of EC behavior and movement. The presence and transport of ECs in aquatic systems are influenced by a complex interplay of physical, chemical, and biological factors.

Physically, the hydrodynamic conditions of water bodies, such as flow rates and turbulence, play a major role in dispersing ECs and affecting their residence times. Temperature is another key factor, influencing both the solubility and reactivity of contaminants, while solar radiation can induce photodegradation of light-sensitive compounds. Rainfall and surface runoff also contribute to the transport of ECs from terrestrial to aquatic systems, impacting their concentration and distribution (Yang et al. 2024). Suspended solids, like sediments and algae, serve as sorbents that may bind to ECs and change their mobility. Chemically, water pH can modulate the ionization state of contaminants and thereby alter their bioavailability and the potential for adsorption to particles or sediments.

Moreover, the levels of dissolved oxygen and nutrients affect microbial communities that can biodegrade ECs and directly impact their persistence in water bodies. These microbial communities are focused on degrading the biological transcending and evolution of ECs, specific bacteria, and algal assemblage composition decaying the fate of pollutants, again via metabolic action. Understanding these factors and their interactions is crucial for predicting EC behavior and devising efficient water practices as a means to reduce possible ecological effects.

10.4 Spatial and Temporal Variations in Contamination Levels: Implications for Environmental Monitoring

The spatial and temporal variations of contamination levels are key determinants of the exposure, retention, and possible environmental effects of pollutants in the environment. Definition of spatial variability: the variation in concentration of contamination and local anthropogenic activities, land carbonation rates vary greatly between different geographical regions, patterns of use, climate conditions, and proximity to sources of pollution. For instance, urban industrial areas will have higher levels of pollutants like heavy metals, inadequate human populations, and are concentrated upon phosphorus fertilizers, and organic chemicals waste disposal infrastructure (Haque 2022).

In contrast, rural and remote areas may show lower levels of pollution but can be susceptible to atmospheric deposition or long-range transport of pollutants, as shown in studies of POPs (Bhardwaj et al. 2018, 2019; Bhardwaj and Jindal 2019; Bhardwaj et al. 2020). Temporal variations, on the other hand, describe the fluctuations of contaminant levels over time, whether by nature, be it seasonal changes or weather patterns, or human activities, for example, agricultural cycles and industrial discharge. Temporal dynamics are of particular importance in aquatic ecosystems, in which contamination levels may spike during the rainy season or runoff, causing episodic contamination events. These epistatic events are often associated with agricultural runoff, wastewater discharge, and stormwater events. When it comes to long-term trends in contamination; however, it would indicate more gradual changes because of shifts in land use practices, pollution control measures, or climate-driven changes in hydrological cycles.

Since spatial and temporal variability is censorious for effective environmental monitoring and risk evaluation it allows the identification of areas with high risk, prediction of contamination trends, and optimization of mitigation strategies. Recent advances in environmental monitoring technologies, including remote sensing, geospatial technologies such as global positioning systems (GPS), and geographic patterns of contamination in space and in time. Combining these tools with spatial–temporal modeling techniques enables more treatment of contaminated sources; accurate prediction of contamination trends; and better decision-making for policy planning and environmental management. Furthermore, these innovations offer meaningful data for immediate and potential environmental implications, facilitating preemptive measures in pollution management and mitigation (Pulicharla et al. 2022).

10.5 Bioaccumulation and Biomagnification in Aquatic Food Webs: Mechanisms and Ecological Implications

The bioaccumulation and biomagnification processes in aquatic ecosystems, which are the overarching amplifiers of dilution, impact the spatial and toxicological orientation of pollutants in a food web source. Bioaccumulation describes the time-related increase in the concentration of a substance in an organism, essentially from the day-to-day consumption of food or water that has been contaminated. This process, in contrast, is more pronounced in organisms found at the bottom of the food chain, such as zooplankton and benthos, among others, which share similar attributes of accumulating several contaminants such as heavy metals, POPs, and pharmaceuticals.

Such contaminants, when once included in the organs, have the potential to remain there for a long duration which is dependent upon the constituent's chemical structure and the organism's metabolic or excretory power. In contrast, bioaccumulation refers to the process of increasing the contaminant at successive trophic levels in the food chain. This is caused by the fact that organisms that are at the top of the food chain consisting of fishes, birds, mammals, etc., take many contaminated preys making them have a higher concentration of pollutants in their bodies than the concentration found in the surrounding environment.

Lipid soluble substances, or biomagnification substances are the substances that are incorporated in fats, and mostly, environmental pollutants have been identified to be polychlorinated biphenyls (PCBs) and Hg which are very ideal (Bhardwaj et al. 2022). The ecological significance of the processes of bioaccumulation and biomagnification remains profound as they result in toxic effects on apex predators disrupting their reproductive health, stunting growth, and in rare circumstances resulting in population decline. Additionally, such mechanisms also pose extraordinary threats to the health of humans, especially those communities that depend on fish and other aquatic fauna for sustenance. It is crucial to follow the processes of bioaccumulation and biomagnification as they are essential for understanding the long-range consequences of environmental contaminants exposure to aquatic environments and for implementing efficient pollution control options.

10.6 Risk Assessment Methodologies

10.6.1 Emerging Pollutants in Aquatic Environments: Screening, Sampling, and Preconcentration Techniques

10.6.1.1 Screening Methods

Screening remains a key technique for identifying unknown pollutants by leveraging compound-specific information available from various databases, research articles, and other sources. An innovative approach was presented, which included solid phase extraction (SPE) and liquid chromatography-high resolution mass spectrometry (LC-HRMS) as both suspect screening and target quantification of pharmaceutical residues in surface water. Similarly, an innovative step for direct injection strategy was developed that incorporated two chromatographic runs and scheduled analysis that facilitated the determination of 154 compounds, 84 precursors, and 70 transformation products in surface water (Canlı et al. 2024). Recent advancements have focused on increasing the sensitivity and throughput of these screening techniques. Notably, novel software algorithms and improved mass spectrometers have enhanced the ability to detect trace levels of ECs even in complex environmental matrices (Yang et al. 2023). Studies have explored the application of high-resolution tandem mass spectrometry (HR-MS/MS) to identify and quantify a wider range of ECs, with increased sensitivity at lower concentrations (Yang et al. 2023).

10.6.1.2 Sampling and Preconcentration

Passive sampling methods have been successfully employed when pollution of this kind must be assessed at minimal volumes in aquatic habitats. For this goal, passive samplers such as polar organic chemical integrative samplers (POCIS) are quite popular. POCIS devices, however, require careful consideration of their longevity and sampling rates. POCIS samplers stored for over nine years showed significantly reduced sampling rates, highlighting the impact of aging on sorption capacity (Wang et al. 2020). Recent innovations have introduced novel microextraction techniques aimed at improving extraction efficiency and preconcentration. Development of an in-situ liquid–liquid microextraction configuration coupled with electroanalysis to detect ECs at trace levels was done (Chormey and Bakırdere 2018).

This method demonstrated higher sensitivity, and a reduced sample preparation time compared to traditional techniques. Additionally, the introduction of a novel approach utilizing magnetic cellulose nanoparticles (MCNPs) coated with ionic liquid for magnetic solid phase extraction (MSPE), which showed high extraction recoveries for common pharmaceuticals like ibuprofen, paracetamol, and diclofenac. Microextraction by packed sorbent (MEPS) has gained prominence due to its simplicity, cost-effectiveness, and minimal solvent use. Studies have demonstrated its effectiveness for concentrating ECs in various matrices, including surface water and wastewater effluents, with recoveries reaching above 90%. Additionally, advancements in polymeric and silica-based sorbents have been integrated into microextraction protocols to improve analyte recoveries and reduce matrix effects (Hashemi and Kaykhaii 2024).

10.6.1.3 Recent Innovations and Green Chemistry Approaches

Improvement and Optimization of Solid Phase Extraction (SPE)

SPE combined with techniques: LC–MS or GC–MS still ranks among the most popular and efficient techniques aimed at the detection of ECs over a wide variety. It is known to be fast, accurate, and reproducible. Nowadays, emphasis is put on the refinement of the SPE procedure, increasing the selectivity and capacity of the adsorbents, and modifying it for the purpose of online monitoring of pollution. Emphasis in recent innovations has been placed on green technologies in the construction of SPE systems, which aim to minimize solvent consumption without sacrificing extraction and preconcentration efficiency. Additionally, new approaches are being designed to integrate SPE into in situ monitoring platforms, such as marine buoys and autonomous gliders, for continuous monitoring of water quality. A key advancement in this field is the development of low-pressure micro-SPE (m-SPE) systems for real time, in situ monitoring of ECs (Tang et al. 2019).

These systems are designed to use smaller volumes of samples and solvents compared to traditional SPE methods. The low-pressure m-SPE technique is particularly advantageous for integration into automated and robotic systems, such as marine gliders and oceanographic buoys, allowing for continuous environmental monitoring. New versions of m-SPE systems are incorporating advanced microfluidic chips to further enhance detection sensitivity and reduce operational costs. Further innovations have been made with SPE cartridge design, the development of a novel sample preparation cartridge that incorporated a mixture of Strata X-AW, Strata X-CW, and Isolute ENV+ in specific ratios, topped with Oasis HLB. This new mixed-bed system enhanced the recovery of a wide range of chemicals, achieving superior efficiency compared to traditional SPE methods.

Researchers incorporated nanostructured adsorbents in SPE cartridges, increasing recovery rates by up to 50% for several pharmaceutical pollutants, while also improving the cartridges; resistance to fouling from complex matrices (Zhou et al. 2024). In a pioneering effort, the introduction of hybrid SPE cartridges made from poly (N-isopropyl acrylamide)-grafted silica NPs was done offering thermoresponsive capabilities. This development represents a significant step toward portable, temperature-sensitive SPE systems, which are ideal for field applications in varying environmental conditions. The research expanded to incorporate biodegradable thermoresponsive materials, making the systems not only more portable but also more environmentally sustainable.

10.6.2 Emerging Methods for Detection and Analysis of Emerging Contaminants (ECs)

The latest expansion in analytical techniques for the detection of ECs has enhanced selectivity and sensitivity, addressing challenges related to matrix interferences in complex environmental samples (Table 10.2). Techniques such as atmospheric pressure chemical ionization (APCI) and atmospheric pressure photoionization (APPI) have been introduced to improve the ionization process, enhancing selectivity and minimizing interference in matrix-rich samples like those from surface waters and wastewater. For instance, a method developed for detecting progestogens in water combines SPE with liquid chromatography coupled to APCI/APPI hybrid mass spectrometry (LC-APCI/APPI-HRPS) (Golovko et al. 2018). This method allows the use of unfiltered water samples in the SPE process, enabling filtration and extraction in a single step, improving workflow efficiency. As concerns over the persistence and bioaccumulation of certain ECs growth, research has increasingly focused on developing alternative, less toxic compounds.

However, a significant obstacle remains in the lack of validated quantitative methods and suitable standards for many of these substances, which hinders comprehensive environmental monitoring. To address this, validation of a quantitative LC–MS method for determining perfluoroether carboxylic acids (PFECAs) in surface, drinking, and wastewater played an important role. The method showed accurate spike recovery and acceptable variability, confirming its suitability for monitoring these challenging pollutants in various water matrices. Another promising innovation in environmental monitoring involves the use of Buckypaper (BP)—a material made from compressed carbon nanotubes (CNTs). Research has explored further modification of BP to enhance its selectivity for a wider range of organic micropollutants, improving the efficiency of extraction (Li et al. 2024). The detection of steroidal estrogens listed on the European WFD watch list due to their endocrine-disrupting properties requires highly sensitive detection methods.

Table 10.2 Detection techniques for pharmaceutical compounds in water and wastewater, covering a wide array of commonly found substance

S. No	Pharmaceuticals compounds	Techniques
1	Aspirin, Codeine, Cyclophosphamide, Galaxolide, Pentoxyfylline, Tonalide, Triclosan, Atenolol, Betaxolol, Bisoprolol, Ciprofloxacin, Clarithromycin, Diclofenac, Carbamazepine, Diazepam, Ibuprofen, Paracetamol, Phenazon	GC–MS (gas chromatography paired with mass spectrometry is effective for volatile and semi-volatile substances without the need for derivatization)
2	Aspirin, Etofibrate, Etofylline, Flurbiprofen, Ketoprofen, Methotrexate, Omeprazole, Sotalol, Estrone, Bezafibrate, Clofibrate, Fenofibrate, Naproxen, Roxithromycin, Salbutamol, Ibuprofen, Paracetamol	GC–MS/MS (derivatization is required for pharmaceuticals with polar functional groups to enhance volatility for GC analysis)
3	17α-Ethinylestradiol, 7β-Estradiol, Acetyl-sulfamethoxazole, Amidotrizoic acid, Iopamidol, Lofepramine, Metronidazole, Nadolol, Penicillin G, Penicillin V, Amoxycillin, Enalapril, Tetracycline, Sulfonamides, Erythromycin, Sulfamethoxazole, Norfloxacin, Simvastatin, Oxytetracycline, Bezafibrate, Clofibrate, Diclofenac, Fenofibrate, Indomethacin, Propranolol, Roxithromycin, Tamoxifen, Virginiamycin, Methotrexate, Ofloxacin, Oxacillin	LC–MS/MS (liquid chromatography combined with mass spectrometry is well suited for polar, nonvolatile pharmaceuticals and enables the detection of substances at trace levels (ng/L))

10.6.3 Risk Quotient (RQ) Method

RQ is routinely utilized for risk assessment for different chemicals (Tang et al. 2019). The RQ is expressed mathematically as RQ = Cexpo/Ceff and is employed by regulatory agencies, e.g., the United States Environmental Protection Agency (USEPA) in matters regarding pesticide-related ecological risk assessments. Meanwhile, the hazard quotient (HQ) is determined by dividing the measured or estimated environmental concentration of a pollutant by its effect concentration, helping to assess potential risks to ecosystems (Tang et al. 2019). To this end, an environmental risk characterization was performed utilizing RQ on submarine sewage outfalls (SSO) derived pollutants. The RQ method was used to conduct an environmental risk assessment concerning pollutants derived from SSO wastes and some ECs have been discovered to pose a significantly high potential risk. The ecotoxicological data used for this assessment was obtained from the database ECOTOXicology Knowledgebase (ECOTOX) controlled by the USEPA. The study reported high ecological risks in some of the SSOs tested. Similarly, a study in Slovenia used the RQ method to assess the environmental risks of common ECs like oxybenzone, estrone, and triclosan, finding moderate to high risks to the environment, with RQs ranging from 0.282 for homosalate benzophenone to 15.5 for estrone (Česen et al. 2019).

10.7 Novel Considerations in Risk Assessment of Emerging Contaminants (ECs)

One challenge in risk assessment is the inconsistency in how field data are presented, as concentrations are often measured based on mass or lipid content, while toxicity data are typically reported by volume or mass (Tang et al. 2019). To address these scientists proposed a fugacity approach to standardize these different units. This method allows for a more consistent comparison of pollutant concentrations across various matrices, such as sediments and biota, and supports more accurate risk assessments for benthic organisms by comparing data to chronic toxicity benchmarks.

Another important consideration is the long-term effects of early-life exposure, particularly with EDCs, which are often overlooked in traditional chemical risk assessments.

Researchers have stressed the importance of adopting rapid, mechanism-based testing methods in line with the 3R principle (Replacement, Reduction, and Refinement) of animal testing (Verderio et al. 2023). For instance, suggestions were made that invertebrate nuclear receptors (NRs) could serve as an alternative to vertebrate NRs, providing a more mechanistic approach to assessing EDCs and other ECs in aquatic organisms (Tang et al. 2019). Despite the growing body of research into the ecological impacts of individual ECs, there remains a significant knowledge gap regarding the toxicity of chemical mixtures and their combined effects. Unlike single-compound toxicity, chemical mixtures can lead to synergistic or antagonistic effects, often referred to as the cocktail effect.

Expanding research on this topic is crucial to developing accurate methods for evaluating the environmental risks posed by ECs. Researchers continued to focus on integrated monitoring approaches for emerging pollutants, incorporating novel biological and chemical methods. New studies have explored the role of artificial intelligence (AI) and machine learning (ML) in predicting the behavior of ECs in complex environmental matrices (Liu et al. 2024). These advanced computational tools have shown promise in improving risk assessments by identifying potential contaminants in previously unexplored regions. Furthermore, in-situ monitoring systems have been enhanced with nanosensor technologies to detect trace pollutants in real time, offering more precise data for environmental management.

10.8 Management and Remediation of Emerging Contaminants (ECs)

The management and remediation of ECs like those described are important for two main reasons. They present significant risks to the environment and to the ecosystem, and drastically to public health and well-being. They also cause significant climate harm and global pollution. To address these challenges, numerous technologies have been developed that have the potential to help in the remediation of ECs (Wang et al. 2024).

10.8.1 Conventional Wastewater Treatment Technologies

Conventional wastewater treatment involves a series of processes, i.e., primary, secondary, and tertiary treatments, utilizing physical, chemical, and biological methods (Gangaraju et al. 2021). During the primary and secondary treatment phases, a significant portion of large particles, suspended solids, and organic materials are eliminated from wastewater. Following these initial treatments, any remaining undesirable substances in the wastewater are addressed in the tertiary treatment stage (Gangaraju et al. 2021).

- **Primary Wastewater Treatment**

Primary wastewater treatment relies on the sedimentation process for solid/liquid (Kesari et al. 2021; Asadi et al. 2021). This primary phase separation reduces the biological oxygen demand (BOD) of the wastewater effluent by 20 to 30% and the total suspended solids (TSS) by nearly 50 to 60% (Kesari et al. 2021). Integrated methods of coagulation and flocculation are often used to allow for effective sedimentation. To allow small particles to merge with each other to form larger particles, a slow mixing process called flocculation is used which helps the particles to settle more easily (Asadi et al. 2021). Coagulants can be classified as metal-based, biopolymer-based, and synthetic coagulants. Natural coagulants like *Stryhnos Potatorum* (Nirmali seed) and *Pisum Sativum* (pea seed) for use in the wastewater generated by the meat food processing industry propose a viable alternative treatment method to chemical coagulants (Pahazri and Fadzilah 2018). A study by Pahazri and Fadzilah found that the parameters of turbidity, TSS, chemical oxygen demand (COD), and oil and grease (O&G) removal by chemical coagulants were like these natural coagulants (Pahazri and Fadzilah 2018). Flocculants can be cationic, anionic, and non-ionic. Biogenic elemental sulfur is a byproduct obtained in the anoxic biodesulfurization of biogas, which can be effectively removed from wastewater using a cationic flocculant called Lizaflock 853 M (Torres-Herrera et al. 2022).

10.8.2 *Secondary Wastewater Treatment*

In the process of secondary wastewater treatment, microorganisms are employed to break down the soluble organic compounds found in wastewater (Sharma et al. 2023). These are broadly of two types, namely, an attached growth system and a suspended growth system. In the former, a bed of substances like gravel and ceramic is present at the bottom of the wastewater tank in a fixed condition. The microorganisms are allowed to grow on this, and the wastewater is allowed to flow on top; thus, finally resulting in the formation of a thick biofilm. This is later removed when it grows too thick (Gangaraju et al. 2021). In suspended growth systems like the activated sludge process wastewater flows around and through the free-floating microorganisms, gathering as biological flocs, which are settled at the bottom. The settled flocs contain microorganisms which are recycled back to treat the wastewater. Some of the environmental implications of the current processes are the need for high energy and chemical use and the production of large volumes of waste. In addition, conventional water decontamination methods lack selectivity toward contaminants. They cannot remove one compound over another, suggesting inefficiency and susceptibility to interference during operation (Pérez et al. 2023).

10.9 Remediation of Emerging Contaminants (ECs)

ECs pose a serious risk to human health and to aquatic ecosystems, thus highlighting the need for effective management and treatment options for these ECs from urban and industrial sources (Zahmatkesh et al. 2022). Generally, conventional wastewater techniques fail to successfully remove trace contaminants, because of this there is a need to identify recent technologies that are effective in doing so (Mahmood et al. 2022).

10.9.1 *Membrane Filtration*

Distinctive materials characterized by specific filtering attributes, such as specific charge, pore size, and hydrophobicity, constitute a membrane that functions as a physical barrier, regulating the types of substances that can pass through to the opposite side. Membrane filtration can be categorized into several types, including ultrafiltration (UF), nano-filtration (NF), microfiltration (MF), forward osmosis (FO), and reverse osmosis (RO) all of which achieve separation through hydraulic pressure techniques. These membrane filtration methods possess the capability to reduce ECs by over 99%; however, their implementation on a larger scale has not yet been applied (Mahmood et al. 2022). UF membranes use a pressure of 2 to 5 bars, and this membrane is said to be microporous with asymmetric characteristics (Kammakakam et al. 2023).

NF has a membrane that is tight and porous and utilizes pressure between the range of 5 to 15 bars (Kammakakam et al. 2023). The particle size varies from 5×10^{-2} to 5×10^{-3} μm (Zahmatkesh et al. 2022) and possesses distinctive characteristics for the removal of pharmaceuticals, including anti-inflammatory medications, sulfonamides, and fluoroquinolone antibiotics, as well as natural hormones such as testosterone, estradiol, and progesterone. (Mahmood et al. 2022). The operating pressure for RO ranges from 15 to 75 bars, with the particle size extending from 5×10^{-3} to 10^{-4} μm. The RO process can separate monovalent ions up to 99.5% of the time, unlike other membrane separation processes based on pressure, which cannot separate small particles like bacteria (Zahmatkesh et al. 2022). Research efforts on the utilization of MF have concentrated on the removal of colloidal particles, dyes, organic matter, and other high molecular weight soluble polyethylene plastics (PEPs) from the waste streams.

MF has been generally employed in combination with other techniques for the removal of complex industrial PEPs. For instance, domestic wastewater contains hormones such as estrone (E1), 17β-estradiol (E2), and 17αethynylestradiol (EE2). Unlike the MF process, UF has a solute rejection regime of above 2 kDa molecular weight for the separation of macro-PEPs. Considering that many of the PEPs fall under the category of macromolecular range, UF has been successful in treating such PEPs from wastewater. NF has a much tighter regime than UF membranes, thus adding several benefits, such as divalent salts and textile wastewater recovery (Dharupaneedi et al. 2019).

10.9.2 Adsorption

Adsorption is a physical process wherein a substance, also called an adsorbate, is attracted to and sticks to the surface of another substance called an adsorbent (Fig. 10.1). Emerging new pollutants like those from wastewater sludge, from the paper industry, silt and soil, and even environmental pharmaceuticals have been found to reduce in amount because of this novel technology. Adsorbents or sorbents can be made up of a wide diversification of materials, including zeolites, mineral matter, organic carbon frameworks, metal–organic frameworks, activated carbon, biochar, hydro-char, CNTs, graphene, carbon composite materials, polymers, and mesoporous clay nanocomposites (Mishra et al. 2023). In a study conducted by Lima et al. (2019) five pharmaceuticals present in hospital effluents were found to have effective removal because of the action of activated carbon. It was observed that for the removal of paracetamol, propranolol, and tetracycline, the removal was complete (100%), and for amoxicillin and sodium diclofenac. It was at least 92.71 ± 3.15% and 91.82 ± 0.95%, respectively, indicating that activated carbon is an adsorbent with high sorption capacity (Lima et al. 2019).

Fig. 10.1 Mechanism of absorption

10.9.3 Nanotechnology and Nanomaterials

Even though traditional wastewater treatments such as are effective in removing traditional contaminants, they oftentimes face significant challenges in terms of the removal of ECs (Rath et al. 2024). Nanomaterials are certain specific materials that show great promise for water purification because of their various specific characteristics (Fig. 10.2). Generally, nanomaterials are materials with at least one dimension that is less than 100 nm (Kreyling et al. 2010). These materials possess unique functionality due to their characteristics of being small in size, and more importantly, volume to specific surface area ratio (Bhardwaj et al. 2023a; Tokas et al. 2024). It has been proposed that materials with a volume to a specific surface area greater than or equal to 60 m^2/cm^3 should be nanomaterials (Kreyling et al. 2010). Nanomaterials exhibit a high specific surface area, providing a high potential for liquid–solid interactions while using a small procedure footprint (Pérez et al. 2023).

10.9.3.1 Metal Oxide Nanoparticles (NPs)

Metal oxide NPs have been found to be effective as a sustainable and environment-friendly wastewater treatment technology (Corsi et al. 2018; Pérez et al. 2023). These metal-NPs possess short intraparticle diffusion distances and high specific surface area and are compressible without significant surface area reduction (Corsi et al. 2018). These kinds of NPs have many surface reaction sites because of which their effective adsorption capacity increases, as well as their rate or reaction, or their kinetics (Pérez et al. 2023). NPs such as magnesium (Mg), gold (Au), and Ag have shown high removal rates of heavy metals (chromium (Cr), nickel (Ni), cadmium (Cd), copper (Cu), Hg, and arsenic (As), etc.) and antimicrobial potential due to their small size, higher surface area-to-volume ratio and unique physical–chemical characteristics. The mode of action of the antimicrobial properties of metal-NPs and metal oxides may lead to cell membrane damage, ROS, and inhibition of biofilm formation (Atacan et al. 2022).

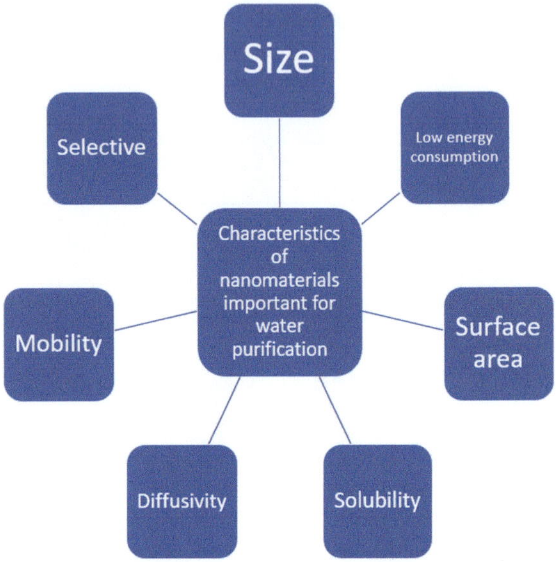

Fig. 10.2 Important nanomaterial characteristics for water purification

10.9.3.2 Carbon Nanotubes (CNTs)

CNTs are efficacious in the removal of contaminants, especially organic materials because of their higher adsorption capacity, high surface area, and variable site interactions in comparison with activated carbon (Sajid et al. 2022). The outer surfaces of specific CNTs have an effective surface area for adsorption. Effects like the hydrophobic effect, pi-pi interactions, covalent bonds, hydrogen bonds, and electrostatic forces between contaminant and CNT interactions allow the CNTs to adsorb numerous materials (Sajid et al. 2022). CNTs allow for pi-pi interactions with organic molecules, especially those containing double bonds. CNTs can increase their capacity for concentrating microbes by increasing their surface area. They are also effective in adsorbing antibiotics through electrostatic forces. They are also effective in adsorbing antibiotics and metal ions through electrostatic forces (Pérez et al. 2023).

10.9.3.3 Graphene

Graphene is a complex of carbon atoms that look like honeycomb—like lattice structures during SP^2 hybridization. It is the structural basis of all forms of graphite, and it has unique characteristics such as a large surface area, and specialized functional groups which make it an efficient material for wastewater treatment (Ali et al. 2019). Chemically modified graphene improves metal ion removal, achieving high sorption values. Graphene oxide (GO) nanosheets effectively adsorb heavy metals like Cd^{2+} and cobalt (Co^{2+}), with adsorption dependent on pH and influenced by humic acid.

Studies show GO's strong potential for large-scale water treatment, particularly for heavy metals like Pb^{2+}, Cd^{2+}, Cu^{2+}, and zinc (Zn^{2+}) (Pérez et al. 2023).

10.9.4 Advanced Oxidation Process (AOP)

AOP is a treatment process that significantly degrades organic and inorganic contaminants in wastewater. These contaminants frequently face ineffective removal when treated with traditional or conventional methods. AOPs, however, use a significantly different molecule for the treatment of highly reactive hydroxyl radicals (those with an OH group), which oxidizes these pollutants into carbon dioxide, inorganic ions, and water (Feng et al. 2023). Per and poly-fluoroalkyl substances (PFAS) are a class of ECs frequently present in pharmaceutical medications and insecticides. Ozone-based oxidation is a type of AOP that is popularly used against these pollutants by using ozone or O_3 to attack the pollutants and thus generate hydroxyl radicals (Liu et al. 2024).

Apart from generating hydroxyl radicals when exposed to UV light, photocatalytic oxidation utilizes semiconductor photocatalysts like TiO_2 that help to degrade organic pollutants and purify water, all the while using the minimum energy required. This method is quick, effective, and safe for the environment when it comes to eliminating stubborn pollutants (Azuazu et al. 2023; Liu et al. 2024). AOP such as photo-Fenton oxidation wherein ferrous ions (Fe^{2+}) react with hydrogen peroxide (H_2O_2) generates hydroxyl radicals in the presence of UV light or other activators (Liu et al. 2024). These techniques are effective in decomposing organic pollutants, removing heavy metals, and purifying water. Fenton and photo-Fenton techniques are inexpensive, have quick response kinetics, and are simple to incorporate into current water treatment infrastructure. In general, AOPs are useful for tackling issues such as disinfection byproducts and developing contaminants, protecting public health, and conserving water resources by decomposing persistent pollutants (Liu et al. 2024).

10.9.5 Phytoremediation

A green technique called phytoremediation uses plants to remove pollutants through several biological and physicochemical processes. Typically, plants absorb pollutants from the soil through their roots, and the microbes that live there help them with microbiological growth and transformation into plants (Bhardwaj et al. 2023b, 2025a). Numerous compounds secreted by roots help to create more active and diverse microbial populations in the rhizosphere (Shi et al. 2023). This plant-based method makes use of several naturally occurring processes, such as phytovolatilization, plant absorption, accumulation, metabolism, and rhizosphere microbial and flora breakdown (Fig. 10.3).

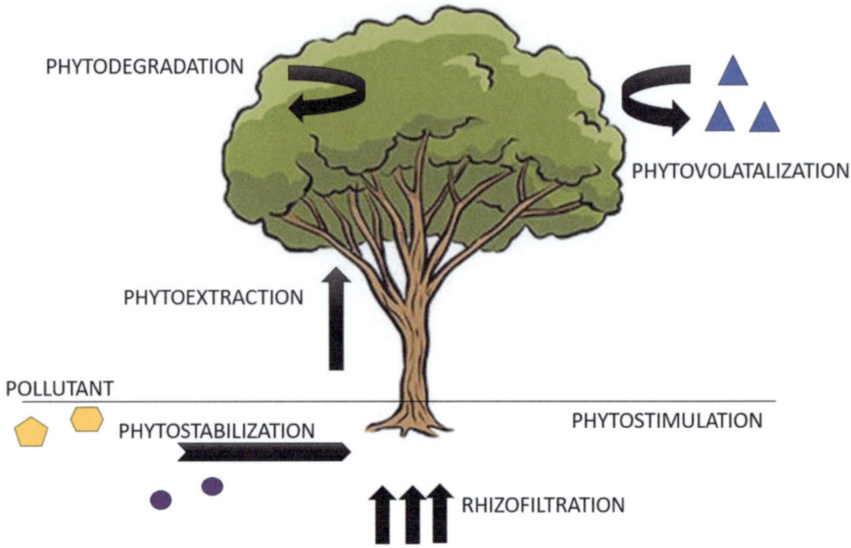

Fig. 10.3 Phytoremediation and its various processes

10.10 Public Policy and Government Regulation

Public policy and government regulation for ECs aim to reduce risks to human health and ecosystems through anticipatory measures. This includes identifying and characterizing contaminants based on toxicity, persistence, and exposure potential as well as promoting research into their effects. Regulatory frameworks emphasize in-depth risk assessments to establish safe thresholds, enforce discharge limitations, and require regular monitoring. Governments encourage pollution prevention through green chemistry, waste reduction, and incentives for cleaner technologies. Adaptive regulations guarantee that policies remain effective as new contaminants are identified. Public awareness campaigns further help to enhance compliance and sustainable practices.

10.11 Future Directions and Conclusions

ECs are hard to remove, build up in nature, and harm ecosystems, making it crucial to find effective solutions. Advanced technologies like filtration, oxidation, and bioremediation using plants, algae, and other microflora, offer hope when used together. However, technology alone is not enough. Future research should focus on the analysis of early detection of toxins and their on-site effects monitoring to understand the hazardous effects of ECs on the ecosystem which will help in understanding its risk assessment and management. Regulatory agencies should implement stringent guidelines for the discharge of ECs into the environment. The maximum permitted concentrations for different contaminants should be detailed by regulatory bodies. There is also a need for collaborations with different international and national governing bodies to develop standardized methods for monitoring and evaluating the environmental impact of ECs. Public awareness campaigns and educational initiatives that encourage the proper disposal of pharmaceuticals and PCPs can significantly reduce the EC. Furthermore, it is important to support incentives for industries to embrace green chemistry methods and develop eco-friendly alternatives to harmful chemicals. Implementing these policy initiatives will require a collaborative effort from governments, industries, and the scientific community to reduce ECs and protect our ecosystems. The authors can make everyone's future safer and healthier by bringing together innovations, stringent regulations, and public awareness.

Acknowledgements The authors sincerely thank Amity University, Noida, for their support and resources that facilitated the completion of this book chapter. Also, they acknowledge editors for their insightful discussions and feedback in completing this book chapter.

References

Ahmad HW, Bibi HA, Chandrasekaran M, Ahmad S, Kyriakopoulos GL (2024) Sustainable wastewater treatment strategies in effective abatement of emerging pollutants. Water (20734441) 16(20)

Alam S, Bhardwaj LK, Mallick R, Rai S (2023) Estimation of heavy metals and fluoride ion in vegetables grown nearby the stretch of River Yamuna, Delhi (NCR), India. Indian J Environ Prot 43(1):64–73

Albaseer SS, Al-Hazmi HE, Kurniawan TA, Xu X, Abdulrahman SA, Ezzati P, Habibzadeh S, Hollert H, Rabiee N, Lima EC, Badawi M (2024) Microplastics in water resources: global pollution circle, possible technological solutions, legislations, and future horizon. Sci Total Environ 173963

Ali I, Mbianda XY, Burakov A, Galunin E, Burakova I, Mkrtchyan E, Tkachev A, Grachev V (2019) Graphene based adsorbents for remediation of noxious pollutants from wastewater. Environ Int 127:160–180

Ali N, Khan MH, Ali M, Ahmad S, Khan A, Nabi G, Ali F, Bououdina M, Kyzas GZ (2023) Insight into microplastics in the aquatic ecosystem: properties, sources, threats and mitigation strategies. Sci Total Environ 169489

Asadi Z, Abdi H, Naji M, Rajabi H, Aminzadeh O, Madadi K (2021) Flocculation process and Increasing sedimentation of total suspended solids in clarifier. J Chem Lett 2(2):82–86

Atacan K, Güy N, Özacar M (2022) Recent advances in photocatalytic coatings for antimicrobial surfaces. Curr Opin Chem Eng 36:100777

Azimzada A, Farner JM, Hadioui M, Liu-Kang C, Jreije I, Tufenkji N, Wilkinson KJ (2020) Release of TiO 2 nanoparticles from painted surfaces in cold climates: characterization using a high sensitivity single-particle ICP-MS. Environ Sci Nano 7(1):139–148

Azuazu IN, Sam K, Campo P, Coulon F (2023) Challenges and opportunities for low-carbon remediation in the Niger Delta: towards sustainable environmental management. Sci Total Environ 165739

Bhardwaj LK, Jindal T (2019) Contamination of lakes in Broknes peninsula, East Antarctica through the pesticides and PAHs. Asian-J Chem 31(7):1574–1580
Bhardwaj LK, Jindal T (2020) Persistent organic pollutants in lakes of Grovnes Peninsula at Larsemann Hill area, East Antarctica. Earth Syst Environ 4(2):349–358
Bhardwaj L, Chauhan A, Ranjan A, Jindal T (2018) Persistent organic pollutants in biotic and abiotic components of Antarctic pristine environment. Earth Systems and Environment 2:35–54
Bhardwaj L, Sharma S, Ranjan A, Jindal T (2019) Persistent organic pollutants in lakes of Broknes peninsula at Larsemann Hills area, East Antarctica. Ecotoxicol 28:589–596
Bhardwaj LK, Rath P, Choudhury M (2023b) A comprehensive review on the classification, uses, sources of nanoparticles (NPs) and their toxicity on health. Aerosol Sci Eng 7(1):69–86
Bhardwaj LK, Rath P, Yadav P, Gupta U (2024) Microplastic contamination, an emerging threat to the freshwater environment: a systematic review. Environ Syst Res 13(1):8
Bhardwaj L, Jindal T (2022) Polar ecotoxicology: sources and toxic effects of pollutants. New Front Environ Toxicol 9–14
Bhardwaj LK, Kumar D, Kumar A (2023b) Phytoremediation potential of Ocimum sanctum: a sustainable approach for remediation of heavy metals. In: Phytoremediation potential of medicinal and aromatic plants. CRC Press, pp 46–57
Bhardwaj LK, Upadhyay D, Jain H, Rath P, Kumar D, Bajpai S (2025a) Recent trends of phytoremediation for the wastewater treatment and resource recovery. In: Biotechnologies for wastewater treatment and resource recovery. Elsevier, pp 135–146
Bhardwaj LK, Rath P, Rehman F, Yadav P, Massahi T, Mittal R, Jain H (2025b) Status of micro/nano-plastic (MNP) pollution in the aquatic system. In: Societal and environmental ramifications of plastic pollution. IGI Global, pp 43–62
Bhardwaj LK (2024) Occurrence of microplastics (MPs) in Antarctica and its impact on the health of organisms. Maritime Technol Res 6(2):265418
Canlı O, Oktem Olgun E, Güzel B, Kaplan M (2024) Sensitive and accurate determination of 168 micropollutants including pharmaceuticals and pesticides in surface water and wastewater samples with direct injection using jet stream ESI LC-MS/MS. Int J Environ Anal Chem 104(8):1778–1804
Česen M, Ahel M, Terzić S, Heath DJ, Heath E (2019) The occurrence of contaminants of emerging concern in Slovenian and Croatian wastewaters and receiving Sava river. Sci Total Environ 650:2446–2453
Chormey DS, Bakırdere S (2018) Principles and recent advancements in microextraction techniques. In: Comprehensive analytical chemistry, vol 81. Elsevier, pp 257–294
Corsi I, Winther-Nielsen M, Sethi R, Punta C, Della Torre C, Libralato G, Lofrano G, Sabatini L, Aiello M, Fiordi L, Cinuzzi F (2018) Ecofriendly nanotechnologies and nanomaterials for environmental applications: Key issue and consensus recommendations for sustainable and ecosafe nanoremediation. Ecotoxicol Environ Saf 154:237–244
Dharupaneedi SP, Nataraj SK, Nadagouda M, Reddy KR, Shukla SS, Aminabhavi TM (2019) Membrane-based separation of potential emerging pollutants. Sep Purif Technol 210:850–866
El-Kalliny AS, Abdel-Wahed MS, El-Zahhar AA, Hamza IA, Gad-Allah TA (2023) Nanomaterials: a review of emerging contaminants with potential health or environmental impact. Discover Nano 18(1):68
Feng H, Yang W, Zhang Y, Ding Y, Chen L, Kang Y, Huang H, Chen R (2023) Electroactive microorganism-assisted remediation of groundwater contamination: advances and challenges. Biores Technol 377:128916
Gangaraju G, Balakrishna K, Uma Devi R, Shah KJ (2021) Introduction to conventional wastewater treatment technologies: limitations and recent advances. In: Advances in wastewater treatment. Materials Research Foundations, pp 1–34. https://doi.org/10.21741/9781644901144-1
Gangopadhyay M, Naithani S, Singh G, Naithani P, Karmakar R (2024) Evaluation of emerging contaminants (ECs) in water resources, spatio-temporal assessment, and environment-friendly remedies: an Indian perspective. In: Climate crisis and sustainable solutions: strategies for

adaptation, mitigation and sustainable development. Springer Nature Singapore, Singapore, pp 303–322

Ghosh A, Tripathy A, Ghosh D (2022) Impact of endocrine disrupting chemicals (EDCs) on reproductive health of human. In: Proceedings of the zoological society, vol 75, no 1. Springer India, New Delhi, pp 16–30

Golovko O, Šauer P, Fedorova G, Kroupová HK, Grabic R (2018) Determination of progestogens in surface and waste water using SPE extraction and LC-APCI/APPI-HRPS. Sci Total Environ 621:1066–1073

Haque SE (2022) Urban water pollution by heavy metals, microplastics, and organic contaminants. In: Current directions in water scarcity research, vol 6. Elsevier, pp 21–43

Hashemi SH, Kaykhaii M (2024) Porous polymer sorbents in micro solid phase extraction: applications, advantages, and challenges. Top Curr Chem 382(4):1–33

Hussain A, Ashique S, Hassan MZ, Afzal O, Asiri YI, Kumar P, Dua K, Webster TJ, Altamimi AS, Altamimi MA (2023) Pharmaceutical contaminants in aquatic systems, conventional and green strategies, recent updates, challenges and policies, and potential outcomes. J Mol Liq 389:122905

Kammakakam I, Lai Z (2023) Next-generation ultrafiltration membranes: a review of material design, properties, recent progress, and challenges. Chemosphere 137669. https://doi.org/10.1016/j.chemosphere.2022.137669

Kaur R, Choudhary D, Bali S, Bandral SS, Singh V, Ahmad MA, Rani N, Singh TG, Chandrasekaran B (2024) Pesticides: an alarming detrimental to health and environment. Sci Total Environ 170113

Kesari KK, Soni R, Jamal QMS, Tripathi P, Lal JA, Jha NK, Siddiqui MH, Kumar P, Tripathi V, Ruokolainen J (2021) Wastewater treatment and reuse: a review of its applications and health implications. Water Air Soil Pollut 232:1–28

Khalikova M, Jireš J, Horáček O, Douša M, Kučera R, Nováková L (2024) What is the role of current mass spectrometry in pharmaceutical analysis? Mass Spectrom Rev 43(3):560–609

Kreyling WG, Semmler-Behnke M, Chaudhry Q (2010) A complementary definition of nanomaterial. Nano Today 5(3):165–168

Li X, Shen X, Jiang W, Xi Y, Li S (2024) Comprehensive review of emerging contaminants: detection technologies, environmental impact, and management strategies. Ecotoxicol Environ Saf 278:116420

Lima DR, Gomes AA, Lima EC, Umpierres CS, Thue PS, Panzenhagen JC, Dotto GL, El-Chaghaby GA, de Alencar WS (2019) Evaluation of efficiency and selectivity in the sorption process assisted by chemometric approaches: removal of emerging contaminants from water. Spectrochim Acta Part Mol Biomol Spectrosc 218:366–373

Liu X, Sathishkumar K, Zhang H, Saxena KK, Zhang F, Naraginiti S, Anbarasu K, Rajendiran R, Aruliah R, Guo X (2024) Frontiers in environmental cleanup: recent advances in remediation of emerging pollutants from soil and water. J Hazard Mater Adv 100461

Mahmood T, Momin S, Ali R, Naeem A, Khan A (2022) Technologies for removal of emerging contaminants from wastewater. Wastewater Treat 159:1–20

Mishra RK, Mentha SS, Misra Y, Dwivedi N (2023) Emerging pollutants of severe environmental concern in water and wastewater: a comprehensive review on current developments and future research. Water-Energy Nexus

Norvill ZN, Shilton A, Guieysse B (2016) Emerging contaminant degradation and removal in algal wastewater treatment ponds: identifying the research gaps. J Hazard Mater 313:291–309

Ortúzar M, Esterhuizen M, Olicón-Hernández DR, González-López J, Aranda E (2022) Pharmaceutical pollution in aquatic environments: a concise review of environmental impacts and bioremediation systems. Front Microbiol 13:869332

Pahazri NF (2018) A study on stryi-icnos potatorum and pisum sativum as natural coagulants for meat food processing wastewater. Doctoral dissertation, Universiti Tun Hussein Onn Malaysia

Pathak VM, Verma VK, Rawat BS, Kaur B, Babu N, Sharma A, Dewali S, Yadav M, Kumari R, Singh S, Mohapatra A (2022) Current status of pesticide effects on environment, human health and it's eco-friendly management as bioremediation: a comprehensive review. Front Microbiol 13:962619

Pérez H, Quintero García OJ, Amezcua-Allieri MA, Rodríguez Vázquez R (2023) Nanotechnology as an efficient and effective alternative for wastewater treatment: an overview. Water Sci Technol 87(12):2971–3001

Pilapitiya PGCNT, Ratnayake AS (2024) The world of plastic waste: a review. Clean Mater 11:100220. https://doi.org/10.1016/j.clema.2024.100220

Pulicharla R, Proulx F, Behmel S, Sérodes JB, Rodriguez MJ (2022) Spatial and temporal variability of contaminants of emerging concern in a drinking water source. RSC Adv 12(32):20876–20885

Rath P, Bhardwaj LK, Yadav P, Bhardwaj AK (2024) A synthesis of biogenic nanoparticles (NPs) for the treatment of wastewater and its application: a review. Biog Wastes-Enabled Nanomater Synth: Appl Environ Sustain 127–148

Rolph CA, Jefferson B, Brookes A, Hassard F, Villa R (2020) Achieving drinking water compliance levels for metaldehyde with an acclimated sand bioreactor. Water Res 184:116084

Sajid M, Asif M, Baig N, Kabeer M, Ihsanullah I, Mohammad AW (2022) Carbon nanotubes-based adsorbents: properties, functionalization, interaction mechanisms, and applications in water purification. J Water Process Eng 47:102815

Samal K, Mahapatra S, Ali MH (2022) Pharmaceutical wastewater as emerging contaminants (EC): treatment technologies, impact on environment and human health. Energy Nexus 6:100076

Sharma M, Agarwal S, Malik R, Kumar G, Dan Bahadur P, Mandal M, Sarkar A, Bantun F, Haque S, Singh P, Srivastava N, Vijai KG (2023) Recent advances in microbial engineering approaches for wastewater treatment: a review. Bioeng 14(1). https://doi.org/10.1080/21655979.2023.2184518

Shi Q, Kaur P, Gan J (2023) Harnessing the potential of phytoremediation for mitigating the risk of emerging contaminants. Curr Opin Environ Sci & Health 32:100448

Tang Y, Yin M, Yang W, Li H, Zhong Y, Mo L, Liang Y, Ma X, Sun X (2019) Emerging pollutants in water environment: occurrence, monitoring, fate, and risk assessment. Water Environ Res 91(10):984–991

Tokas R, Bhardwaj LK, Kumar N, Jindal T (2024) Nanotechnology for sustainable development and future: a review. In: Green and sustainable approaches using wastes for the production of multifunctional nanomaterials, pp 221–233

Tong X, Mohapatra S, Zhang J, Tran NH, You L, He Y, Gin KYH (2022) Source, fate, transport and modelling of selected emerging contaminants in the aquatic environment: current status and future perspectives. Water Res 217:118418

Torres-Herrera S, González-Cortés JJ, Almenglo F, Yeste MP, Ramírez M, Cantero D (2022) Optimization of biogenic sulfur flocculation from an anoxic desulfurization bioreactor using response surface methodology. Fuel 323:124367

Ullah Z, Peng L, Lodhi AF, Kakar MU, Mehboob MZ, Iqbal I (2024) The threat of microplastics and microbial degradation potential; a current perspective. Sci Total Environ 177045

Verderio P, Lecchi M, Ciniselli CM, Shishmani B, Apolone G, Manenti G (2023) 3Rs principle and legislative decrees to achieve high standard of animal research. Animals 13(2):277

Wang L, Liu R, Liu X, Gao H (2020) Sampling rate of polar organic chemical integrative sampler (POCIS): influence factors and calibration methods. Appl Sci 10(16):5548

Wang F, Xiang L, Leung KSY, Elsner M, Zhang Y, Guo Y, Pan B, Sun H, An T, Ying G, Brooks BW (2024) Emerging contaminants: a one health perspective. Innovation 5(4):100612

Włodarczyk R, Kwarciak-Kozłowska A (2021) Nanoparticles from the cosmetics and medical industries in legal and environmental aspects. Sustain 13(11):5805

Yang Y, Li C, Yang L, Zheng M, Liu G (2023) Application of non-target screening by high-resolution mass spectrometry to identification and control of new contaminants: implications for sustainable industrial development. Sustain Horiz 5:100049

Yang W, Shi M, Zhao T, Xu Z, Chu W (2024) Unseen streams tracing emerging contaminants from stormwater to surface water: a brief review. J Environ Sci

Zahmatkesh S, Bokhari A, Karimian M, Zahra MMA, Sillanpää M, Panchal H, Alrubaie AJ, Rezakhani Y (2022) A comprehensive review of various approaches for treatment of tertiary wastewater with emerging contaminants: what do we know? Environ Monit Assess 194(12):884

Zhang X, Liu X, Zhan C (2024) Emerging contaminants in aquatic ecosystems: sources, effects, and mitigation approaches. Int J Aquac 14

Zhou J, Chang Y, Tian H, Qu JH, Wang Q, Wang J, Jiang Z (2024) Recent advances in accelerating solid-phase extraction. J Chromatogr Open 100156

Chapter 11
Regulatory Framework and Policy Responses

Anamika Shrivastava and Harshita Jain

Abstract Emerging contaminants (ECs), not commonly monitored, are synthetic or naturally occurring chemicals or agents of biological origin that have the potential to enter the environment and cause adverse effects on human health or the ecosystem. They originate from several anthropogenic activities such as chemical leaching, domestic wastes, improper handling, and disposal of chemicals and have caused pollution of varying degrees to different spheres of the environment. They have been classified into various groups such as plasticizers, surfactants, pharmaceuticals, and personal care products (PPCPs), nanomaterials. The ECs have been reported to be a serious issue in the aquatic ecosystem with severe adverse impacts and risen into a major threat to water distribution system. As a result of the increasing concentration of ECs, there is a need for prioritization of its regular assessment and monitoring. Various policies and regulatory frameworks have been developed regarding the levels of contaminants in the environment which combine source-related measures along with environmental quality standards. However, the total emission of pollutants into the environment remains on a constant rise. Currently, the protection of the aquatic ecosystem from ECs requires the implementation of more stringent policies, standards, and frameworks for its control. Moreover, the standards released by the Bureau of Indian Standards (BIS) and World Health Organization (WHO) also need to be updated from time to time, especially with respect to ECs which are possible with recent data and research in the area. The present chapter focuses on the existing laws and standards to control emerging pollutants in the aquatic ecosystem, their shortcomings, and suggestions for their strengthening. Additionally, the role and responsibilities of the regulatory bodies are also discussed to explore the future course of action for the prevention and control of ECs' release.

Keywords Emerging contaminants (ECs) · Policies and laws · Environment · Human health

---

A. Shrivastava (✉) · H. Jain
Amity Institute of Environmental Sciences (AIES), Amity University, Noida, Uttar Pradesh, India
e-mail: ashrivastava1@amity.edu

H. Jain et al. (eds.), *Emerging Contaminants in the Aquatic Environment*, Advances in Wastewater Research, https://doi.org/10.1007/978-981-96-8548-6_11

List of Abbreviations

AI	Artificial intelligence
BIS	Bureau of Indian Standards
CECs	Contaminants of emerging concern
DDT	Dichlorodiphenyltrichloroethane
ECs	Emerging contaminants
EEA	European Environment Agency
EIPPPs	Eco-industrial parks and public-private partnerships
EPA	Environmental Protection Agency
EPR	Extended producer responsibility
EU	European Union
Hg	Mercury
IMO	International Maritime Organization
ML	Machine learning
MPs	Microplastics
NGOs	Non-governmental organizations
OECD	Organization for Economic Co-operation and Development
PCBs	Polychlorinated biphenyls
PCPs	Persistent chlorinated pesticides
PFAS	Per- and polyfluoroalkyl substances
PFOA	Perfluorooctanoic acid
POPs	Persistent organic pollutants
PPCPs	Pharmaceuticals and personal care products
PPP	Public-private partnership
PUB	Public Utilities Board
SDGs	Sustainable Development Goals
UN	United Nations
UNEP	United Nations Environment Program
UNIDO	United Nations Industrial Development Organization
WFD	Water Framework Directive
WHO	World Health Organization
WWF	World Wide Fund for Nature

11.1 Introduction

New-generation compounds or relatively recent substances that have later been held to have deleterious impacts on populations and the environment are known as ECs or contaminants of emerging concern (CEC) (Boahen et al., 2023). There is a plethora of them and many of them are by-products of modern industrial and consumer processes: medications, personal care products (PCPs), industrial chemicals, pesticides, microplastics (MPs), nanomaterials, and endocrine disruptors (Puri et al., 2023). They commonly enter the environment through agricultural run-off, industrial discharge, effluents, wastewater discharge, and atmospheric deposition wherein they accumulate in soil, water, and air (Meng et al., 2021; Boahen et al., 2023).

Some ECs are present in very low concentrations: parts per billion or trillion, and many are parts per trillion, which implies severe consequences. For instance, MPs and nanoplastics disrupt food chains, and per- and polyfluoroalkyl substances (PFAS) also called “forever chemicals” are persistent and bioaccumulate (Geissen et al., 2015; Wilkinson et al., 2017). Due to their low levels of occurrence as well as their functions within complex environmental structures, the management of ECs is always very challenging. The impacts of ECs are still under study, while these are unlike many general pollutants, not yet fully controlled or often closely analyzed (Gavrilescu et al., 2015; Puri et al., 2023).

In this regard, the need for such contaminants to be captured in environmental legislations and preventive approaches when our knowledge of them is advanced cannot be downplayed as far as the health of biological diversity, public health, and environmental conservation are concerned. ECs pose significant challenges to the environment due to their potential harm to human health and ecosystems. Advances in technology have improved our ability to detect and monitor these contaminants, but effective treatment methods remain a challenge (Geissen et al., 2015; Wilkinson et al., 2017). Continued research and innovation in treatment and remediation technologies are crucial for addressing EC issues in environmental management and policy. The safety of our soil and water, as well as the preservation of our ecosystem, depends on ongoing study and innovation in this area (Meng et al., 2021). To protect the health and welfare of present and future generations, environmental managers and legislators must move quickly to address new contamination problems (Wilkinson et al., 2017).

With increased regulatory standards adopted together with capital investments in sustainable ways, they strive to minimize the proliferation of ECs in their environment. Social research work has to involve efforts from scientists, policymakers, and industrialists in devising proper ways of combating this increasing phenomenon (Boahen et al., 2023). Hence, everyone ought to embrace the proper health of our planet and ensure we do not allow these chemicals to wreak havoc on people’s health and our environment further. The situation of developing toxins in the environment is well controlled, they have to work hard to come up with stricter measures and provide funding for sustainable processes (Dulio et al., 2018; Puri et al., 2023). Finding practical answers to address this expanding problem requires cooperation between scientists, legislators, and business executives. Prioritizing the well-being of our planet and taking proactive steps to stop additional harm from these dangerous pollutants are essential.

11.1.1 Importance of Regulatory and Policy Measures

Due to their characteristics, ECs offer challenging problems that require decisions on the regulatory and policy levels. They frequently resist the implementation of standardization environmental managerial processes because of their characteristics and low visibility (Puri et al. 2023). Some of the hazards that have been identified in this category are MPs, body products, drugs, and PFAS are very risky to the environment, wildlife, and society. In this way, regulation means suggesting how or what should be monitored, controlled, and managed to reduce such risks, setting up acceptable confines and prioritizing some of them while putting mechanisms in the provision of regulating their source (Wilkinson et al., 2017).

In a similar manner, they give impetus to investments in research and development of advanced detection and treatment technologies that help reduce the environmental burdens of the ECs such as nanofiltration, ozonation, and bioremediation (Wilkinson et al., 2017; Meng et al., 2021). In addition, to ensure consumers engage in responsible behaviors and the policy deployments are successful, complementary frameworks that would support the interaction of stakeholders, and raise public awareness of the policies are very crucial (Gavrilescu et al., 2015; Geissen et al., 2015). Protections of water quality, soil integrity, and human welfare could be at long term, and in some cases, permanent risk from unchecked environmental accumulation of ECs due to inadequate policies and regulation. Therefore, to ensure the unsustainability of the dangers of ECs, flexible and progressive policies are required and necessary to maintain ecological sustainability and protect subsequent generations (Wilkinson et al., 2017). Awareness and control strategies of ECs as a policy and legal instrument have significant impacts on conserving human and environmental health. Such measures assist in setting upper limits to pollutants in water, air, and the soil and ensure that polluters; be they industrial goliaths or average citizens, are made to answer for their actions (Meng et al., 2021). By implementing and adhering to these regulations, we are able to minimize the spread of contaminants across our environment to promote a clean and safe environment for everyone in society.

As pointed out by Schellenberg et al. (2020), regulatory and policy measures can only play limited roles in dealing with ECs or even prevent or control the contamination since there are always cracks in the system or lack of enforcement or implementation of the measures apart from contestations that result from how new knowledge of the contaminants emerge. Besides, there is the danger of over-reliance on regulations which hinders the development of new and sustainable solutions to the new EC challenges (Boahen et al., 2023). For instance, due to contaminants in drinking water supplies such as perfluorooctanoic acid (PFOA), the regulatory agencies set lower acceptable rates of the chemical (Eriksen and Perrez, 2014). Conducting also established measures to ensure that manufacturers dealing in PFOA-containing products ensured limited pollution by the chemical (Meng et al., 2021). But in one case this absence of regulation of PFAS in drinking water when scientific literature shows that they are toxic to humans. This, therefore, brings to light the idea that there is a need to enhance the enforcement of eighty measures of ECs in an effort to prevent them from evolving into typical threats to the population's health.

11.1.2 Global Challenges in Regulating Emerging Contaminants (ECs)

There are significant global challenges in the management of Education and Legal frameworks that relate to ECs due to their high variability, abundance, and scientific departure. Due to the fact that each of the categories of ECs is detected, monitored, and eliminated with different techniques, the regulation of each category is complicated. Besides, there is poor information about its toxicity and its effects on the environment, biological systems, and human beings which is an obstacle to formulating evidence-based rules and regulation (Johnson and Bell, 2023). Moreover, the ability to detect these ECs at trace levels also varies significantly across countries, primarily because techniques such as high-resolution mass spectrometry, and other modern analytical tools used in analyzing the samples remain expensive plus are not easily accessed in the low-middle income countries (Meng et al., 2021).

Some of the global challenges that affect the regulation of ECs include first, the pace at which scientific research on new contaminants is continuously changing, second, the lack of policy harmony that is between international borders, and third, industry pressure that exerts an influence in policy formulation (Meng et al., 2021). In view of this, regulatory agencies must remain active and responsive to address new contaminants as pointed out by Low et al. (2024). These analyses show that work to address the risks associated with these substances requires efficient cooperation between scientists, policymakers, and industry representatives (Boahen et al., 2023). Another factor that cannot go unnoticed is that monitoring and regulation of ECs should look at the implications of the contaminants bearing in mind the consequences on health as well as the surroundings. The problem of accountability and transparency in the regulation process is also raised to enhance the protection of the public interest over industry revenues. With the implementation of multi-disciplinary collaboration, we can develop a stronger and more adaptive regulatory system capable of protecting our community from the threats posed by new contaminants in our environment.

For instance, thereof we have PFAS which according to recent findings has been credited for diseases such as cancer, and reproductive disorders not forgetting pollution of water resources (Boahen et al., 2023). To carry out the research scientists, industry participants, and the general public need to collaborate to develop more effective monitoring strategies and apply more rigorous standards regulating the usage and disposal of products containing PFAS (Meng et al., 2021). Thus, collaborations with all sorts of stakeholders are necessary, although stringent regulation alone might not suffice to solve the problem of PFAS contamination (Puri et al., 2023). Moreover, there are possibilities of difficulties to facilitate greater control on regulations compliance because of the sophistication of these chemicals and their immense application in products.

Unfortunately, the final issue is the lack of public participation and awareness; consumers remain unaware of how they pollute the EC due to inadequate disposal of devices like electronics, plastics, etc., or disposal of medications (Puri et al., 2023). These problems can be solved only with the help of a unified approach all over the world and more funding for research and technologies, as well as the support of strong international laws to provide reasonable and non-prejudiced regulating of ECs to protect the people's health and the environment.

11.2 International Frameworks and Conventions

Again, to some extent, international cooperation in other environments, and social and economic problems are greatly facilitated through international frameworks and conventions. Through legal, policy, and operational guidelines, these accords enable nations to solve pressing issues such as pollution, climate change, loss of species diversity, and sustainable development (Mohapatra et al., 2023). The Stockholm Convention addresses the problem of persistent organic pollutants (POPs), and the Basel Convention seeks to limit hazardous wastes that can impact the health of human beings and the environment (Sharma et al., 2014). These frameworks are important for promoting commonality, synthesizing activities internationally, and ensuring accountability in the achievement of set goals (Aidonojie and Anani, 2024). International conventions can bring joint actions of governments, organizations, and other interested members and provide working mechanisms for discussing issues that cannot be solved by one state only (Khalid et al., 2018; Puri et al., 2023). However, success rates depend on their sound implementation, adequate funding, and the political will of the member countries to honor their commitments and adapt to new global challenges (Hendry, 2017; Mohapatra et al., 2023).

Such cooperation can also give the assurance that all countries are on the right track in combating this problem and they can pool resources and study information in the process (Puri et al. 2023). In addition, inviting input from different global agencies and scholars shall improve the efficiency of further monitoring and prevention activities which positively reflects the necessity of achieving better sustainability and security in the world (Johnson and Bell, 2023; Ebinghaus 2023). Sometimes this may also prove disadvantageous since proper standards and targets may be set by different countries and it may always be challenging to have international collaboration in such a way that a host of targets will suit all nations in agreement. Moreover, to overcome these challenges countries need to find institutionalized ways of reporting and coordinating. This can include status meetings and sharing of data with all the appropriate stakeholders as well as the creation of specific adherence to best practices for data gathering and assessment (Hendry, 2017). This way it becomes easy for countries to coordinate the implementation process of the policies and ensure that everyone is moving in the direction of a common goal or agenda. In addition, it also creates opportunities to share ideas related to better practices and the effective and efficient solutions of ECs to achieving long-term goals.

11.2.1 Key Conventions on Emerging Contaminants (ECs)

International responses to other pressing environmental and health matters concerning hazardous substances and waste have indispensable intergovernmental treaties such as the Stockholm, Basel, and Minamata Accords (Bassett, 2016; Khalid et al., 2018; Šebková, 2023). The Stockholm Convention on POPs aims to eliminate or restrict the use and production of dangerous toxic substances that accumulate in ecosystems and are hazardous to human and animal health: for example, dioxins, polychlorinated biphenyls (PCBs), and dichlorodiphenyltrichloroethane (DDT) (Selin, 2014; Bassett, 2016). Basel Convention controls the transboundary movement of hazardous wastes effectively as avoiding dumping of wastes in developing nations also encapsulates the regulation of the export/import of wastes and aims at minimizing the generation and disposal of hazardous wastes (Kummer, 1999).

The use of mercury (Hg) is reflected in goods, processes, and emissions and has been regulated by the Minamata Convention on Hg which was named for Hg poisoning devastation in Japan (Khalid et al., 2018). Altogether, these treaties are useful for exacting cooperation, capacity, and responsibility in managing and dealing with pollutant and hazardous waste. This chapter will argue that their implementation is key to attaining sustainable development and the preservation of the environment for future generations (Low et al., 2024). They have been developed to solve environmental problems and control the exportation of dangerous materials around the world (Low et al., 2024). These conventions aspire to facilitate collaboration between countries to prevent health risks and adverse impact of pollutants on the natural world (Malisch et al., 2023).

More so by following the agreements and developing practices predicting the future health of the planet for future generations (Low et al., 2024). There is more that needs to be done in order to improve the state of our planet and work together as humanity to combat climate change. Such cooperation would improve the current environment and make it even healthier for future generations as would be seen from the following highlights by Bassett (2016). There are several conventions that need to be followed by all the countries in the world and chemists and engineers need to work toward the eradication of problems caused by pollution and toxic substances which leads to loss of biodiversity and someday is not so far when maybe the whole world will turn into a desert so it's very important for every country to take some steps toward ensuring health of the environment.

It can be done by compliance with policies that assemble enhanced restrictions and sanctions, ensuring effective sustainable policies, and putting into practice efficient green technologies (Malisch et al., 2023). Besides, cooperative efforts among state actors, business actors, and non-state actors are critical in managing the environmental challenges globally (Bassett, 2016). It is high time that people pay attention to the conditions of their environment and devote much attention to the inhabitants of the Earth (Low et al., 2024). Hence, while commitment and financial support for multidisciplinary collaboration in green technologies and policies are useful ways of approaching counter-environmental issues, overemphasizing stringent laws may not effectively guarantee compliance/implementation (Malisch et al., 2023). Moreover, the choice of the ecological approach directly interferes with the economic needs of the Earth and the well-being of all living creatures, which can be barely compatible with politics (Bassett, 2016).

11.2.2 Role of the United Nations Environment Program (UNEP) and Contributions of Other International Organizations

United Nations Environmental Program (UNEP) fulfills the important function of exercising leadership, fostering cooperation, and driving forward the international environmental protection regime in addressing the challenges that new ECs pose (UNEP, 2011). UNEP develops, and enhances, scientific assessments, and serves as an implementing agency for instruments regulating pollutants and hazardous substances, including the Stockholm, Basel, and Minamata Conventions of the United Nations (UN) (2011). Through programs like the Global Chemicals Outlook and partnerships such as the Global Environment Facility, UNEP supports countries in identifying emerging pollutants, developing regulatory frameworks, and implementing sustainable development initiatives (Yakubu, 2017).

EC global initiatives are coordinated by the UNEP (Yakubu, 2017). The UNEP facilitates countries' efforts to mitigate the impact of these chemicals on human health and the environment through research, policy development, and capacity-building initiatives. These bodies also offer invaluable support, as do global organizations such as the WHO and the Organization for Economic Cooperation and Development (OECD) (UNEP, 2011). All these organizations collaborate to shape sustainable solutions to address the ECs, whose impact on human health and ecosystems is growing. Through partnerships and knowledge and best practice sharing, these organizations can more effectively handle the complex issues posed by ECs (Puri et al., 2023).

Encouraging information sharing and cooperation, organizations also assist governments and industries in their decision-making processes, seeking to ensure that these contaminants are not released into nature (Yakubu, 2017). Their approach also includes steps for raising awareness and education with the aim of enabling members of the public to act themselves, both as protectors of themselves from ECs and in their efforts on behalf of protection against these classes of compounds (UNEP, 2011). The combined initiatives are making headway toward a cleaner, healthier Earth for generations to come. An annual report published in 2007 by the European Environment Agency (EEA) reported that "Member States are supporting these ambitions and efforts within their countries" as "some Member States have already implemented measures to address ECs such as MPs through research development, policy, and regulation."

Through the sharing of data and best practices, they can develop effective regulations and guidelines to minimize these contaminants' effects on aquatic ecosystems. UNEP's work is also supported by a large number of other international organizations. For instance, the International Maritime Organization (IMO) works against pollution that has chemicals and MPs in waters, while the WHO assesses toxins' effects on health (Najam, 2003; UNEP, 2011; Puri et al., 2023). The United Nations Industrial Development Organization (UNIDO) supports cleaner industrial technology (Biermann, 2007). These organizations collaborate to close knowledge gaps, encourage capacity-building, and offer financial and technical assistance, empowering nations to successfully manage new pollutants and reduce the hazards they pose to ecosystems and public health (Stephan and Zelli, 2018).

11.2.3 Limitations and Gaps in International Agreements

ECs have been addressed in international treaties; however, those are not cohesive, and certain elements are missing, such as reliable reporting and databases between states and general principles of how to control and manage these contaminants (Sanganyado, 2022). Additionally, there is a worrying deficiency of implementation and enforcement features in many treaties which leads to pollution and the continuing degradation of people's health and that of the ecosystem (Puri et al., 2023). If these weaknesses and shortfalls are not complemented, there would be a tangible challenge for the global effort to deal with the threats posed by the ECs which are on the rise. It is of great significance that the countries work to develop standards that are not only clear but also applicable with regard to ECs (Meng et al., 2021). This would require improvement in the level of confidentiality as well as communication among states in order to capture the needed accurate data for implementation.

In addition, there should be more robust accountability measures that can help trace the sources of pollution among countries and identify potential measures to minimize the risk to people's health and the environment (Aidonojie and Anani, 2024). It is only through these interventions that the international community will be able to thwart the menace posed to ECs and safeguard the world for future generations. If these strategies are collaborated on and implemented, it will go a long way in cleaning and making the world a safer place for many. Furthermore, new compounds appear more quickly than international standards can adjust; many international accords fail to take into consideration the changing nature of the emerging pollutants (Meng et al., 2021).

Establishing precise guidelines is made more difficult by the scientific ambiguity around the effects of these pollutants on the environment and human health (Wilkinson et al., 2017). Moreover, monitoring, data gathering, and the enforcement of international agreements are complicated by the differences in resource availability and capacity between developed and poor countries (Aidonojie and Anani, 2024).

The world community runs the risk of failing to adequately handle the escalating threat of ECs in the absence of greater collaboration, more inclusive regulations, and flexible frameworks.

11.3 National-Level Policies in Emerging Contaminants (ECs)

Protecting public health and the environment (Talib and Randhir, 2016) requires national-level policies related to ECs in India. Thus, the government needs to incorporate monitoring and controlling these contaminants in order to stop the negative impacts of these contaminants on the water sources, the environment as well as human health (Naidu et al., 2016). India should also collaborate with international organizations and experts to stay abreast of the latest research and technology for addressing ECs effectively. India can ensure future generations have a safe and sustainable environment while enacting strong policies and enforcement measures (Noguera-Oviedo and Aga, 2016). Furthermore, meaningful decisions regarding the sources and impacts of ECs will also require more investment in research and development to further understand these sources and impacts.

For instance, public awareness campaigns and education programs can also help create responsible behavior and curb the use of harmful substances (Naidu et al., 2016; Schellenberg et al., 2020). India can take a lead in environmental protection, and prove to other countries what to do, by taking a proactive approach to address the rise of ECs. Although investment in research and education programs is important, it might not be highly enough to tackle the issue of ECs until regulations and enforcement mechanisms are realigned to tackle it (Wilkinson et al., 2017). Further, public awareness campaigns to change behavior commonly are ineffective, and more specific actions may be required to conserve for future generations a safe and sustainable environment.

11.3.1 Public versus Private Sector Roles in Regulation of Emerging Contaminants (ECs)

Typically, roles for the public sector involve the setting of and enforcing regulatory standards for the protection of public health and the environment in the works of EC regulations. At the same time, the Environmental Protection Agency (EPA) and other public sector agencies lead research, monitor contaminants, and write and enforce regulations to deal with new threats (Shaikh and Birajdar, 2024). Private sector roles, on the other hand, will involve developing and putting into practice methods to detect and remediate contaminants and working with regulation agencies to ensure that standards are being met. Each of the roles played by public and private sector entities in addressing the challenges posed by ECs stands in supporting human health and the environment (Richardson, 2009; Sánchez-Ocampo et al., 2022).

Partnering in synergy, between the public and private sectors can superpower each other's capabilities of addressing new contaminants successfully (Shaikh and Birajdar, 2024). Together, this collaboration guarantees that total and powerful answers are made to watch over our ecological surroundings and well-being from consistently developing perils of arising contaminants (Sánchez-Ocampo et al., 2022). The EPA might give guidance for testing methods and regulatory standards for a new chemical that is an EC, and a company may use that guidance to create a next-generation filtration system to remove the contaminant from drinking water supplies (Milano and Zugliani, 2019). Sharing knowledge and resources, both sectors can jointly face the challenges introduced by new contaminants and protect public health and environmental safety (Sánchez-Ocampo et al., 2022). Collaboration between private companies and the public sector can produce inventive solutions, but profit-driven motives degrade the effectiveness and safety of filtration systems (Shaikh and Birajdar, 2024).

11.3.2 Policy Gaps in Low-Income Countries

ECs in low-income countries have the potential to create policy gaps since governments are not in a position to fund such services. These countries usually fail to equip themselves with adequate resources and infrastructure for monitoring these contaminants, thus exposing vulnerable groups to more risks of contact (Sanganyado, 2022). It is very important to close these policy loopholes so as to guarantee the health of the persons in those areas and probably the sustainability of their assets (Gwenzi et al., 2022). Further, as Taheran et al. (2018) pointed out, there is a need to expand funding, monitoring, and research to identify new emerging compounds to determine the possible extent of dangers that they pose and come up with ways of mitigating them carefully.

Finally, with the help of proper actions for the challenges mentioned above, we can achieve a better health and future. Thus, the relationships between governments, scholars, and industries can play a significant role in solving the problem of using water for PFAS-containing products and the development of treatment technology and legislation (Sanganyado, 2022a). Both organizations and citizens will be able to collaborate, using data and research, to stop environments from becoming further polluted, and to shield communities from the effects of new substances. Monitoring and research also receive additional funding with improved early identification of the threats hence improved response times (Taheran et al., 2018). As much as collaboration and enhanced funding can support combating environmental issues, it is also crucial to promote dependency in preventing and combating pollution (Gwenzi et al., 2022). The large-scale implication of education and awareness to the consumers can go a long way in making the desired changes toward minimizing waste and its impacts on the environment.

Moreover, most low-income countries depend on donor funding, which is unpredictable and uncoordinated (Sanganyado, 2022). In addition, many developing countries are dependent on aid which is unpredictable and often arbitrary (Sanganyado, 2022). Inadequate legislative frameworks for pollution control may arise from national governments prioritizing more pressing developmental issues above environmental and public health issues (Taheran et al., 2018). Additionally, poor enforcement of any current restrictions is a result of low stakeholder engagement and public understanding of the dangers posed by developing pollutants (Xabadia et al., 2021). There is also often a lack of collaboration between government, business, and non-governmental organizations, leading to fragmented reactions to ECs (Sanganyado, 2022). The scope and complexity of emerging pollutants may not be adequately addressed by capacity-building programs, and environmental policies in low-income nations may not be in line with international norms (Xabadia et al., 2021). The inability of low-income nations to close these policy gaps hinders their capacity to reduce the long-term health and environmental hazards brought on by these pollutants.

11.4 Regulatory Challenges

The implication of policies for ECs is fraught with some challenges by virtue of the fact that they are ECs that are relatively complex and remain heavily regulated they include (Milano and Zugliani, 2019). Therefore, any policymaker has a challenge in how he/she can update his/her working knowledge on the latest research and technological advancement in order to tackle the challenges arising from the emergence of contaminants (Rathi et al., 2021). Perhaps, initiating policies for the public's healthy and safe environment can be done with successful partnerships between government agencies, industries, and environmentalists (Petrie et al., 2015). As for these policies, their constant monitoring and assessment are required ultimately to overcome potential difficulties that may arise and to make the necessary improvements that may ensure the implementation of these policies (Rathi et al., 2021). It is thus clear that when these stakeholders are in coordination, regulation can always meet the required objective of reflecting change.

Currently, one may not notice incomplete aspects in a policy but when in communication with change agents, they may be pointed out as critical issues that would erupt in the future. When the government is involved, the proactivity of communities and incentives for cooperation, the health and well-being of a community, as well as the environment for future generations, is enhanced (Petrie et al., 2015). For instance, in the consideration of new emissions regulations for industrial factories, a routine assessment can assist in determining areas of weaknesses or noncompliance that are likely to occur (Rathi et al., 2021). Sometimes potential problems can be identified when stakeholders like environmental groups, industries, and government agencies are involved in the evaluation process (Milano and Zugliani, 2019). It can, therefore, result in enhanced airflow and the health status of inhabitants of areas close to these factories.

Although engaging the stakeholders in the assessment process could assist in the discovery of the noncompliance concerns, the method, in turn, causes disruption and controversy that slow down the key enchants of the essential rules (Parida et al., 2021). It is crucial for policymakers to stay informed about the latest research and technological advancements in order to effectively address the risks associated with ECs (Rathi et al., 2021). Collaboration among government, industry, and environmental advocates is needed for the development of comprehensive and effective policies that can protect public health and the environment (Petrie et al., 2015). Furthermore, ongoing monitoring and evaluation are essential to address challenges and make adjustments to ensure success (Rathi et al., 2021).

Regular communication and collaboration can help identify potential gaps in current policies and address them before they become major issues. A proactive and cooperative approach can help policymakers to protect the health and well-being of communities and also protect the environment for future generations (Petrie et al., 2015). An example is the case of making new emissions regulations for factories and continuous evaluations can help in identifying the loopholes and compliance issues that may arise (Rathi et al., 2021). By involving stakeholders such as environmental groups, industry representatives, and government agencies in the evaluation process, potential problems can be addressed promptly to ensure the effectiveness of the regulations (Milano and Zugliani, 2019). This proactive approach can lead to improved air quality and health outcomes for communities living near these factories. While involving stakeholders in the evaluation process may help identify compliance issues, it can also lead to delays and conflicts that hinder the timely implementation of necessary regulations (Parida et al., 2021).

11.4.1 Lack of Standardized Monitoring Protocols

First, there is no specific monitoring pattern associated with the new chemical substances thus making it difficult to determine accurately the level of emergence of ECs and their risks to the environment and human health. This lack of consistency in the monitoring protocols also poses a challenge when developing policies and procedures to reduce or contain these contaminants (Wilkinson et al., 2017). The lack of standard protocols to monitor these contaminants, therefore, makes it hard for researchers, policymakers, and regulators to decide on the best approaches to eliminating or controlling these contaminants. With little guidance as to how ECs can or should be monitored, there could potentially be many that remain undiscovered or at least, under-identified and the consequences of their appearances remain unclear for ecosystems and public health in the long run (Gavrilescu et al., 2015).

Further, a lack of identical protocols may hamper the comparability and reproducibility of data, as well as directly hinder study comparisons and meta-analysis due to the differences in the methods used and locales (Ghestem et al., 2021). For future directions, the paper insists on stakeholders combining efforts in standardizing monitoring protocols to embrace ECs in order to promote environmental and human safety (Wilkinson et al., 2017). For instance, the contamination of MP may be challenging for researchers who lack well-defined monitoring protocols to fairly quantify the level of pollution in a specific ecosystem (Gavrilescu et al., 2015). Such situations could lead to reduced timings of adoption of other relevant precautionary measures to avoid further losses to wildlife and human beings.

A detailed counterexample to this situation could be seen in the case of the Flint water crisis, where a lack of standardized monitoring protocols resulted in high levels of lead contamination going undetected for years (Ghestem et al., 2021). Such failure to properly supervise the problem proved to be fatal for the inhabitants' health and proved the significance of consistent and prompt control measures when viewing the problem from the safety of public health and the environment's aspects (Petrie et al., 2015). Of course, delays in the application of mitigation strategies are always a potential issue, but it is crucial to carry out a careful monitoring and evaluation process in order not to pay an excess of attention to interventions that are not effective and suitable for the considered situation (Ghestem et al., 2021). Entering the implementation stage without careful data collection and analysis may lead to deterioration of the situation or inefficient use of resources.

11.4.2 Challenges in Risk Assessment for Emerging Contaminants (ECs)

Some of the drawbacks highlighted include the inability to accurately predict toxicity levels and, the dearth of information whereby the impacts of these ECs on the environment and human health are hard to determine in advance are some of the drawbacks in risk assessment (Noguera-Oviedo and Aga, 2016). Furthermore, risk assessors could find it harder to react to new pollutants arising because of the swiftness at which technology is progressing, and the constant launch of new chemicals (Lin, 2020). These emerging pollutants are different and constantly evolving, and for a number of reasons, this poses challenges for risk assessment. One challenge is that there is limited scientific knowledge as to the effects of most new pollutants on the environment, their toxicity, or their degree of biodegradation.

This is due to the fact that the new generation emergent compounds encompass MPs, medicines, and PFAS that exhibit a steep chemical profile that hinders their ability to predict how they act in the environment unlike the traditional pollutants, which include pesticides or heavy metals (Noguera-Oviedo and Aga, 2016). Such a duality makes it more challenging to assess the threats that threaten ecosystems and people's health (Philip et al., 2018). The absence of established testing procedures and analytical instruments for identifying and measuring these pollutants in environmental media, including soil, water, and air, is another major obstacle (Lin, 2020).

Current monitoring methods are frequently insufficient or nonexistent, and many new pollutants reside in trace amounts that make detection challenging (Noguera-Oviedo and Aga, 2016). It is difficult to determine the degree of pollution and exposure levels when there is a lack of trustworthy monitoring data, especially in isolated or underdeveloped areas. Furthermore, there may be unidentified synergistic or cumulative effects from newly discovered pollutants. When several pollutants are present in the environment, their combined toxicological effects may be different from those of the individual compounds (Philip et al., 2018). Risk assessments frequently ignore the complex relationships that may occur in contaminated environments in favor of concentrating on individual contaminants (Noguera-Oviedo and Aga, 2016).

Furthermore, because these substances are constantly changing, new compounds keep appearing, making risk assessment frameworks more difficult to use. These frameworks are already complicated by a lack of historical data (Lin, 2020). Regulatory decision-making is further complicated by the lack of clarity about the effects of new pollutants on the environment and human health (Vishal et al., 2021). Without enough information, regulators and policymakers may find it difficult to set acceptable risk levels or safe exposure limits (Philip et al., 2018). In addition, it is essential but often not easy to integrate toxicologists, environmental scientists, and chemical and public health personnel due to bureaucratic barriers and lack of sponsorship.

Nonetheless, this case shows that regulatory agencies and scientists need to generate new approaches that will help to assess and mitigate the risks associated with these contaminants (Lin, 2020). In light of rapidly evolving circumstances, it becomes easier to prevent the harm such compounds may cause to public health and the environment (Philip et al., 2018). This can be made possible through periodic water, air, and soil analysis in order to determine the qualities of the existing water air, and soil and carrying out risk analyses on new chemicals before they are let loose in the environment. Regulators, scientists, and industry associations might still be able to work together and uncover other hazards and efficiently develop safeguards to minimize the adverse effects of the hazard (Lin, 2020).

Hence, the approaches enumerated above can help minimize the risk factors that enhance safety and health for people of the current and future generations to enjoy whenever they are out in the fields (Vishal et al. 2021). However, to regulate the new chemical substances and for the protection of human health and the environment considerable aspects must not be neglected, and that is monitoring and risk assessments, yet significant significance should be given to the economic consequences of the new chemistry regulation, because extreme measures will hamper innovations and economic development (Lin, 2020; Vishal et al. 2021). Furthermore, the regulators and scientists as well as the industry may not reach a consensus on the best way to deal with the risks that may be present.

11.4.3 Resource Constraints and Enforcement Issues

The effective implementation of policies addressing emerging pollutants is severely hampered by resource limitations and enforcement concerns. In controlling these substances, many nations—especially those with low and moderate incomes—face significant financial and technological constraints (Fayemiwo et al., 2022). It is difficult to determine the level of pollution or monitor its progression in the absence of sophisticated monitoring infrastructure, such as labs that can identify traces of toxins (Ionela et al., 2021). Furthermore, a lack of financing for scientific study restricts our knowledge of the long-term effects of new pollutants on the environment and human health, depriving politicians of the information they need to develop strong legislation (Richardson and Kimura, 2019).

The inability of many governments to enforce compliance because of a lack of resources for monitoring and sanctions makes it difficult to enforce current policies (Fayemiwo et al., 2022; Richardson and Kimura, 2019). While industries may put short-term financial gain ahead of environmental responsibility, enforcement is further complicated by fragmented regulatory frameworks and varying regulations across borders (Ionela et al. 2021). For both local and international governments, tackling new pollutants continues to be a major challenge in the absence of sufficient financial and technical resources as well as effective enforcement mechanisms (Richardson and Kimura, 2019).

11.5 Policy Responses and Innovations

ECs in the aquatic environment have become an emerging concern to ecosystems and human health. Traditional regulatory approaches tend to miss the mark in dealing with the complex, changing nature of these pollutants. Thus, innovative policy responses are needed to bridge the gap. This section discusses four critical strategies: Polluter Pays Principle (PPP), Extended Producer Responsibility (EPR), Economic Instruments, and Public-Private Partnerships (EIPPPs) (Fig. 11.1). Together, these mechanisms provide unique contributions to the regulatory landscape with pathways to mitigate the adverse consequences of ECs.

11.5.1 Polluter Pays Principle (PPP)

The PPP is one of the most common principles of environmental legislation. This implies a polluter incurs the cost of ceasing, averting, and eliminating pollution. This gives an incentive for polluters to reduce their environmental impact in part by internalizing the cost of pollution into operational expenses. Depending on the region, PPP was implemented in different ways in order to deal with ECs. Say industrial

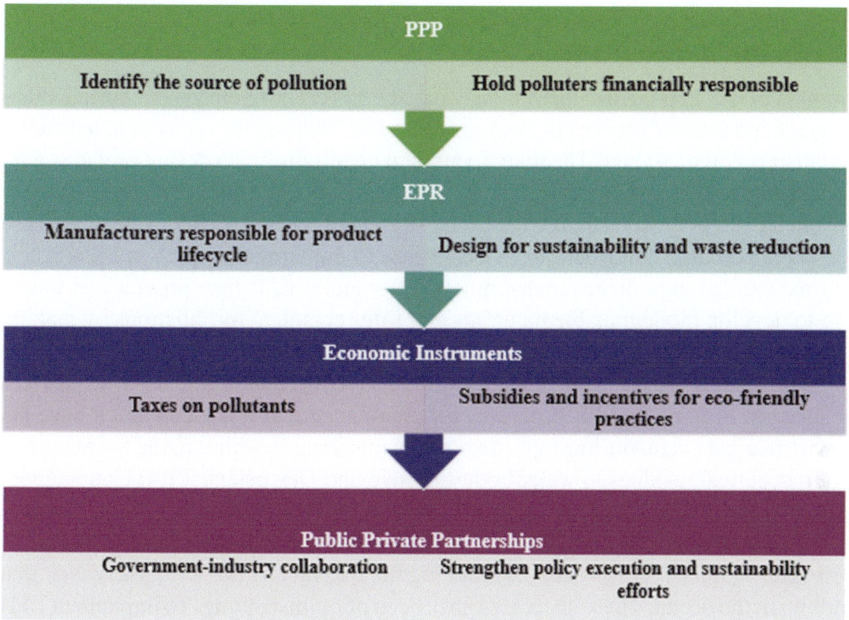

Fig. 11.1 Policy responses and innovations

facilities that discharge wastewater with pharmaceutical residues or synthetic chemicals may be assessed levies based on what's being discharged. In such cases, these funds are refocused to support environmental restoration projects or the strengthening of wastewater treatment infrastructure (Khan, 2015). Although PPP has been applied to ECs, there are challenges in conducting this work.

However, many ECs are not regulated according to established criteria. In addition, their background sources (including household products and agricultural runoffs) make it difficult to identify responsible parties. To do this, some jurisdictions have implemented innovative strategies, for instance, by levying fees on the sale of products that release ECs into the environment encouraging better product design and environmentally more benign alternatives. The European Union (EU) case studies show the success of PPP in water pollution. For instance, the European Water Framework Directive (WFD) uses PPP including industrial acceptance of the costs pertaining to meeting environmental quality standards. Successes notwithstanding, enforcement is an important problem. To improve the effectiveness of PPP for ECs, stronger penalties for noncompliance with the techniques need to be established, alongside improved monitoring (Aragão, 2022).

11.5.2 Extended Producer Responsibility (EPR)

Another policy approach in the war on ECs is the EPR. The company has taken EPR as an extended form of producers' responsibility that involves the collection, recycling, and disposal of a product. The main focus is to get producers to design products with a more minimal environmental impact. In EC, EPR could be directed at the pharma, personal care, or plastics industries (Ramasubramanian et al., 2023). However, these sectors make large contributions to the release of contaminants such as MPs, active pharmaceutical ingredients, and synthetic fragrances. EPR then provides an incentive to develop biodegradable materials and safer chemical formulations by making producers accountable for how they manage the end-of-life of their products.

EPR schemes have been implemented in several countries to varying degrees of success. In Canada, EPR programs for pharmaceutical waste collection have cut down to a great extent the improper disposal of medications, minimizing the release of pharmaceutical residues to water bodies (Sauvé and Desrosiers, 2014; Compagnoni et al., 2024). Much like Japan's EPR system for plastic packaging, which has led to reduced plastic pollution, including MPs, in its marine environment, Oregon's beverage container EPR system is reducing the quantity of plastic nursery tape that ends up in the ocean. These successes have been notwithstanding. To implement EPR properly, strong regulatory frameworks, adequate infrastructure for waste collection and processing, and active consumer participation are needed. Further, it may also be resource-intensive to monitor and enforce compliance. To overcome these problems, governments may offer producers financial and technical support, as well as consumer participation incentives provided by deposit-refund schemes (Pouikli, 2020).

11.5.3 Economic Instruments: Taxes, Subsidies, and Incentives

Concerning the use of economic levers like taxes, subsidies or incentives for changing behavior and enforcing environmental regulation newcomers to the field of environmental enforcement will soon find out the strength of this tool. In an effort to reduce the release of ECs into the environment, these instruments could be designed to encourage industries and consumers to practice in ways that reduce the release of ECs from the environment. Environmental taxes put a dollar bill on the head of someone doing something associated with pollution, making it financially harder to do (Hibiki, 2024). An example is a tax on the use of non-biodegradable microbeads in some PCPs in order to encourage manufacturers to find alternatives that are less damaging. Also, levying farm inputs such as pesticides and fertilizers that are responsible for runoff, which often carries contaminants like pesticides and nitrates into water bodies can reduce that runoff. They can also have a profound consequence in encouraging the application of cutting-edge technologies for pollution control.

Governments can for example subsidize the installation of advanced wastewater treatment systems capable of EC removal, e.g., pharmaceuticals and industrial chemicals. They reduce the financial cost for industries and municipalities in accelerating technologies to become cleaner. Financial incentives are also possible such as tax breaks or grants to industries to adopt sustainable practices. Take for example companies engaging in research and development for environmentally friendly products and companies benefiting from circular principles. They also promote innovation and competitiveness and reduce the environmental impact of their operation (Garella, 2021). Economic instruments, however, can be flexible and cost-effective, as long as they are carefully designed and implemented. So, policymakers must consider the bearing of the appropriate tax rates, availability of alternative technologies, and socio-economic impact across various stakeholders. Further, these instruments must be continuously monitored and evaluated in order that they achieve their intended environmental ends.

11.5.4 Public-Private Partnerships in Policy Implementation

The new work and a new emerging concept that has been developed is Public-Private Partnerships where knowledge from both the public and private sectors is used to meet multifaceted environmental issues like the ECs. Public-Private Partnerships can harness the advantages of both the public and the private sector to promote environmentally sound, innovative, technical, and managerial approaches to pollution control and environmental management in most fields of economic activity (Casprini and Palumbo, 2022). The most important strength of Public-Private Partnerships is their capacity to make use of inputs from multiple sources, such as resources, expertise, and technology.

Partnerships between government agencies and private companies can create state-of-the-art wastewater treatment facilities able to treat ECs. Such partnerships can also help build monitoring systems to relay real-time data on contamination levels such that interventions are timely. The Public Utilities Board (PUB) in Singapore works with private firms to develop and operate more advanced water treatment technology under its new water initiative. The program has been key to reducing the country's reliance on freshwater sources and attenuating the effects of new contaminants. In Sweden, similar partnerships between municipal authorities and private industries have also created success with pharmaceutical take-back programs.

They stop drugs from ending up in the water where they shouldn't, thereby cutting down the number in the aquatic habitat. However, due to their potential, Public-Private Partnerships have problems, ranging from differences in their objectives, risk-sharing arrangements, and governance structure. Issues must be addressed with clear legal frameworks and contractual agreements (Wen Meng and Li, 2024). As well as, they are important to foster trust and effective communication with their partners. As for the future, we believe that Public-Private Partnerships can be pivotal to scaling up solutions for ECs. For example, partnerships could work to design decentralized treatment systems for rural and peri-urban areas that are frequently without centralized infrastructure. In addition, Public-Private Partnerships can work to raise public awareness of the dangers of ECs and why responsibility with regard to disposal is important.

11.6 Stakeholder Engagement

11.6.1 Role of Non-governmental Organizations (NGOs) and Civil Society in Policy Advocacy

However, NGOs and civil society are very important in influencing robust policies and regulatory measures on ECs. These entities are between the public and the public policy givers; they are amplifiers of marginalized voices and guarantors of critical environmental agendas (Göbel Ottolini and Schulze, 2021). This means that NGOs conduct policy advocacy by conducting research, publishing reports, or lobbying for legislative changes. For instance, previously, internationally peers Worldwide Fund (WWF) and Greenpeace have been pivotal in reminding people about the fact that water bodies exist with MPs and pharmaceutical residues. These groups also provide scientific data and case studies to support the formulation of evidence-based policies.

The law civil society organization is also a watchdog role, ensuring activities are in accord with laws; and industries and governments are made to account for any environmental violations. Through campaigns, petitions, and legal actions, it advocates tighter enforcement of environmental regulations and ensures compliance with international standards (Alam, 2023). Finally, NGOs help to create multi-stakeholder collaborations, involving scientists, policymakers, and industry leaders to respond to new contaminants. They develop knowledge-sharing, innovation, and holistic solutions. For one, international initiatives like the Plastic Pollution Coalition are global coalitions hard at work across sectors to decrease plastic waste and its pollutant byproducts.

11.6.2 Community Participation in Regulatory Processes

Effective environmental governance is based on community involvement. However, communities especially when those that are directly affected by pollution, have a wealth of knowledge and information that can greatly aid regulatory processes. Through their participation, policies are socially inclusive and context-specific (Wang and Yu, 2022). Community participation can be by means of public consultations, stakeholder meetings, and participatory workshops. Through these platforms, community members are able to voice their concerns, offer feedback on proposed regulations, and essentially co-create solutions. For example, before approving

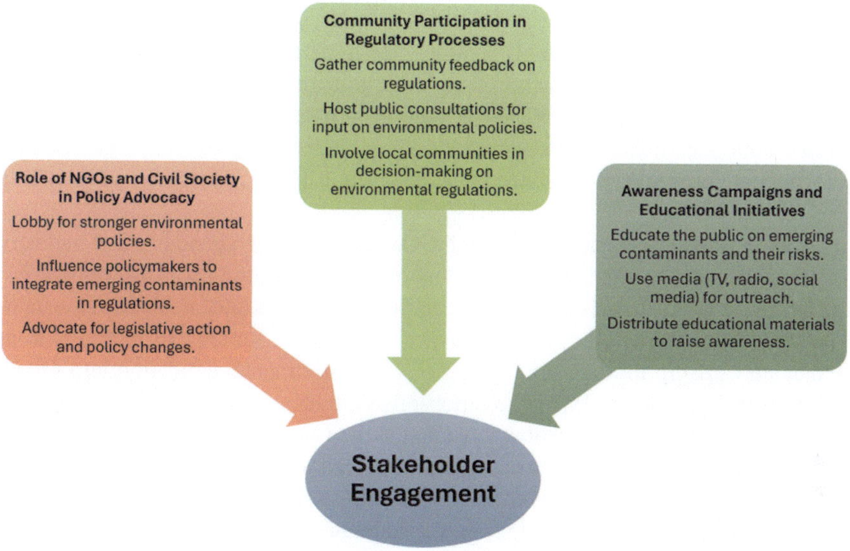

Fig. 11.2 Stakeholder engagement in regulatory processes

industrial projects, public hearings are sometimes held to evaluate the project's environmental impact in many countries.

Particularly, indigenous and local knowledge should be included in regulatory frameworks. Traditional resource management practices and sustainable ideologies are deeply ingrained in Indigenous understanding of their landscapes, making them a vital resource for governments and external organizations. This knowledge can be recognized and integrated into the development of policies aimed at ECs. Strong regulatory enforcement can be enormously strengthened by empowering communities to monitor and report pollution levels (Bekele Kebede and Kuma, 2023). Many regions have been successful in utilizing citizen science initiatives in which local residents receive training on how to collect and analyze environmental data. For instance, community network-based water quality monitoring programs in India and Africa have empowered community groups to play a pivotal role in gathering the needed data to inform regulatory agencies, thereby triggering immediate action, and optimizing water management (Fig. 11.2).

11.6.3 Awareness Campaigns and Educational Initiatives

Increasing awareness about the risks of ECs can help trigger behavioral change and support the development of regulatory measures that are accepted by the public. Awareness campaigns and educational initiatives can bridge the knowledge gap and help individuals become more informed. Public education campaigns against ECs can be targeted at informing the public about sources, impacts, and strategies for their mitigation (de Sousa, 2023). These campaigns employ a staff of consultants who take a multimedia approach in their work and are able to reach a large number of diverse audiences in and beyond the areas of impact through social media, public service announcements, and community events. For example, campaigns to alert citizens that unsafe drug disposal can create various health risks drew many to drug take-back programs.

Such initiatives typically involve NGOs, government agencies, and industry players. For instance, the UNEP has championed plastic pollution global campaigns, spreading the word about how it wounds marine ecosystems and encouraging sustainable replacements for all types of plastics. Long-term strategy to build environmental stewardship through the integration of environmental education into school and university curricula. Water pollution-related educational programs, as well as ECs, can arm students with knowledge and skills to overcome these challenges. It is also very useful to educate a specific group such as farmers, industrial workers, and local government officials through workshops, seminars, and field visits. Waste management, pollution prevention, and sustainable resource use are best demonstrated through these programs (Krishnakumar et al., 2022). Agricultural runoff is one of the leading sources of ECs; for example, training farmers to use fertilizers and pesticides responsibly would reduce agricultural runoff.

11.7 Future Directions in Regulatory Frameworks

11.7.1 Integration of Science-Based Evidence in Policymaking

EC regulation requires science-based policymaking. The more we know about ECs scientifically, the more we need to update our policies based on the evidence we do know about their sources, paths, and impacts. The gap between scientific research and policy implementation is one of the critical challenges in the existing regulations framework (Kaylor et al., 2024). It is imperative that governments and regulatory bodies create a mechanism to speed up the translation of scientific findings into actionable policies. Scientific advisory panels that supply evidence-based advice for policy makers as well as dynamic regulatory frameworks that permit standard and guideline updating with new research can achieve this. As an example, recent research into the toxicity of nanomaterials has prompted the reworking of safety tolerances in some countries. Other categories of ECs should be regulated globally in similar approaches should therefore be adopted. Interdisciplinary research is critical for ECs (Haddaway et al., 2016). Toxicologists, environmental scientists, and chemists can extend the understanding of ECs and their impacts on the environment.

11.7.2 The Role of Big Data and Artificial Intelligence (AI) in Policy Development

Artificial intelligence (AI) has come to improve the regulatory frameworks. These technologies can enrich the monitoring, analysis, and prediction of EC trends so that more informed and proactive policymaking is possible. Vast amounts of environmental data from different sources including satellite imagery, sensors, and citizen science initiatives are aggregated by big data platforms (Iandolo et al., 2021). This data can then be analyzed by AI algorithms to find pollution hot spots, predict contaminant behavior, and evaluate risk in real time. For example, machine learning (ML) models can estimate the transport of pharmaceutical residues in aquatic systems and thereby assist regulators in directing their mitigation measures. As with monitoring station placement decisions, predictive analytics can determine the locations of monitoring stations that will collect the most data while minimizing manpower and cost. The other way AI can be helpful is to simulate how the different policy scenarios would play out and let policymakers figure out what works for specific regulatory strategies (Alotaibi and Nassif, 2024). These simulations can help decision-makers allocate resources, prioritize enforcement actions, and design intervention measures.

11.7.3 Strengthening Global Cooperation and Harmonization

Pollutants are a transboundary issue, where most water-borne contaminants can readily move river, ocean, and atmospheric boundaries and ensure that pollutants emanating from one country do not remain within those boundaries. For effective regulation, therefore, robust international cooperation and harmonization of standards are needed. There are frameworks for managing hazardous substances, based on international agreements, like the Basel Convention and the Stockholm Convention (Awewomom et al., 2024). Yet these agreements need to be extended to also include a greater breadth of ECs. UNEP and WHO can be a relevant factor in consolidating international actions through synchronized recommendations and data exchange within countries. Many ECs are not well controlled in most developing countries due to the fact that; developing countries generally have the least technical and institutional capability to control them. It should, however, direct its efforts to activities such as training, technology development, and financial assistance for the developing world. This will go a long way in putting more order in a just and efficient international response to ECs.

11.7.4 Linking Regulations to Sustainable Development Goals (SDGs)

Achievement of related SDGs, underwater quality, health, environmental sustainability, and the ability to regulate ECs are intrinsically linked (Fet, 2024). SDG 6 focuses on ensuring the availability and management of water and sanitation sustainably for everyone. To achieve this goal, controlling the ECs is critical because they threaten water quality and aquatic ecosystems. Specific SDG 6 targets and indicators should be specified within regulatory frameworks. Management of ECs is closely associated with SDG12 (economic, social, and environmental aspects) that require sustainable production and consumption patterns. Reducing the release of ECs into the environment through policies that support the circular economy and its efforts to minimize and reduce waste such as EPR. Also, within focus under SDG 14 is protecting marine ecosystems from pollution. ECs play an important role as targets of regulatory measures, including plastics (MPs) and POPs, to protect aquatic biodiversity and ensure the health of marine ecosystems.

11.8 Conclusion

Due to anthropogenic activities, ECs are increasing as a threat to aquatic ecosystems and human health subjected to chemical leaching, domestic waste, and improper disposal practices. Relatively new contaminants, from plasticizers and surfactants to pharmaceuticals, PCPs, and nanomaterials, are increasingly present and constitute risks to water quality and distribution system material compatibility. EC concentrations, however, exhibit rising concentrations though there exist policies and regulatory frameworks for regulating this release despite the fact high EC concentrations point toward the need for a more robust and dynamic approach to regulations like the ones stipulated by the BIS and WHO. However, standards such as these are always evolving so rapidly that we need to update them for the rapid evolution of industrial processes and the discovery of new contaminants.

Integration of scientific advancements, improved quality standards monitoring, and enforcement of environmental quality standards need to strengthen the regulatory framework. It points out important gaps in current policies and suggests solutions that can be taken to fill in those gaps. For these efforts to succeed, regulatory bodies will need to play a part that will have to adapt to a rapidly changing environment in order to better accommodate mounting challenges. Besides this, the development of proper strategies in order to address EC pollution, which includes prevention and mitigation, demands a joint collaboration between various stakeholders like governments, industry, and civil society. Because no one approach can be used to protect aquatic ecosystems from the adverse impacts of these ECs. It involves more stringent regulations, more innovative policy instruments, and also sustainability. With a focus on improving research, enhancing governance, and international cooperation, the global community will be making the move toward a more resilient and sustainable management of ECs to ensure the protection of environmental and public health for future generations.

Acknowledgment The authors sincerely express gratitude to Amity University and the Amity Institute of Environmental Sciences (AIES) for their unwavering support, encouragement, and the invaluable academic environment that has fostered this study. The resources, guidance, and collaborative opportunities provided have significantly contributed to the development of this chapter.

References

Aidonojie PA, Anani OA (2024) Legislative framework in addressing emergent pollutants and ecological impacts. In: Emergent pollutants in freshwater plankton communities. CRC Press, pp 163–176

Alam M (2023) Environmental education and non-governmental organizations. The Palgrave encyclopedia of urban and regional futures. Springer International Publishing, Cham, pp 495–502

Alotaibi E, Nassif N (2024) Artificial intelligence in environmental monitoring: in-depth analysis. Discov Artif Intell 4(1):84

Aragão A (2022) Polluter-pays principle. Encyclopedia of contemporary constitutionalism. Springer International Publishing, Cham, pp 1–24

Awewomom J, Dzeble F, Takyi YD, Ashie WB, Ettey ENYO, Afua PE, Sackey LN, Opoku F, Akoto O (2024) Addressing global environmental pollution using environmental control techniques: a focus on environmental policy and preventive environmental management. Discov Environ 2(1):8

Bassett A (2016) A more nuanced approach to environmental hazards?: a critical review of the existence, priorities and scope of the Minamata Convention on Mercury. Doctoral dissertation, University of British Columbia

Bekele Y, Kebede B, Kuma T (2023) Assessing the role of community participation in integrated watershed management in Dandi Lake watershed Dandi district, West Showa, Oromia, Ethopia. Appl Water Sci 13(11):207

Biermann F (2007) Reforming global environmental governance: from UNEP towards a world environment organization. In: Global environmental governance: perspectives on the current debate. pp 103–123

Boahen E, Owusu L, xAdjei-Anim SO (2023) A review of emerging environmental contaminants of global concern

Casprini E, Palumbo R (2022) Reaping the benefits of digital transformation through public-private partnership: a service ecosystem view applied to healthcare. Glob Public Policy Gov 2(4):453–476

Compagnoni M, Grazzi M, Pieri F, Tomasi C (2024) Extended producer responsibility and trade flows in waste: the case of batteries. Environ Resour Econ 1–34

de Sousa FDB (2023) Consumer awareness of plastic: an overview of different research areas. Circ Econ Sustain 3(4):2083–2107

Dulio V, van Bavel B, Brorström-Lundén E, Harmsen J, Hollender J, Schlabach M, Slobodnik J, Thomas K, Koschorreck J (2018) Emerging pollutants in the EU: 10 years of NORMAN in support of environmental policies and regulations. Environ Sci Eur 30:1–13

Ebinghaus R, Barbaro E, Nash SB, de Avila C, De Wit CA, Dulio V, Felden J, Franco A, Gandrass J, Grotti M, Herata H (2023) Berlin statement on legacy and emerging contaminants in polar regions. Chemosphere 327:138530

Eriksen HH, Perrez FX (2014) The Minamata Convention: a comprehensive response to a global problem. Rev Eur, Comp & Int Environ Law 23(2):195–210

Fayemiwo O, Carden K (2022) Emerging contaminants: approaches for policy and regulatory responses in low-income countries. In: Emerging contaminants in the terrestrial-aquatic-atmosphere continuum. Elsevier, pp 343−352

Fet AM (2024) Linking sustainable good health and well-being to the sustainable development goals: an historical perspective. In: Towards sustainable good health and well-being: the role of health literacy. Springer Nature Switzerland, Cham, pp 3−20

Garella PG (2021) The effects of taxes and subsidies on environmental qualities in a differentiated duopoly. Lett Spat Resour Sci 14(2):197–209

Gavrilescu M, Demnerová K, Aamand J, Agathos S, Fava F (2015) Emerging pollutants in the environment: present and future challenges in biomonitoring, ecological risks and bioremediation. New Biotechnol 32(1):147–156

Geissen V, Mol H, Klumpp E, Umlauf G, Nadal M, van der Ploeg M, van de Zee SE, Ritsema CJ (2015) Emerging pollutants in the environment: a challenge for water resource management. Int Soil Water Conserv Res 3(1):57–65

Ghestem JP, Togola A, Kuczyńska A (2021) Proposal and recommendations for monitoring emerging pollutants in EU groundwater

Göbel C, Ottolini L, Schulze A (2021) Science as a lever: the roles and power of civil society organisations in citizen science. Sci Citiz Sci 331

Gwenzi W, Kanda A, Muhoyi E, Mukura TJ (2022) Emerging contaminants in the terrestrial-aquatic-atmosphere continuum: a global perspective. In: Emerging contaminants in the terrestrial-aquatic-atmosphere continuum. Elsevier, pp 17−25

Haddaway NR, Bernes C, Jonsson BG, Hedlund K (2016) The benefits of systematic mapping to evidence-based environmental management. Ambio 45:613–620

Hendry S (2017) Contaminants of emerging concern. In: Routledge handbook of water law and policy. Routledge, pp 95−108

Hibiki A (2024) Choice of policy instruments: tax, regulation or subsidy. In: Introduction to environmental economics and policy in Japan. Springer Nature Singapore, Singapore, pp 19−27

Iandolo F, Loia F, Fulco I, Nespoli C, Caputo F (2021) Combining big data and artificial intelligence for managing collective knowledge in unpredictable environment—insights from the Chinese case in facing COVID-19. J Knowl Econ 12(4):1982–1996

Johnson C, Bell SJ (2023) Linking emerging contaminants to production and consumption practices. Wiley Interdiscip Rev: Water 10(1):e1615

Kaylor SD, Dalton RM, Greaver T, Herrick JD, Leath E, Novak K, Ridley CE (2024) Emerging scientific approaches for identifying ecologically adverse effects of air pollution. Environ Manag 74(5):835–845

Khalid S, Shahid M, Natasha BI, Sarwar T, Shah AH, Niazi NK (2018) A review of environmental contamination and health risk assessment of wastewater use for crop irrigation with a focus on low and high-income countries. Int J Environ Res Public Health 15(5):895

Khan MR (2015) Polluter-pays-principle: the cardinal instrument for addressing climate change. Laws 4(3):638–653

Krishnakumar S, Singh DSH, Godson PS, Thanga SG (2022) Emerging pollutants: impact on environment, management, and challenges. Environ Sci Pollut Res 29(48):72309–72311

Kummer K (1999) International management of hazardous wastes: the Basel convention and related legal rules. Oxford University Press, USA

Lin X, Xu J, Keller AA, He L, Gu Y, Zheng W, Sun D, Lu Z, Huang J, Huang X, Li G (2020) Occurrence and risk assessment of emerging contaminants in a water reclamation and ecological reuse project. Sci Total Environ 744:140977

Low ML, Karim J, Abu Tahrim N, Tan MY, Lim KS, Yeap SP, Mijan NA, Ramalingam M (2024) Advancing sustainable water quality monitoring and remediation in malaysia: innovative analytical solutions for detecting and removing emerging contaminants. ACS ES&T Water

Malisch R, Malisch K, van Leeuwen FR, Moy G, Tritscher A, Witt A, Alvarez J (2023) Overview of WHO-and UNEP-coordinated human milk studies and their link to the Stockholm Convention on persistent organic pollutants. Persistent organic pollutants in human milk. Springer International Publishing, Cham, pp 29–61
Meng Y, Liu W, Fiedler H, Zhang J, Wei X, Liu X, Peng M, Zhang T (2021) Fate and risk assessment of emerging contaminants in reclaimed water production processes. Front Environ Sci & Eng 15:1–16
Milano E, Zugliani N (2019) Capturing commitment in informal, soft law instruments: a case study on the Basel committee. J Int Econ Law 22(2):163–176
Mohapatra P, Nøklebye E, Arora T, Basu AR (2023) International regulatory frameworks and best practices in management of POPs. In: Managing persistent organic pollutants in India: case studies on Vapi and Surat, Gujarat. Springer Nature Switzerland, Cham, pp 127–151
Naidu R, Jit J, Kennedy B, Arias V (2016) Emerging contaminant uncertainties and policy: the chicken or the egg conundrum. Chemosphere 154:385–390
Najam A (2003) The case against a new international environmental organization. Glob Gov 9:367
Noguera-Oviedo K, Aga DS (2016) Lessons learned from more than two decades of research on emerging contaminants in the environment. J Hazard Mater 316:242–251
Parida VK, Saidulu D, Majumder A, Srivastava A, Gupta B, Gupta AK (2021) Emerging contaminants in wastewater: a critical review on occurrence, existing legislations, risk assessment, and sustainable treatment alternatives. J Environ Chem Eng 9(5):105966
Petrie B, Barden R, Kasprzyk-Hordern B (2015) A review on emerging contaminants in wastewaters and the environment: current knowledge, understudied areas and recommendations for future monitoring. Water Res 72:3–27
Philip JM, Aravind UK, Aravindakumar CT (2018) Emerging contaminants in Indian environmental matrices–a review. Chemosphere 190:307–326
Pouikli K (2020) Concretising the role of extended producer responsibility in European Union waste law and policy through the lens of the circular economy. In: ERA forum, vol 20, no 4. Berlin/Heidelberg: Springer Berlin Heidelberg, pp 491–508
Puri M, Gandhi K, Kumar MS (2023) Emerging environmental contaminants: a global perspective on policies and regulations. J Environ Manag 332:117344
Ramasubramanian B, Tan J, Chellappan V, Ramakrishna S (2023) Recent advances in extended producer responsibility initiatives for plastic waste management in Germany and UK. Mater Circ Econ 5(1):6
Rathi BS, Kumar PS, Show PL (2021) A review on effective removal of emerging contaminants from aquatic systems: current trends and scope for further research. J Hazard Mater 409:124413
Richardson SD (2009) Water analysis: emerging contaminants and current issues. Anal Chem 81(12):4645–4677
Richardson SD, Kimura SY (2019) Water analysis: emerging contaminants and current issues. Anal Chem 92(1):473–505
Sánchez-Ocampo EM, Téllez-López AM, Islas-Flores H, Gómez-Oliván LM (2022) Environmental laws and politics, the relevance of implementing regulation of the presence of emerging pollutants in Mexico: a systematic review. Water Emerg Contam & Nanoplastics 1(2)
Sanganyado E (2022) Policies and regulations for the emerging pollutants in freshwater ecosystems: challenges and opportunities. Emerg Freshw Pollut 361–372
Sauvé S, Desrosiers M (2014) A review of what is an emerging contaminant. Chem Cent J 8:1–7
Schellenberg T, Subramanian V, Ganeshan G, Tompkins D, Pradeep R (2020) Wastewater discharge standards in the evolving context of urban sustainability–the case of India. Front Environ Sci 8:30
Šebková K (2023) The Stockholm convention, global monitoring plan and its implementation in regional and global monitoring reports. Persistent organic pollutants in human milk. Springer International Publishing, Cham, pp 63–73
Selin H (2014) Global environmental law and treaty-making on hazardous substances: the Minamata convention and mercury abatement. Glob Environ Polit 14(1):1–19

Shaikh M, Birajdar F (2024) Groundwater and public health: exploring the connections and challenges. Int J Innov Sci Res Trends Innov 9:1351–1361

Sharma BM, Bharat GK, Tayal S, Nizzetto L, Larssen T (2014) The legal framework to manage chemical pollution in India and the lesson from the persistent organic pollutants (POPs). Sci Total Environ 490:733–747

Stephan HR, Zelli F (2018) International Organizations and the global environment. In: Politics of the environment. Routledge, pp 52–70

Taheran M, Naghdi M, Brar SK, Verma M, Surampalli RY (2018) Emerging contaminants: here today, there tomorrow! Environ Nanotechnol, Monit & Manag 10:122–126

Talib A, Randhir TO (2016) Managing emerging contaminants: status, impacts, and watershed-wide strategies. Expo Health 8:143–158

United Nations Environment Programme (2011) Division of early warning and assessment. In: UNEP year book 2011: emerging issues in our global environment. UNEP/Earthprint

Vasilachi IC, Asiminicesei DM, Fertu DI, Gavrilescu M (2021) Occurrence and fate of emerging pollutants in water environment and options for their removal. Water 13(2):181

Wang B, Yu G (2022) Emerging contaminant control: from science to action. Front Environ Sci & Eng 16(6):81

Wen X, Meng F, Li S (2024) Analyzing the effect of public private partnership mode on sewage treatment in China. Sci Rep 14(1):9531

Wilkinson J, Hooda PS, Barker J, Barton S, Swinden J (2017) Occurrence, fate and transformation of emerging contaminants in water: an overarching review of the field. Environ Pollut 231:954–970

Xabadia A, Ortuzar I, Martinez Y, Esteban E (2021) Contaminants of emerging concern: a review of biological and economic principles to guide water management policies. Int Rev Environ Resour Econ (ART-2021-125430)

Yakubu OH (2017) Addressing environmental health problems in Ogoniland through implementation of United Nations environment program recommendations: environmental management strategies. Environ 4(2):28

Chapter 12
Emerging Contaminants in Mumbai's Untreated Sewage to Rhine's Industrial Runoff—A Comparative Case Study in India and Europe

Niketa Bhati, Laxmi Kant Bhardwaj, and Harshita Jain

Abstract The issue of emerging contaminants (ECs) in aquatic systems is expanding globally. This chapter looks at this subject using a comparative case study of Mumbai, India, and the Rhine River Basin in Europe to consider the differences between the effects of untreated wastewater in Mumbai and industrial runoff in the Rhine on ECs and the ecology. The objective of this chapter is to pull from these case studies to propose effective EC pollution reduction methods in both developing and developed countries in the world. Mumbai is an Indian metropolis with a dense population and much of its effluent ends up in the aquatic systems untreated. This effluent could include a mix of residential and industrial ECs which will be influenced by their ecological footprint. The European Rhine River Basin has a long industrial history which could influence the industrial runoff. This case study will pick apart the types of ECs, we find at each location. The ECs in Mumbai consist of pharmaceuticals, personal-care products (PCPs), and home chemicals while in the Rhine it majorly consists of heavy metals, persistent organic pollutants (POPs), and industrial by-products. This chapter will compare Mumbai and Rhine to consider the options that can effectively reduce the ECs. It will also provide viable solutions such as improved wastewater treatment and disposal systems for Mumbai and stricter regulations on industrial discharges along with continuous clean-ups for Rhine. Further, it will highlight the need for international cooperation on this environmental issue and its role in tackling it.

Keywords Emerging contaminants (ECs) · Wastewater · Water pollution · Aquatic system · Mumbai · Rhine river basin · Environment

---

N. Bhati · H. Jain (✉)
Amity Institute of Environmental Sciences, Amity University, Noida, Uttar Pradesh, India
e-mail: hjain@amity.edu

L. K. Bhardwaj
Amity Institute of Environmental Toxicology, Safety and Management, Amity University, Noida, Uttar Pradesh, India

H. Jain et al. (eds.), *Emerging Contaminants in the Aquatic Environment*, Advances in Wastewater Research, https://doi.org/10.1007/978-981-96-8548-6_12

List of Abbreviations

AOPs	Advanced oxidation processes
DDT	Dichlorodiphenyltrichloroethane
ECs	Emerging contaminants
GC-MS	Gas chromatographic-mass spectrometry
H_2O_2	Hydrogen peroxide
HMMM	Hexa methoxymethyl melamine
HPLC–ESI–MS/MS	High-performance liquid chromatography-electrospray ionization tandem mass spectrometry
MBR	Membrane bioreactor
MPs	Microplastics
PCBs	Polychlorinated biphenyls
PCPs	Personal care products
PFOA	Perfluorooctanoic acid
PFOS	Perfluorooctane sulfonate
POPs	Persistent organic pollutants
UV	Ultraviolet
WFDIED	Water framework directive and industrial emissions directive

12.1 Introduction

ECs are an increasing threat to the aquatic ecosystems of the world (Kumar et al. 2022a). These contaminants, which include pharmaceuticals, PCPs, and industrial chemicals, among other synthetic compounds, have been found in waters worldwide, thereby generating a wake-up call for their potential environmental and human health effects (Ávila and García 2015). Most importantly, compared with traditional pollutants, ECs originate from common human behavior and are not fully removed by conventional wastewater treatment steps, hence their contribution to persistence in the environment (Rout et al. 2021). The general presence of ECs in aquatic ecosystems now demands a better understanding of their sources, impacts, and mitigation possibilities.

ECs are unique because they span a wide range of substances with varying chemical properties and ecological impacts. For instance, the presence of pharmaceuticals and PCPs, found mainly in urban wastewater, leads to the disruption of hormonal systems even at low concentrations (Ebele et al. 2017). Industrial chemicals comprise POPs and heavy metals that have a high possibility of coming from industrial site runoff, which thus accumulates in the food chain, leading to long-term disruption of ecosystems and human health (Masindi and Muedi 2018; Agarwal and Jain 2024; Alam and Markandey 2020; Alam et al. 2023). This diversity and complexity of ECs necessitate the tailoring of their monitoring and management needs toward the environment (Fig. 12.1).

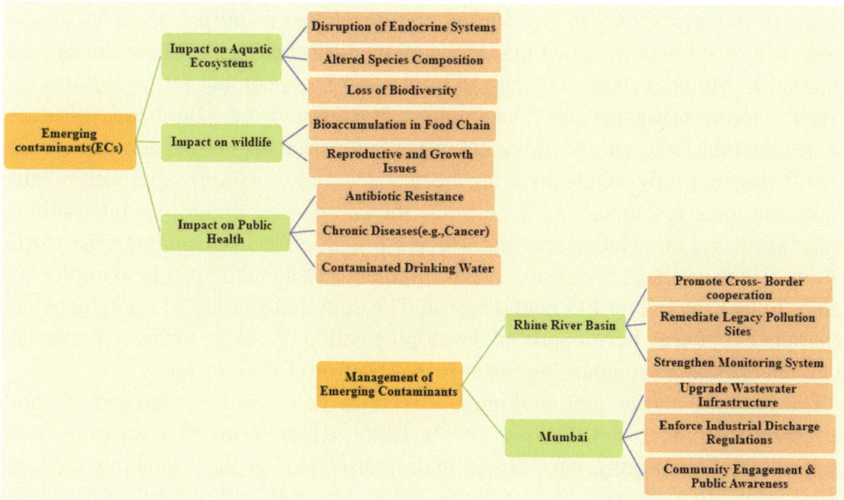

Fig. 12.1 Impact of emerging contaminants (ECs) and their management

The issues of ECs are addressed in this chapter by discussing two completely different geographical and socio-economic contexts, Mumbai in India and the Rhine River Basin in Europe. The City of Mumbai, generally known as a teeming city with a highly dense population, has problems associated with untreated sewage and, thus, boasts an entry of numerous ECs into its water supply systems (Praja Foundation 2021). The high rate of urbanization and industrialization in the city indicates that both residential and industrial effluents are discharged into the nearby water bodies. These effluents often carry a mixture of pharmaceuticals, PCPs, and household chemicals, which, with inadequate wastewater treatment infrastructure, enter aquatic systems largely untreated. This situation threatens the local ecology and the health of millions depending on these water sources for daily needs.

On the other hand, the Rhine River Basin, with its long industrial history, faces industrial runoff as the leading origin of ECs (Uehlinger et al. 2009). The Rhine is one of the most important waterways in Europe and accommodates an intensive network of industries along its course (Leuven et al. 2009). They include heavy metals, POPs, and other byproducts of an industry that have been known for decades to be the results of industrial endeavors that cause river pollution. Changes in policies and new technologies in wastewater treatment have brought big improvements in water quality but the legacy of industrial pollution is also now experienced on the Rhine in terms of its ecological health (Water Quality 2020). Understanding the types and impacts of ECs in this region would be very helpful for long-term challenges and opportunities for remediation in industrialized areas.

These varied sources and contexts give an overview of the different ECs and their varying impact in different regions. This chapter will draw upon a comparative analysis between Mumbai and the Rhine on the peculiar challenges and similarities related to managing ECs in developing and developed countries. It draws lessons from each case to come up with effective strategies for reducing EC pollution. The situation in Mumbai shows urgency in improving the treatment infrastructures and proper disposal of wastewater. The situation at Rhine, on the other hand, illustrates the need to take stringent measures in terms of industrial discharges and those under remediation presently. While invoking the need for context-specific approaches while respecting local resources and conditions, the comparison underlines international cooperation and knowledge sharing. There are joint efforts from around the world, putting scientific research and policy development together with practical implementation, to be effective in EC management. The common sharing of knowledge and experiences between developed and developing nations would contribute immensely to global efforts in safeguarding water resources from EC pollution.

The possible causes and ecological effects of ECs will be examined for both the case of Mumbai and the case of the Rhine River Basin by comparing their scenarios and suggesting targeted mitigation strategies in the following sections. This comparative analysis, unlike the traditional review, will not only expand our understanding of EC pollution but also provide valuable information for policy-making and management practices that can help protect aquatic ecosystems and public health around the world. By comparing the two contrasting regions, a potential common challenge for innovative solutions to contribute toward a holistic approach to the management of ECs in aquatic systems can be envisaged (Fig. 12.2).

12.2 Mumbai, India: A Megacity's Challenge with Untreated Wastewater

Mumbai is one of the biggest, and most populous cities in India, and with all these population-related factors comes enormous management difficulties in tackling its wastewater. The rapid expansion of urbanization and a growing population have therefore surpassed the infrastructure growth of the city's systems of wastewater, meaning that half of the wastewater discharge that directly goes into the environment has not been properly treated. Thus, untreated wastewater poses tremendous threats to the environment and public health and favors the rapid proliferation of ECs in the region's water bodies (Singh et al. 2018). The urban expanse of the city has a high population density with sprawling industrial activities and sprawling informal settlements, and all these factors converge to overwhelm the waste-water burden. The city produces about 2,700 million liters of wastewater every day, but only around 70% is treated in some form (Maharashtra Pollution Control Board 2005). The untreated one mainly carries a cocktail of ECs and finds its way into rivers, creeks, and the Arabian Sea. Poor sanitation facilities, especially within slums where there are very low levels of coverage imply that sanitation access to households is appalling, and infrastructure provision is not keeping up with population growth.

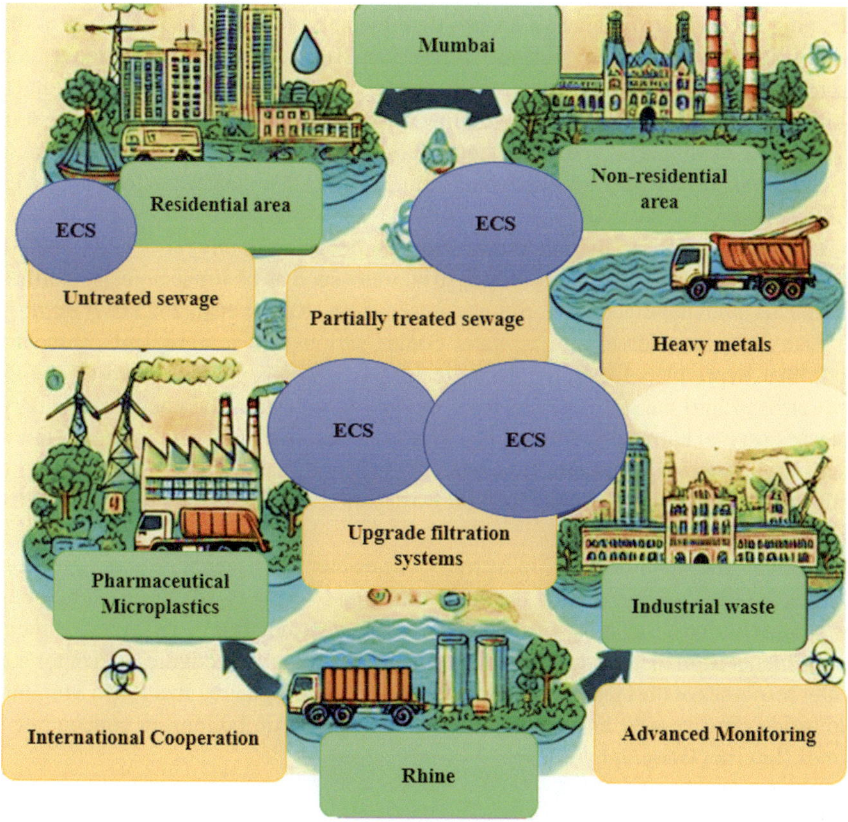

Fig. 12.2 Comparative analysis of emerging contaminant (EC) pollution in Mumbai and the Rhine River Basin

12.2.1 Potential Sources of Emerging Contaminants (ECs) in Mumbai's Wastewater

There are several sources that contribute to constituting ECs in Mumbai's wastewater and these come from various sources which can be industrial, commercial, or residential. The difference and levels of contaminant outputs from each sector make the problem of ECs a complex one to be addressed. Sources of ECs in the residential sector include pharmaceuticals and PCPs, as well as household chemicals (Kumar et al. 2022a). The reliance on pharmaceuticals is one of the major modes of pollution, and of these, many pharmaceuticals unused or expired are flushed down the household drains. This disposal method adds massive quantities of active pharmaceuticals to the wastewater (Stringer 2011; Munshi et al. 2023).

PCPs which consist of shampoos, conditioners, lotions, and soaps are used on a daily basis by millions of people and these products also include the household cleaning agents that contain surfactants, disinfectants, and solvents contribute to the wide range of ECs found in domestic wastewater (Kumar et al. 2022a). Moreover, the commercial area in Mumbai also greatly contributes to this. The different hospitals, clinics, and pharmacies pour large volumes of medical waste into the environment in the form of medicines and disinfectants (Sarkar et al. 2024). Hotels and restaurants produce wastewater with a cocktail of cleaning agents, detergents, and food additives (Shukla and Chatterjee 2011).

ECs are abundant in the effluent of Mumbai because the city has a highly active industrial sector. In fact, all major industrial units such as pharmaceuticals, textiles, chemicals, and manufacturing units are present here (Kumar et al. 2022a). A number of such factories release into the water bodies various active ingredients, solvents, and other byproducts from pharmaceuticals. Dyes, softeners, and bleaching agents are released into the environment from the textile industry. The chemical industry releases several harmful chemicals like POPs and heavy metals with improper treatment into the sewage system. It is thus a resultant effect of the presence of these industries, causing highly contaminated wastewater with a complex mixture of ECs (Kumar et al. 2022a).

Further, the treatment of industrial wastewater is inadequate (Singare and Dhabarde 2014). Very few of the industrial plants possess advanced techniques of treatment and also, most of the time they neglect environmental rules. Hence, the effluents from these industrial zones bypass the treatment activities and enter the water resources of the city (Saravanane et al. 2014). The high urbanization in Mumbai alongside the riverside and creekside lends itself to toxic buildup in concentrated zones (Internet Geography 2023).

12.2.2 Ecological Impacts of Emerging Contaminants (ECs) in Mumbai's Water Bodies

Untreated wastewater contribution to Mumbai's eutrophication and contamination can have wide ecological impacts. Untreated effluents in water bodies increase the levels of ECs, which in turn there are various harmful effects on aquatic life (Mishra et al. 2023; Bhardwaj et al. 2025a). Antibiotics, artificial hormones, and active pharmaceutical ingredients interfere with the hormonal activities of fish and other wildlife and affect the endocrine and reproductive systems of aquatic organisms (Yuan et al. 2023). Overturf et al. (2015) described the exposure effect of these ECs on sex ratios, fertility rate, and growth patterns. Heavy metals and POPs are, however, another major threat. As part of the food chain, however, they bioaccumulate and concentrations are relatively higher at higher trophic levels (Alharbi et al. 2018).

Some heavy metals that are extremely toxic even at low concentrations can injure the neurological system, damage various organs, and fail reproduction in aquatic organisms (Tchounwou et al. 2012; Alhelou et al. 2019; Bozkir et al. 2010; Brack et al. 2018; Brunn et al. 2023). They are also industrial chemicals that are persistent in the environment and remain for prolonged periods unaltered. It is a chronic toxicity that their presence in aquatic systems causes interfacing with other species and the entire food chain. Thus, deterioration in water quality from ECs leads to a loss of biodiversity within such ecosystems. Species intolerant of polluted conditions might decline or even become extinct, thus lowering biodiversity (Fa 2012).

This loss of species might further disturb ecological balance and impair ecosystem services like purification of water, provision of habitats, and nutrient cycling (Adla et al. 2022). An increase in robust and pollution-tolerant species has been known to further alter community structures and functions, often weakening ecosystem resilience. The waters of Mumbai pose great health risks to local populations because of the presence of ECs. Many of these residents drink, cook, and bathe in these waters, thus exposing them directly to the effects of contaminated water. Health problems observed among town residents are the result of exposure to these water-borne ECs, ranging from gastrointestinal infections and diseases related to the skin to long-term chronic diseases such as cancer and neurological disorders (Water-related Illnesses 2024). Contaminated drinking water bearing trace levels of pharmaceuticals may present serious challenges in combating antibiotic-resistant bacteria among the resident population (Madhogaria et al. 2024). The situation is worse in the informal settlements, where little access to clean water and proper sanitation is seen. In these communities, the residents rely on the local water bodies for their daily needs, which makes them highly susceptible to the effects of untreated wastewater. The inability to have access to these facilities results in the instant release of wastewater, sometimes mixed with human excreta, into the nearby water bodies, hence worsening the situation of contamination (Hutton and Chase 2018). Children, pregnant women, and the aged are more vulnerable to health problems since they are more exposed to infection and chronic conditions. Apart from the risk of health-related problems to humans, pollution of the water bodies affects economic use. ECs have adversely affected fisheries, which are otherwise a prime source of livelihood for many people in Mumbai. Accumulation of toxic substances within tissues of the fish and deaths to the fish stock can render the waterbody contaminated. The pattern is unsustainable, resulting in declining fish stocks. In doing so, the livelihoods of local fisher folks are put at risk, and in some cases, communities that then depend heavily upon fish as a primary protein source are in danger of hunger.

12.3 Rhine River Basin, Europe: A Legacy of Industrial Activity

Based upon the long history of industrial activity in the Rhine River Basin, one of the biggest waterways in Europe, with a long imprint on environmental health (Uehlinger et al. 2009). It is about 1,230 km long, with its sources in the Swiss Alps, flowing through Germany and France into the North Sea at the bottom of France and the Netherlands. A basin area of about 185,000 square kilometers, and some of the most densely populated and industrialized regions of Europe (Ngo et al. 2023). The Rhine has been the body of commerce and industry in forms of transport of goods and raw materials for which cities such as Basel, Strasbourg, Mannheim, and Rotterdam have developed at the banks of the river using it as an element of economic development (Spaulding 2015). The strategic importance of the route contributes to it being a commercial zone, hence many ports, factories, and industrial complexes are situated along its course and this process of industrialization has brought key economic benefits to the region, fostering growth and development. At the same time, high industrial intensity has brought a lot of pollutants to the Rhine, destroying the quality of water and ecological status (Meybeck 2002).

12.3.1 Types of Emerging Contaminants (ECs) Likely Present in the Rhine

The Rhine River Basin is highly industrialized; therefore, many ECs find their way into the basin. The ECs include POPs, PCPs, industrial contaminants, and microplastics (MPs) of major ecological and public health concern (Wilkinson et al. 2017). These toxic organic chemical compounds can resist degradation which accumulates in the environment and in living organisms (Guo et al. 2019). POPs detected in the Rhine River include polychlorinated biphenyls (PCBs), dioxin, and other pesticides such as dichlorodiphenyltrichloroethane (DDT) (Wilhelm et al. 2007).

POPs are produced from many sources, accumulated mainly from industrial processes, chemical manufacture, and agricultural activities (Das et al. 2024; Besten et al. 2000; Duester et al. 2024; Gaurkhede et al. 2021; Jain 2024; Krauss et al. 2019; Kumar et al. 2024). Although some POPs are outlawed and others are controlled, this pollutant stays in the environment because it is resistant to natural degradation processes and can be biomagnified in the food chain. It may also be remobilized from the impacted river system through flooding and dredging, which will contaminate the environment once more (Alharbi et al. 2018).

PCPs comprise a large group of chemical substances used in medications, cosmetics, and MPs which are small fragments of plastic less than 5 mm are another important emerging pollutant in the Rhine (Tanwar et al. 2014). Most of the particles arise from larger pieces of plastic debris, synthetic textile fibers, and PCPs containing microbeads. MPs are widespread in the aquatic environment and become a threat to aquatic life due to their ingestion and possible release of toxic substances that may have adsorbed on their surface (Ali et al. 2023). MPs on the Rhine have been traced both to urban and industrial activities along the river course.

Used in an automobile context, coatings, and plastics this is commonly called Hexa Methoxymethyl Melamine (HMMM). It has also been identified in the Rhine River near heavy industries (Dsikowitzky and Schwarzbauer 2015). The wastewater from automotive manufacturing facilities contains this compound, which is discharged into the river. Several segments of the Rhine have detected HMMM. Concentrations indicate that the contents came from industrial origins (Schwarzbauer and Ricking 2010). The compound is shown to induce acute toxic effects in aquatic animals, such as daphnia, and also demonstrates genotoxic potential (Dsikowitzky and Schwarzbauer 2015). The fluctuation of HMMM concentrations along the Rhine implies degradation or partitioning of the compound, thus making it cumbersome to monitor. In this case, the presence of HMMM in the Rhine needs thorough water quality monitoring and enhanced wastewater treatment for the prevention of its release into the environment.

Lipophilic organic pollutants are marked by their capability to dissolve in fats and accumulate in the environment and living organisms (Miniero et al. 2015). Analysis by techniques of gas chromatographic-mass spectrometry (GC–MS) has already been able to identify a wide variety of anthropogenic compounds in the Rhine, including industrial chemical and pharmaceuticals as well as other substances that quite often possess significant physiological and ecotoxicological properties (Ying et al. 2022). Of particular interest in the water samples of the Rhine are the varieties of brominated compounds detected. Among them are first-time discoveries in river waters-composition including mono- and di-brominated (methoxyphenyl) propionic acids and hydroxymethylacetophenones. The compounds are detrimental since they may pose adverse effects on aquatic life. Examples include brominated compounds, which have been reported to be persistent and bioaccumulative (Schwarzbauer and Heim 2005). Additionally, they interfere with endocrine functions and can cause other adverse biological impacts. Perfluorooctane sulfonate (PFOS) and perfluorooctanoic acid (PFOA) are two ECs being found in the Rhine River, although under grave environmental and health challenges due to their persistence in environment and bioaccumulation abilities and inherent toxicology. These compounds occur broadly in the environment in different compartments of the system, such as water bodies, drinking water, and fish (Quinete et al. 2009). Concentrations of PFOS in water samples and PFOS accumulation in fish liver have been studied and, by inference, PFOS has been shown to be widely distributed and likely to bioaccumulate. Rhine River PFOA and PFOS concentration range varies between 21 and 48 ng/L and 5 to 18 ng/L, respectively (Kurwadkar et al. 2022). These contaminants are produced at industrial processes, some are in agricultural runoff, and some are from treated sewage sludge, ending up in fertilizer. PFOS and PFOA are harmful to the environment and human health and are analyzed by high-performance liquid chromatography-electrospray ionization tandem mass spectrometry (HPLC–ESI–MS/MS). The ongoing research reveals a need for further monitoring and much more improvement in the methodology aimed at pollution management of these dangerous compounds in the Rhine River.

12.3.2 Ecological Consequences of Emerging Contaminants (ECs) in the Rhine Basin

POPs are a class of chemicals that have shown resistance to decomposition and persist in the environment, as well as in living organisms. In aquatic life, these pollutants have been proven to interfere with endocrine systems, undeniably impair reproduction, and significantly contribute to high mortality rates of fish and other wildlife. On their way up the food chain, these POPs bioaccumulate, placing higher trophic levels at severe risk, with man who consumes the polluted fish ultimately suffering (Alharbi et al. 2018). The Rhine River contains a vast amount of MPs from the degradation of larger litter, synthetic textiles, and PCPs and these tiny MP fragments are ingested by aquatic life, sometimes causing physical obstructions and injuries (Mani et al. 2015). Besides that, these MPs have the ability to adsorb and translocate other toxic contaminants such as heavy metals and organic toxins, further worsening their ecological damage (Mohite and Joya 2021; Namieśnik et al. 2014; Negi and Chandel 2021; Philip et al. 2018; Rahman et al. 2009; Raut and Kashyap 2021; Samal et al. 2022; Schiff 2017). Ingestion of MPs can alter feeding behaviors in fish and invertebrates and lead to reduced growth and reproductive success (Mani et al. 2015; Bhardwaj et al. 2025b).

Strong ecological implications are associated with the presence of mono- and di-brominated (methoxyphenyl) propionic acids and hydroxymethyl acetophenones in the Rhine River (Ruff et al. 2015). For the first time since the discovery of brominated substances as environmental pollutants in the river, these compounds showed properties that could be particularly ominous. They are capable of surviving in a non-processed environment, degrading slowly, and having a very high tendency to bioaccumulate in aquatic organisms (Weijs et al. 2015). Ecologically, these pollutants may interfere with the endocrine functions of aquatic organisms, leading to reproductive problems, development problems, and disturbances to their growth patterns.

The fact that these are found in the river infers that such compounds are stable in water and that there is a probability that they can be adsorbed to particulate matter which interferes with a great range of organisms in the river ecosystem, from microorganisms up to larger fish and invertebrates (Windsor et al. 2018). The spatial distribution and intensity of emission sources of these brominated compounds indicate that industrial activities along the Rhine make a major contribution to the presence of these compounds (Yuan et al. 2022). The stability and bioaccumulative nature of these pollutants therefore imply long-term ecological impacts that require continuous monitoring and regulatory measures to mitigate their effects. For the Rhine River, an existing, major threat has been identified from the ecological perspective is HMMM, which is applied heavily in the automotive industry in coatings and plastics. HMMM has been detected in the river itself as there are attached wastewater discharge origins from the automotive manufacturing industries. The substance was found in various sections of the Rhine River, especially in portions where significant industrial operations occur. HMMM exhibits acute toxic effects on aquatic organisms like daphnia. Daphnia is a small planktonic crustacean very essential to any freshwater ecosystem as it has its place in the ecological balance as one of the primary food items taken by most species of fish. Its exposition to HMMM brings instant harm and mortality. HMMM also has been found to have genotoxic potential, implying interference with genetic information within cells that leads to mutations. It is possible to affect aquatic species in the long term by genotoxicity, which subsequently affects the reproduction of such species as population decline in sensitive species. Changes in concentration along the rivers have added a layer of complexity in monitoring and assessment that is very difficult to readily assess for proper mitigation strategies (Dsikowitzky and Schwarzbauer 2015).

HMMM in the Rhine River calls for comprehensive water quality monitoring and better methods of wastewater treatment with a view to reducing their effects on the environment. Such measures are necessary to protect the aquatic ecosystem and continue having a healthy and diverse river for a long time. Both PFOS and PFOA are POPs, and both have a significant impact on the ecological network of the Rhine River (Cornelsen et al. 2021). The term highly resistant to degradation in the environment is given to these perfluorinated compounds. This, therefore, grows to bioaccumulation of the ecosystem of a river. Most of these pollutants enter the Rhine through the main gateways: industrial discharges, agricultural runoff, and effluents from wastewater treatment plants (WWTPs). Once in the aquatic environment, they are capable of attaching to sediments and also bioaccumulating in the tissue of aquatic organisms, including fish (Cornelsen et al. 2021).

Contamination by PFOS and PFOA has serious environmental implications. Perfluorinated compounds were detected in water, sediment, and biota from the Rhine River, and were ubiquitous. PFOS and PFOA disrupt aquatic animal endocrine functions that in turn affect their reproduction, development, and health (Ma et al. 2023). It has been reported that PFOS levels in fish of the Rhine River are high, especially within the liver tissue, which was normally established to impair the proper functioning of the livers and increase the susceptibility of the fish to diseases (Rüdel et al. 2022).

12.4 Case Comparisons: Mumbai versus Rhine

Recent comparative studies of ECs in Mumbai, India, and the Rhine River Basin, Europe, reveal the following conclusions. The socio-economic and environmental contexts are remarkably different, and this also influences the presence and impact of such pollutants. The sources of ECs that are mainly present in Mumbai are untreated or partially treated sewage arising from domestic, commercial, and industrial activities (Mathew and Kanmani 2020). Effluent from industries and sewage containing pharmaceuticals and PCPs contains large amounts of contaminants. On the contrary, contamination in the Rhine River Basin is mainly caused by old industries and present industrial discharges. The Rhine has undergone reforms in regulatory standards and treatment technology but still bears the marks of POPs, MPs, and other specialist pollutants (Internationale Kommission zum Schutz des Rheins et al. 2020). Types and concentrations of contaminants vary in the various regions, reflecting their unique pollution profile. For instance, pharmaceuticals and PCPs are present at higher concentrations relative to slums and industrially polluted areas in Mumbai (Mishra et al. 2023). Both legacy persistent pollutants such as PCBs and DDT and newer substances including HMMM, PFOS, and PFOA contaminate the Rhine River Basin as well (Göckener et al. 2021). Concentrations vary with proximity to industrial activities and the efficiency of wastewater treatment (Table 12.1).

The ecological and health effects caused by the pollutants are serious in both regions. In Mumbai, contamination alters the endocrine functions of aquatic life, leads to the bioaccumulation of heavy metals, and results in biodiversity loss. The health impacts of related conditions include water-borne diseases and antibiotic-resistant bacteria, mostly for informal settlers who lack proper sanitation (Kumar et al. 2022b). Some impacts observed in the Rhine River Basin are endocrine disruption, reproductive problems among aquatic life, and POP bioaccumulation of MPs also,

Table 12.1 Challenge and management strategies in Mumbai and the Rhine River Basin

Aspect	Mumbai	Rhine River Basin
Context	Rapid urbanization, dense population, and industrialization	Extensive industrial history, significant commercial and industrial activities
Sources of ECs	Untreated residential, commercial, and industrial wastewater	Industrial discharges, agricultural runoff, and wastewater treatment plants
Types of ECs	Pharmaceuticals, PCPs, household chemicals, heavy metals, and POPs	POPs, MPs, pharmaceuticals, PCPs, and industrial chemicals
Wastewater treatment	Inadequate infrastructure, and low treatment coverage (70%)	Advanced treatment technologies are in place, but some legacy pollutants remain
Key challenges	High volume of untreated wastewater, and inadequate sanitation facilities in informal settlements	Historical industrial contamination, ongoing discharges of various ECs, and new pollutants
Ecological impacts	Disruption of aquatic ecosystems, bioaccumulation of toxins, reduced biodiversity	Endocrine disruption, bioaccumulation of POPs, MPs ingestion, and long-term ecological damage
Health risks	Exposure to waterborne diseases, chronic health conditions, and antibiotic resistance	Risks from contaminated drinking water, accumulation of toxins in fish, and potential health issues for humans
Mitigation strategies	Enhancing wastewater treatment infrastructure, improving disposal practices, and addressing informal settlement challenges	Strengthening regulations, advancing remediation technologies, and enhancing monitoring
Future research focus	Development of low-cost treatment solutions, effective sanitation for informal settlements, and community engagement strategies	Novel remediation methods for persistent pollutants, advanced monitoring techniques, and cross-border collaboration

the pollutants affect the food chains and pose health risks to humans through ingestion of fishes and contaminated drinking water. Approach-specific control measures are needed to address the problem (Kurwadkar et al. 2022). In Mumbai, the deplorable condition of wastewater treatment infrastructure needs to be upgraded to handle the diverse load of ECs, and investments made in high-tech equipment and sanitation infrastructure are required. A more comprehensive regulatory framework should evolve to ensure that industries conform not only to waste disposal but also to proper waste management. Moreover, public awareness and community involvement are necessary for proper management. In the Rhine River Basin, it is important to keep investing in modern treatment technologies and robust monitoring systems to address, not just historic pollution, but also priority chemicals. Cooperation between countries in the basin to harmonize regulations and share knowledge would also address the management of pollution problems.

12.4.1 *Mitigating Emerging Contaminants (EC) Pollution: Tailored Approaches for Different Challenges*

Depending on the type and impact of those pollutants in their relevant zones, EC control measures should include source-specific control measures. Thus, all mitigation techniques should take care of the individual problems of ECs, which vary in their socio-economic and environmental contexts. For Mumbai, the situation regarding EC pollution control consists of the fact that in this city, urbanization levels are high, saturation levels are high, and infrastructure for treatment is still under development. Upgradation of wastewater treatment facilities is a major strategy with Mumbai because these facilities are supposed to effectively remove a wider spectrum of contaminants (Starkl et al. 2018). Upgradation of existing treatment plants with advanced technologies like membrane bioreactors and advanced oxidation processes (AOPs) would be able to handle pharmaceuticals, PCPs, and industrial chemicals more efficiently than the currently adopted methods (Kanafin et al. 2023). Decentralized treatment for informal settlements and small-scale industries is pertinent to improving local pollution sources. Appropriate disposal practices applied by the source including pharmaceuticals and other household chemicals should be encouraged. Public education along with campaigns for awareness programs is recommended to promote responsible behavior and develop a better waste management practice (Hasan 2004) (Fig. 12.3).

The Rhine River Basin has a historic record of industrial development and therefore calls for more alternative strategies. Although there has been steady progress with regard to new technologies in wastewater treatment, much of the existing legacy of industrial pollutants, including POPs, MPs, and perfluorinated compounds, demands an integrated long-term management strategy. Strengthening regulations and enforcement of industrial discharges is necessary so that no more additional pollution can be allowed. This can be achieved through monitoring and controlling the release of

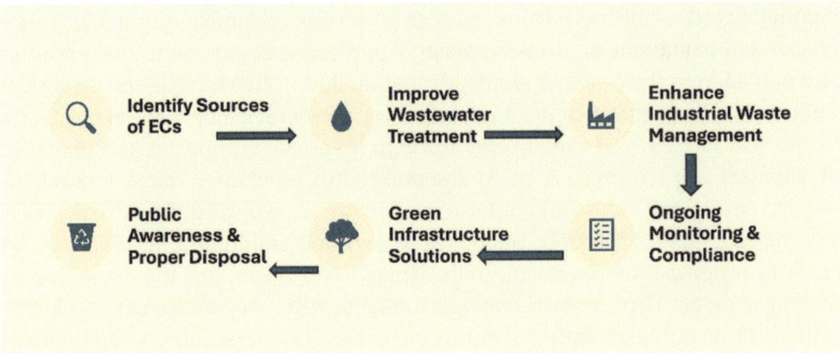

Fig. 12.3 Process steps to treat emerging contaminants (ECs) in Mumbai wastewater

hazardous substances besides full compliance with very stringent discharge standards (Wang et al. 2021). Hence, remedial actions against historical pollution, such as dredging of sediments and remediation of contaminated sites, are required. Since the basin is transboundary in nature, international cooperation is necessary. The shared River Rhine among these countries can aid collaboration which would lead toward harmonized regulations, joint monitoring programs, and best practices sharing (Sadoff and Grey 2005) (Fig. 12.4).

Innovative solutions should be brought about that could reduce the EC pollution in both regions. In Mumbai, the use of green infrastructure solutions, including constructed wetlands and vegetated swales, would provide increased capacity for treatment but also decrease runoff contamination into waters (Stefanakis 2019). In the Rhine River Basin, the R&D in understanding and managing emerging pollutants will positively impact the effectiveness of mitigation strategies, including new types of MPs and novel industrial chemicals (Ghosh et al. 2023). Finally, there is a multidisciplinary need for technological development, regulatory strictures, and public engagement in the management of ECs, all of which depend on specific regional necessities. If both Mumbai and the Rhine River Basin can implement context-specific solutions and improve international cooperation, then there will be improvement in the management of ECs and, therefore, the water resources and ecosystems in these places.

12.5 Strategies for Mumbai: Wastewater Treatment and Disposal Practices

Mumbai has suffered in recent decades due to the rapid growth of urbanization and population, now overburdening the wastewater treatment infrastructure and causing an increase in the pollution of the city water bodies with contaminants. Work, therefore, needs to address the issue holistically through a combination of

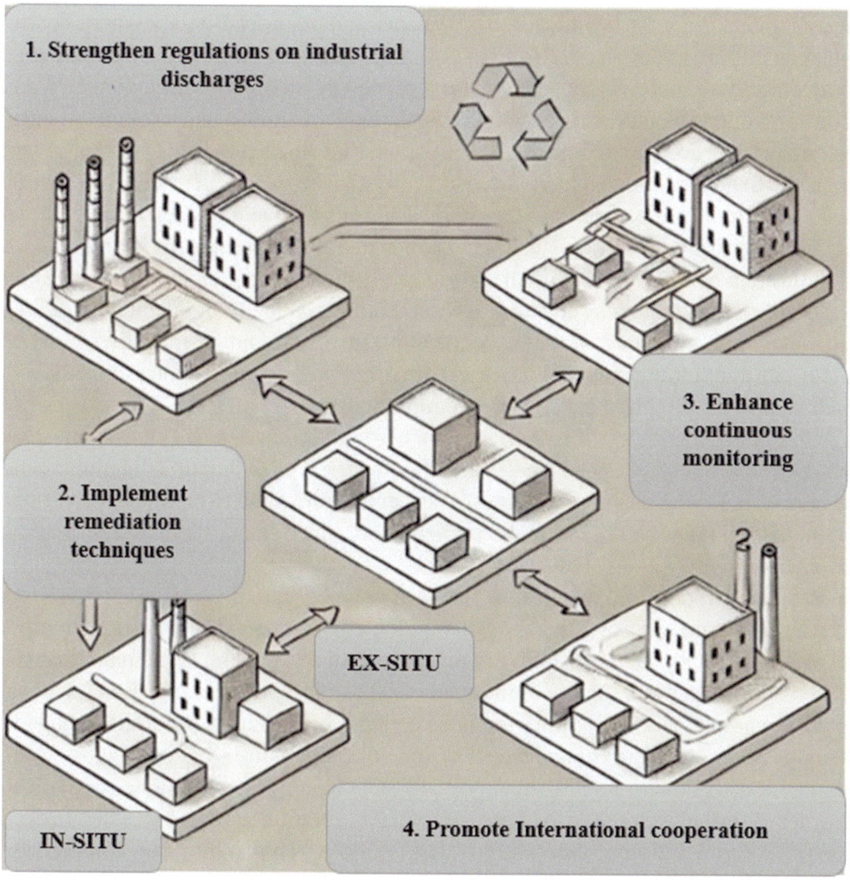

Fig. 12.4 Process steps to treat emerging contaminants (ECs) in Rhine River Basin

remedial measures, both in terms of treating pollutants and handling them for safe disposal. The conventional treatment facilities available in Mumbai lack the capacity to remove a wide range of ECs such as pharmaceuticals, PCPs, and industrial chemicals. Investments in advanced treatment technologies will bridge this gap. Membrane bioreactors (MBRs) are promising technologies that work by combining biological treatment with membrane filtration to achieve a higher contaminant removal efficiency (Jijingi et al. 2024). In addition, there are AOPs, like ozonation and ultraviolet/hydrogen peroxide (UV/H_2O_2) treatment, that can be used as add-on secondary treatments to existing plants to improve the removal of such persistent pollutants (Rekhate and Srivastava 2020). Such technologies aim at counterbalancing the complexity of wastewater and ensure that the water produced is sufficiently safe (Schulte-Wülwer-Leidig et al. 2018; Shanker 2024).

Simultaneously with the upgrading of infrastructure, decentralized and community-based solutions may play a crucial role in addressing wastewater management, such as in informal settlements and industrial zones where central mechanisms are not in place (Ansari et al. 2023). Constructed wetlands, for example, treat water with biologically active natural processes involving action by plants and microorganisms (Kushwaha et al. 2023). Such relatively inexpensive solutions can be integrated into urban planning and solve the smaller flows of wastewater. Important areas are in not-centralized facilities, especially on-site treatment systems like septic tanks and small-scale anaerobic digesters. These minimize the loads imposed upon the municipal plants and enhance local water quality. A significant source of ECs in the waters of Mumbai and deserve stricter regulations and implementation. Strict standards of discharge may ensure that several industrial facilities do not over-pollute the waters and thus prevent major degradation of water quality. Industries must be under constant monitoring and checked for compliance to ensure their adherence to such standards (Singh et al. 2023). Promoting cleaner production in industries can also help curb the generation of hazardous waste in industries. By employing toxic materials and substances at a small scale in industries, so that their salination would be minimized, integrating recycling streams, and conforming to treatment processes, the industries may be able to make tremendous reductions in their overall load on the environment (Zhang et al. 2024).

Public awareness and responsible disposal practices are integral components of a relevant program. Alerting the residents on the environmental impacts of improper disposal of drugs and household chemicals is usually enough to actuate changes in behavior (Rogowska and Zimmermann 2022). The problem of public disposal requires campaigns highlighting the need to hand over unused or expired drugs to designated disposal programs instead of infecting the entire system of wastewater. Increased use of environment-friendly PCPs and cleaning agents can be encouraged in order to further reduce the load of contaminants entering the system (Salonen et al. 2024). The integration of green infrastructure into urban planning can complement efforts that treat wastewater. Sustainable design can further prevent contamination and provide an integrated effect of land use on the water quality by planning sufficient sewer system facilities and encouraging treatment facilities as part of new developments (Arrington et al. 2023). The municipal wastewater management of Mumbai can be significantly enhanced by applying these measures, reducing the impact of ECs on its water bodies. The combination of leading-edge treatment technologies, decentralized solutions, strengthened regulatory frameworks, public awareness, and green infrastructure can collectively improve the city's capacities to address the challenges posed by wastewater and ensure protection for both aquatic ecosystems and public health.

12.6 Strategies for the Rhine: Regulation, Remediation, and Ongoing Monitoring

The application of highly robust regulatory frameworks is an essential part of the approach. European legislation like the Water Framework Directive and the Industrial Emissions Directive (WFDIED) provides an excellent basis for regulating water pollution (Carvalho et al. 2019). These regulations not only demand a decrease in industrial sources of pollutant discharges but also call for very strict quality standards for surface waters. Such regulations have to be enforced effectively, whether by regular inspection and severe punishment or deterrents for non-compliance. Since modern scientific understanding is in constant evolution, regulations are required to be updated in compliance with ECs, including MPs and novel industrial chemicals, to provide ultimate control over potential pollutants (Aguilar-Aguilar et al. 2023). Equally important in this regard is the effective remediation of contaminated sites along the Rhine in order to minimize the long-term effects of the pollution. Contamination of sediments and water bodies is addressed using both *in-situ* and *ex-situ* methods. *In-situ* techniques include monitored natural attenuation, in which natural processes degrade pollutants, supplemented by active methods that include bioremediation and chemical treatment. The clean-up at hotspots requires remediation approaches *ex-situ*, such as dredging contaminated sediments for treatment or disposal (Azubuike et al. 2016). Better targeting of remediation efforts based on comprehensive risk assessments helps to make better resource use and deals with the worst contamination problems that require urgent attention (Tong et al. 2022; Varjani and Chaithanya Sudha 2018).

Continuous monitoring is important in uncovering and managing emerging contamination. Wide-ranging and systematic monitoring should be conducted, and routine sampling of water, sediment, and biota should monitor contaminant levels and trends. Advanced analytical techniques including high-resolution mass spectrometry must be included to identify and quantify as wide a range of contaminants as possible, including non-regulated but potentially hazardous ones (Li et al. 2015). Thus, while monitoring continues in the field, research projects meant to understand the behavior, fate, and effects of ECs continue to prevail. Collaborating the research institutions, the governmental bodies, and the industries, would take their efforts to a higher level with management strategies being implemented based on the most updated knowledge from the sciences involved. More importantly, public involvement and awareness play an equally important role in EC management. Citizen education on the sources of contaminants and proper practices of waste disposal should be encouraged to promote a culture of responsibility toward the environment. Involving communities in monitoring programs and decision-making processes can also enhance the effectiveness of water management approaches because the concerns and opinions of the parties affected by it are addressed (Debrah et al. 2021).

12.7 Discussion and Future Prospects

ECs in both Mumbai and the Rhine River Basin present different yet interlinked challenges, thus necessitating focused, innovative solutions. In Mumbai's case, the issue to be addressed is deficiencies in its infrastructure for wastewater treatment infrastructure that allow such a broad array of contaminants—ranging from pharmaceuticals and PCPs to industrial chemicals—to freely enter water bodies primarily untreated. The diversified residential, commercial, and industrial sources of ECs in the rapidly growing city demand a major overhaul of the present wastewater management systems. Future research in Mumbai should be focused on newer advanced treatment technologies capable of removing a wide range of ECs (Singh et al. 2023). Some of the promising solutions include MBRs, AOPs, and integrated treatment systems that can improve efficiency in contaminant removal.

At the same time, it should explore if there is efficiency in decentralized treatment systems for wastewater systems especially when the population outpaces infrastructure growth in informal settlements (Kanafin et al. 2023). It should also look at community-based activities and public education and incentives-based programs on environmentally friendly products to considerably contribute to source EC reduction. This is, however, different from the Rhine River because of its history and current industrial development. Legacy pollution and new ECs still afflict the ecological health of this river, though major strides have been made in water quality through tighter regulations and the implementation of technological innovations. More attention in the future for Rhine research should be paid to improving techniques for detecting and quantifying new ECs. Better analytical techniques include high-resolution mass spectrometry and more advanced chromatographic techniques, which will allow for better identification of such pollutants and tracing them across time (Ayala-Cabrera et al. 2021).

Research is also needed to determine the persistence and bioaccumulation of contaminants, including POPs and MPs, that become important agents threatening aquatic life and human health. Innovative remediation technologies, including bioremediation and phytoremediation, are potential steps toward mitigation of Rhine contamination. The study on applying plants, microbes, and other natural processes to eliminate or neutralize contaminants might well provide some innovative avenues for improving water quality (Liu et al. 2023). Crucially, transboundary issues regarding pollution become heavily reliant on international coordination for it is a major European waterway. Cross-border policy frameworks, as well as collaborative remediation projects, can improve the effectiveness of regional and global efforts to manage ECs by reinforcing sound cross-border coordination and development paths. Future research must focus on the long-term effects of ECs on both ecological systems and human health. Identification of chronic effects associated with low-level exposure to pharmaceuticals, MPs, and many other pollutants will be critical when making risk assessments and appropriate strategies in the mitigation of those risks. Longitudinal studies, tracking the impact of health effects of ECs on aquatic organisms and humans, can be very informative regarding the effectiveness of the current management practices and future policy decisions (Wantzen et al. 2022).

12.8 Conclusion

This comparison of Mumbai and the Rhine River Basin shows how they managed rising toxins in their aquatic systems. Emerging toxins threaten water quality, ecological health, and human well-being in both locations, which have differing socioeconomic and industrial environments. Mumbai freely discharges treated wastewater into its aquatic body. Improvements in public health protection require a high-tech treatment of pollutants through advanced technologies and improvements in the community's infrastructure base. Instead, the pollutants emitted from past and ongoing industrial activities along the Rhine River Basin include a complex mixture of POPs, MPs, and perfluorinated chemicals. While strict regulations and technological advances have made progress, industrial pollution, and new contaminants present new challenges to monitor, innovate, and cooperate internationally. Mumbai will improve wastewater treatment and disposal and manage informal settlements. These are essential to reducing EC pollution and protecting local ecosystems. The Rhine needs improved monitoring, cleanup, and coordination across borders in the management of legacy and emerging pollution. Future studies will concentrate on new treatment methods, continuous monitoring and improvement, and long-term effects of ECs on the environment and health. International cooperation and sustainable industrial practices are necessary for dealing with new pollutants worldwide.

Acknowledgements The authors sincerely express gratitude to Amity University and the Amity Institute of Environmental Sciences (AIES) for their unwavering support, encouragement, and the invaluable academic environment that has fostered this study. The resources, guidance, and collaborative opportunities provided have significantly contributed to the development of this chapter.

References

Adla K, Dejan K, Neira D, Dragana Š (2022) Degradation of ecosystems and loss of ecosystem services. In: One health. Academic Press, pp 281–327

Agarwal S, Jain H (2024) Investigating the potential of floral waste as a vermicompost and dual-functional biosorbent for sustainable environmental management. Water Air Soil Pollut 235(5):1–18

Aguilar-Aguilar A, de Leon-Martinez LD, Forgionny A, Soto NYA, Mendoza SR, Zarate-Guzman AI (2023) A systematic review on the current situation of emerging pollutants in Mexico: a perspective on policies, regulation, detection, and elimination in water and wastewater. Sci Total Environ 905:167426

Alam S, Bhardwaj LK, Mallick R, Rai S (2023) Estimation of heavy metals and fluoride ion in vegetables grown nearby the stretch of River Yamuna, Delhi (NCR), India. Indian J Environ Prot 43(1):64–73

Alam SM, Markandey K (2020) Consequences of unplanned growth: a case study of metropolitan Hyderabad. In: Urban and regional planning and development: 20th century forms and 21st century transformations. pp 203–219

Alharbi OM, Khattab RA, Ali I (2018) Health and environmental effects of persistent organic pollutants. J Mol Liq 263:442–453

Alhelou R, Seiwert B, Reemtsma T (2019) Hexamethoxymethylmelamine–a precursor of persistent and mobile contaminants in municipal wastewater and the water cycle. Water Res 165:114973

Ali N, Khan MH, Ali M, Ahmad S, Khan A, Nabi G, Ali F, Bououdina M, Kyzas GZ (2024) Insight into microplastics in the aquatic ecosystem: properties, sources, threats and mitigation strategies. Sci Total Environ 913:169489

Ansari A, Dutt D, Kumar V (2024) Catalyzing paradigm shifts in global waste management: a case study of Saharanpur Smart city. Waste Manage Bull 2(1):29–38

Arrington DA, Harris RJ, Layman CA, Gomes DG (2023) Leveraging green infrastructure for efficient treatment of reclaimed water. Sci Total Environ 872:162232

Ávila C, García J (2015) Pharmaceuticals and personal care products (PPCPs) in the environment and their removal from wastewater through constructed wetlands. In: Comprehensive analytical chemistry, vol 67. Elsevier, pp 195–244

Ayala-Cabrera JF, Santos FJ, Moyano E (2021) Recent advances in analytical methodologies based on mass spectrometry for the environmental analysis of halogenated organic contaminants. Trends Environ Anal Chem 30:e00122

Azubuike CC, Chikere CB, Okpokwasili GC (2016) Bioremediation techniques–classification based on site of application: principles, advantages, limitations and prospects. World J Microbiol Biotechnol 32(11):180

Bhardwaj LK, Upadhyay D, Jain H, Rath P, Kumar D, Bajpai S (2025a) Recent trends of phytoremediation for the wastewater treatment and resource recovery. In: Biotechnologies for wastewater treatment and resource recovery. Elsevier, pp 135–146

Bhardwaj LK, Rath P, Rehman F, Yadav P, Massahi T, Mittal R, Jain H (2025b) Status of micro/nano-plastic (MNP) pollution in the aquatic system. In: Societal and environmental ramifications of plastic pollution. IGI Global, pp 43–62

Bozkir ED, Leentvaar J, Frijters I, Hofstra MA (2010) Impacts of national and international actors on river basin management–case of river Rhine. In: The global dimensions of change in River Basins. p 125

Brack W, Escher BI, Müller E, Schmitt-Jansen M, Schulze T, Slobodnik J, Hollert H (2018) Towards a holistic and solution-oriented monitoring of chemical status of European water bodies: how to support the EU strategy for a non-toxic environment? Environ Sci Eur 30:1–11

Brunn H, Arnold G, Körner W, Rippen G, Steinhäuser KG, Valentin I (2023) PFAS: forever chemicals—persistent, bioaccumulative and mobile. Reviewing the status and the need for their phase out and remediation of contaminated sites. Environ Sci Eur 35(1):1–50

Carvalho L, Mackay EB, Cardoso AC, Baattrup-Pedersen A, Birk S, Blackstock KL, Borics G, Borja A, Feld CK, Ferreira MT, Globevnik L (2019) Protecting and restoring Europe's waters: an analysis of the future development needs of the water framework directive. Sci Total Environ 658:1228–1238

Cornelsen M, Weber R, Panglisch S (2021) Minimizing the environmental impact of PFAS by using specialized coagulants for the treatment of PFAS polluted waters and for the decontamination of firefighting equipment. Emerg Contam 7:63–76

Das S, Parida VK, Tiwary CS, Gupta AK, Chowdhury S (2024) Emerging contaminants in the aquatic environment: fate, occurrence, impacts, and toxicity. In: Bioremediation of emerging contaminants in water, vol 1. American Chemical Society, pp 1–32

Debrah JK, Vidal DG, Dinis MAP (2021) Raising awareness on solid waste management through formal education for sustainability: a developing countries evidence review. Recycling 6(1):6

Dsikowitzky L, Schwarzbauer J (2015) Hexa (methoxymethyl) melamine: an emerging contaminant in German rivers. Water Environ Res 87(5):461–469

Den Besten PJ, Postma JF, Wegener JWM, Keidel H, Klink A, Mol J, Van de Guchte C (2000) Biological and chemical monitoring after pilot remediations in the delta of the rivers Rhine and Meuse. Aquat Ecosyst Health Manage 3(3):317–334

Duester L, Livrozet N, Poturalski S, Stoetter T, Gerloff AL, Heintz MD (2024) The ICPR measuring programme chemistry and its monitoring approach–a look back and a glimpse of the future. Water Int 49(3–4):446–454

Ebele AJ, Abdallah MAE, Harrad S (2017) Pharmaceuticals and personal care products (PPCPs) in the freshwater aquatic environment. Emerg Contam 3(1):1–16

Fa JE (2012) The environment and endangered species/species extinctions

Gaurkhede N, Adane VS, Khonde S (2021) Identification of interruptions in Urban Drainage systems and their sustainable solutions for alleviating flood risk in Mumbai, an Indian megacity. IDRiM J 11(1):108–130

Ghosh S, Sinha JK, Ghosh S, Vashisth K, Han S, Bhaskar R (2023) Microplastics as an emerging threat to the global environment and human health. Sustainability 15(14):10821

Göckener B, Fliedner A, Rüdel H, Badry A, Koschorreck J (2021) Long-term trends of per-and polyfluoroalkyl substances (PFAS) in suspended particular matter from German rivers using the direct total oxidizable precursor (dTOP) assay. Environ Sci Technol 56(1):208–217

Guo W, Pan B, Sakkiah S, Yavas G, Ge W, Zou W, Tong W, Hong H (2019) Persistent organic pollutants in food: contamination sources, health effects and detection methods. Int J Environ Res Public Health 16(22):4361

Hasan SE (2004) Public awareness is key to successful waste management. J Environ Sci Health, Part A 39(2):483–492

Hutton G, Chase C (2018) Water supply, sanitation, and hygiene

Internationale Kommission zum Schutz des Rheins, Commission Internationale pour la Protection du Rhin, Internationale Commissie ter Bescherming van de Rijn, International Commission for the Protection of the Rhine (2020) Assessment Rhine 2020. In: Assessment Rhine 2020. https://www.iksr.org/fileadmin/user_upload/DKDM/Dokumente/Broschueren/EN/bro_En_Assessment_%E2%80%9CRhine_2020%E2%80%9D.pdf

Internet Geography (2023) What challenges have been caused by urban growth in Mumbai? Internet Geography. https://www.internetgeography.net/topics/what-challenges-have-been-caused-by-urban-growth-in-mumbai/

Jain H (2024) From pollution to progress: groundbreaking advances in clean technology unveiled. Innov Green Dev 3(2):100143

Jijingi HE, Yazdi SK, Abakar YA, Etim E (2024) Evaluation of membrane bioreactor (MBR) technology for industrial wastewater treatment and its application in developing countries: a review. Case Stud Chem Environ Eng 10:100886

Kanafin YN, Satayeva A, Abdirova P, Inglezakis VJ, Arkhangelsky E, Poulopoulos SG (2023) Membrane bioreactor and advanced oxidation processes for combined treatment of synthetic wastewater containing naproxen, bisphenol A, and sulfamethoxazole. J Water Process Eng 55:104250

Krauss M, Hug C, Bloch R, Schulze T, Brack W (2019) Prioritising site-specific micropollutants in surface water from LC-HRMS non-target screening data using a rarity score. Environ Sci Eur 31(1):1–12

Kumar D, Ram S, Srivastav AL (2022a) Urban water quality and wastewater discharges in mega and metro cities of India—Challenges, impacts, and management: an overview. Curr Dir Water Scarcity Res 6:223–244

Kumar R, Qureshi M, Vishwakarma DK, Al-Ansari N, Kuriqi A, Elbeltagi A, Saraswat A (2022b) A review on emerging water contaminants and the application of sustainable removal technologies. Case Stud Chem Environ Eng 6:100219

Kumar G, Kumar S, Paul T, Pal P, Shukla SP, Kumar K, Jha AK, Pradeep S (2024) Ecotoxicological risk assessment of triclosan, an emerging pollutant in a riverine and estuarine ecosystems: a comparative study. Mar Pollut Bull 205:116667

Kurwadkar S, Dane J, Kanel SR, Nadagouda MN, Cawdrey RW, Ambade B, Struckhoff GC, Wilkin R (2022) Per-and polyfluoroalkyl substances in water and wastewater: a critical review of their global occurrence and distribution. Sci Total Environ 809:151003

Kushwaha A, Talukdar S, Mohanan VP, Lata S, Gupta M, Goswami L, Kim BS (2023) Bio-treatment of the swine wastewater and resource recovery: a sustainable approach towards circular bioeconomy. In: Bio-based materials and waste for energy generation and resource management. Elsevier, pp 299–329

Leuven RS, van der Velde G, Baijens I, Snijders J, van der Zwart C, Lenders HR, bij de Vaate A (2009) The River Rhine: a global highway for dispersal of aquatic invasive species. Biol Invasions 11(9):1989–2008

Li E, Endter-Wada J, Li S (2015) Characterizing and contextualizing the water challenges of megacities. JAWRA J Am Water Resour Assoc 51(3):589–613

Li X, Shen X, Jiang W, Xi Y, Li S (2024) Comprehensive review of emerging contaminants: detection technologies, environmental impact, and management strategies. Ecotoxicol Environ Saf 278:116420

Liu CJ, Deng SG, Hu CY, Gao P, Khan E, Yu CP, Ma LQ (2023) Applications of bioremediation and phytoremediation in contaminated soils and waters: CREST publications during 2018–2022. Crit Rev Environ Sci Technol 53(6):723–732

Ma T, Wu P, Wang L, Li Q, Li X, Luo Y (2023) Toxicity of per-and polyfluoroalkyl substances to aquatic vertebrates. Front Environ Sci 11:1101100

Madhogaria B, Banerjee S, Kundu A, Dhak P (2024) Efficacy of new generation biosorbents for the sustainable treatment of antibiotic residues and antibiotic resistance genes from polluted waste effluent. Infect Med 3(1):100092

Maharashtra Pollution Control Board (2005) Report on environment status of Mumbai region. In: Maharashtra Pollution Control Board. https://www.mpcb.gov.in/sites/default/files/focus-area-reports-documents/envreportmumbai.pdf

Mani T, Hauk A, Walter U, Burkhardt-Holm P (2015) Microplastics profile along the Rhine River. Sci Rep 5(1). https://doi.org/10.1038/srep17988

Masindi V, Muedi KL (2018) Environmental contamination by heavy metals. Heavy Met 10(4):115–133

Mathew RA, Kanmani S (2020) A review on emerging contaminants in indian waters and their treatment technologies. Nat Environ Pollut Technol 19(2):549–562

Meybeck M (2002) Riverine quality at the Anthropocene: propositions for global space and time analysis, illustrated by the Seine River. Aquat Sci 64(4):376–393

Miniero R, Iamiceli A, De Felip E (2015) Persistent organic pollutants. Elsevier eBooks. https://doi.org/10.1016/b978-0-12-409548-9.09496-3

Mishra RK, Mentha SS, Misra Y, Dwivedi N (2023) Emerging pollutants of severe environmental concern in water and wastewater: a comprehensive review on current developments and future research. Water-Energy Nexus 6:74–95

Mohite RA, Joya LK (2021) A case study report on waste water management & sewage disposal in Mumbai suburban region. J Emerg Technol Innov Res 8:2349–5162

Munshi R, Dhiman A, Maurya M (2023) A cross-sectional survey to assess the drug disposal practices of unused and expired medicines among lay public visiting a tertiary care hospital in an urban metropolis. Indian J Physiol Pharmacol 67(1):29–35

Namieśnik J, Wolska L, Czernych R, Gałęzowska G, Cieszyńska M (2014) Classifying the transformation products (TPs) of emerging contaminants (ECs) for prioritizing research into their impact on the environment and human health. Transform Prod Emerg Contam Environ 1–64

Negi R, Chandel MK (2021) Analysing water-energy-GHG nexus in a wastewater treatment plant of Mumbai Metropolitan Region, India. Environ Res 196:110931

Ngo H, Bomers A, Augustijn DC, Ranasinghe R, Filatova T, van der Meulen B, Herget J, Hulscher SJ (2023) Reconstruction of the 1374 Rhine river flood event around Cologne region using 1D-2D coupled hydraulic modelling approach. J Hydrol 617:129039

Overturf MD, Anderson JC, Pandelides Z, Beyger L, Holdway DA (2015) Pharmaceuticals and personal care products: a critical review of the impacts on fish reproduction. Crit Rev Toxicol 45(6):469–491

Philip JM, Aravind UK, Aravindakumar CT (2018) Emerging contaminants in Indian environmental matrices–a review. Chemosphere 190:307–326

Praja Foundation (2021) Status of civic issues in Mumbai. https://www.praja.org/praja_docs/praja_downloads/Report%20on%20Status%20of%20Civic%20Issues%20in%20Mumbai%202021.pdf

Quinete N, Wu Q, Zhang T, Yun SH, Moreira I, Kannan K (2009) Specific profiles of perfluorinated compounds in surface and drinking waters and accumulation in mussels, fish, and dolphins from southeastern Brazil. Chemosphere 77(6):863–869

Rahman MM, Kim HD, Lee WK (2009) Properties of crosslinked waterborne polyurethane adhesives with modified melamine: effect of curing time, temperature, and HMMM content. Fibers Polym 10:6–13

Raut AN, Kashyap SR (2021) Assessment of water reuse potential of wastewater treatment systems used in Mumbai, India. Int Res J Adv Sci Hub 3(1S):12–19

Rekhate CV, Srivastava JK (2020) Recent advances in ozone-based advanced oxidation processes for treatment of wastewater-a review. Chem Eng J Adv 3:100031

Rogowska J, Zimmermann A (2022) Household pharmaceutical waste disposal as a global problem—a review. Int J Environ Res Public Health 19(23):15798

Rout PR, Zhang TC, Bhunia P, Surampalli RY (2021) Treatment technologies for emerging contaminants in wastewater treatment plants: a review. Sci Total Environ 753:141990

Rüdel H, Radermacher G, Fliedner A, Lohmann N, Koschorreck J, Duffek A (2022) Tissue concentrations of per-and polyfluoroalkyl substances (PFAS) in German freshwater fish: derivation of fillet-to-whole fish conversion factors and assessment of potential risks. Chemosphere 292(133483):10–1016

Ruff M, Mueller MS, Loos M, Singer HP (2015) Quantitative target and systematic non-target analysis of polar organic micro-pollutants along the river Rhine using high-resolution mass-spectrometry–identification of unknown sources and compounds. Water Res 87:145–154

Sadoff CW, Grey D (2005) Cooperation on international rivers: a continuum for securing and sharing benefits. Water Int 30(4):420–427

Salonen H, Salthammer T, Castagnoli E, Täubel M, Morawska L (2024) Cleaning products: their chemistry, effects on indoor air quality, and implications for human health. Environ Int 190:108836

Samal K, Mahapatra S, Ali MH (2022) Pharmaceutical wastewater as emerging contaminants (EC): treatment technologies, impact on environment and human health. Energy Nexus 6:100076

Saravanane R, Ranade VV, Bhandari VM, Rao AS (2014) Urban wastewater treatment for recycling and reuse in industrial applications: Indian scenario. In: Industrial wastewater treatment, recycling and reuse. Elsevier Inc., pp 283–322

Sarkar P, Basak PR, Kumar K, Karmakar S, Kumar S (2024) Biomedical waste: types, current status, management and impact on aquatic body – a comprehensive review | chronicle of aquatic science. https://chronicleofaquaticscience.in/biomedical-waste-types-current-status-management-and-impact-on-aquatic-body-a-comprehensive-review

Schiff JS (2017) The evolution of Rhine river governance: historical lessons for modern transboundary water management. Water Hist 9:279–294

Schulte-Wülwer-Leidig A, Gangi L, Stötter T, Braun M, Schmid-Breton A (2018) Transboundary cooperation and sustainable development in the Rhine Basin. Achiev Chall Integr River Basin Manag 123–147

Schwarzbauer J, Heim S (2005) Lipophilic organic contaminants in the Rhine river, Germany. Water Res 39(19):4735–4748

Schwarzbauer J, Ricking M (2010) Non-target screening analysis of river water as compound-related base for monitoring measures. Environ Sci Pollut Res 17(4):934–947

Shanker U (2024) Modern applications and current status of ECs in the industries. In: Occurrence, distribution and toxic effects of emerging contaminants. p 237

Shukla TN, Chatterjee C (2011) Issues, practices, policy, enforcement and implementation of waste management in India. Kushagra Int Manage Rev 1(1):107

Singare PU, Dhabarde SS (2014) Industrial pollution scenario due to discharge of waste water effluents along Dombivali Industrial Belt of Mumbai, India–a physico-chemical study. Interdisc Environ Rev 15(1):20–35

Singh BJ, Chakraborty A, Sehgal R (2023) A systematic review of industrial wastewater management: evaluating challenges and enablers. J Environ Manage 348:119230

Singh A, Kazmi A, Starkl M, Sayanekar S, Herlekar M (2018) Sewage management challenges in mega cities in India: a case study of Mumbai. Desalin Water Treat 116:329–341

Spaulding RM (2015) Rhine river commerce and the Continental system. Revisiting Napoleon's continental system: local, regional and European experiences. Palgrave Macmillan UK, London, pp 114–132

Starkl M, Anthony J, Aymerich E, Brunner N, Chubilleau C, Das S, Ghangrekar MM, Kazmi AA, Philip L, Singh A (2018) Interpreting best available technologies more flexibly: a policy perspective for municipal wastewater management in India and other developing countries. Environ Impact Assess Rev 71:132–141

Stefanakis AI (2019) The role of constructed wetlands as green infrastructure for sustainable urban water management. Sustainability 11(24):6981

Stringer R (2011) Medical waste and human rights. Presentación ante el Relator Especial de la Comisión de Derechos Humanos de las Naciones Unidas, Salud Sin Daño

Tanwar S, Di Carro M, Ianni C, Magi E (2014) Occurrence of PCPs in natural waters from Europe. In: Personal care products in the aquatic environment. Springer International Publishing, Cham, pp 37–71

Tchounwou PB, Yedjou CG, Patlolla AK, Sutton DJ (2012) Heavy metal toxicity and the environment. In: Molecular, clinical and environmental toxicology: volume 3: environmental toxicology, pp 133–164

Tong X, Mohapatra S, Zhang J, Tran NH, You L, He Y, Gin KYH (2022) Source, fate, transport and modelling of selected emerging contaminants in the aquatic environment: current status and future perspectives. Water Res 217:118418

Uehlinger U, Wantzen KM, Leuven RSEW, Arndt H (2009) The Rhine River Basin. https://d-nb.info/1088795714/34

Varjani SJ, Chaithanya Sudha M (2018) Treatment technologies for emerging organic contaminants removal from wastewater. Water Remediat 91–115.

Wang M, Liao G, Li Y (2021) The relationship between environmental regulation, pollution and corporate environmental responsibility. Int J Environ Res Public Health 18(15):8018

Wantzen KM, Uehlinger U, Van der Velde G, Leuven RS, Schmitt L, Beisel JN (2022) The Rhine river basin. In: Rivers of Europe. Elsevier, pp 333–391

Water Quality (2020) https://www.iksr.org/en/topics/water-quality

Water-related Illnesses (2024) National Institute of Environmental Health Sciences. https://www.niehs.nih.gov/research/programs/climatechange/health_impacts/waterborne_diseases

Weijs L, Dirtu AC, Malarvannan G, Covaci A (2015) Bioaccumulation and biotransformation of brominated flame retardants. In: Comprehensive analytical chemistry, vol 67. Elsevier, pp 433–491

Wilhelm M, Ewers U, Wittsiepe J, Fürst P, Hölzer J, Eberwein G, Angerer J, Marczynski B, Ranft U (2007) Human biomonitoring studies in North Rhine-Westphalia, Germany. Int J Hyg Environ Health 210(3-4):307–318

Wilkinson J, Hooda PS, Barker J, Barton S, Swinden J (2017) Occurrence, fate and transformation of emerging contaminants in water: an overarching review of the field. Environ Pollut 231:954–970

Windsor FM, Ormerod SJ, Tyler CR (2018) Endocrine disruption in aquatic systems: up-scaling research to address ecological consequences. Biol Rev 93(1):626–641

Ying Y, van Kollenburg G, van den Doel A, Brekelmans S, Zhou H, Stroomberg G, Jansen J (2022) Evaluating anthropogenic origin of unidentified volatile chemicals in the river Rhine. Water Air Soil Pollut 233(7):241

Yuan M, Faggio C, Perugini M, Aliko V, Wang Y (2023) Pharmaceuticals, personal care products and endocrine disrupting chemicals: the physiological consequences of exposure to pollutants in aquatic animals. Front Physiol 14:1145052

Yuan B, Rüdel H, de Wit CA, Koschorreck J (2022) Identifying emerging environmental concerns from long-chain chlorinated paraffins towards German ecosystems. J Hazard Mater 424:127607

Zhang Z, Chen Z, Zhang J, Liu Y, Chen L, Yang M, Osman AI, Farghali M, Liu E, Hassan D, Ihara I (2024) Municipal solid waste management challenges in developing regions: a comprehensive review and future perspectives for Asia and Africa. Sci Total Environ 930:172794

MIX
Papier aus verantwortungsvollen Quellen
Paper from responsible sources
FSC® C105338

If you have any concerns about our products,
you can contact us on
ProductSafety@springernature.com

In case Publisher is established outside the EU,
the EU authorized representative is:
Springer Nature Customer Service Center GmbH
Europaplatz 3, 69115 Heidelberg, Germany

Printed by Libri Plureos GmbH
in Hamburg, Germany